APPLIED SCIENCE OF
INSTRUMENTATION

A Technician's Reference to Calibration and Process Control

Applied Science of Instrumentation: A Technician's Reference to Calibration and Process Control is intended to be an educational resource for the user and contains procedures commonly practiced in industry and the trade. Specific procedures vary with each task and must be performed by a qualified person. For maximum safety, always refer to specific manufacturer recommendations, insurance regulations, specific job site and plant procedures, applicable federal, state, and local regulations, and any authority having jurisdiction. The *electrical training ALLIANCE* assumes no responsibility or liability in connection with this material or its use by any individual or organization.

© 2017 *electrical training ALLIANCE*

This material is for the exclusive use by the IBEW-NECA JATCs and programs approved by the *electrical training ALLIANCE*. Possession and/or use by others is strictly prohibited as this proprietary material is for exclusive use by the *electrical training ALLIANCE* and programs approved by the *electrical training ALLIANCE*.

All rights reserved. No part of this material shall be reproduced, stored in a retrieval system, or transmitted by any means whether electronic, mechanical, photocopying, recording, or otherwise without the express written permission of the *electrical training ALLIANCE*.

1 2 3 4 5 6 7 8 9 – 17 – 9 8 7 6 5 4 3 2

Printed in the United States of America

M39522

Contents

Chapter 1 — INTRODUCTION TO INSTRUMENTATION ... 1
Purpose of Instrumentation ... 2
Typical Working Environment ... 3
Instrumentation and Control Safety ... 3
Math in Instrument and Control Applications ... 9
Instrumentation Vocabulary ... 10
DC Theory ... 12

Chapter 2 — FUNDAMENTALS OF PROCESS AND CONTROL SYSTEMS ... 19
Historical Background ... 20
Defining Control Systems ... 21
Defining Process Variables ... 23
Control Signals ... 24
Signal Types ... 27
Operator Interface ... 28
Components of a Loop ... 30

Chapter 3 — INSTRUMENTATION SYMBOLS AND DIAGRAMS ... 41
Instrument Symbols and Identifiers ... 42
Instrumentation Documents ... 53

Chapter 4 — CALIBRATION PROCEDURE AND DOCUMENTATION ... 61
Calibration ... 62
As-Found Test ... 72
Error Correction Procedures ... 75
As-Left Test ... 77
Specification Sheets ... 80

Chapter 5 — PRINCIPLES OF PRESSURE ... 87
Pressure Fundamentals ... 88
Pressure Scales ... 91
Pressure Devices and Their Functions ... 96
Pressure Units of Measure ... 105
Multivariable Relationships ... 106

Contents

Chapter 6   PRINCIPLES OF LEVEL 111
Level Fundamentals 112
Methods of Level Measurement 117
Density Compensation 134
Environmental Concerns 138

Chapter 7   PRINCIPLES OF FLOW 141
Flow Fundamentals 142
Volumetric Flow Equation 144
Law of Conservation of Energy 145
Flow Characteristics 148
Methods of Flow Measurement 151
Environmental Considerations 165

Chapter 8   PRINCIPLES OF TEMPERATURE 169
Purpose of Temperature Measurement 170
Methods of Temperature Measurement 173
Thermowell Basics 188

Chapter 9   PRINCIPLES OF SMART INSTRUMENT COMMUNICATION AND CALIBRATION 191
Basics of Smart Instruments 192
Types and Methods of Communication 192
Smart Instrument Communicators 194
Systems and Applications 198
Using the Communicator to Calibrate a Smart Instrument 201

Chapter 10   CONTROL VALVES, ACTUATORS, AND ACCESSORIES 209
Control Valve Assembly Components 210
Principles of Pneumatics 210
Types Of Control Valves 211
Types of Actuators 218
Types of Accessories 225

Contents

Chapter 11   ANALYTICAL MEASUREMENT 241
Analytical Measurement 242
Conductivity 242
pH Measurement 243
Oxidation Reduction Potential 246
Silica Analyzers 247
Continuous Emission Monitoring Systems 247
Combustion Analysis 248
Gas Chromatography 251
Oxygen 253
Density 254
Flammable Gas Detectors 257
Hydrogen Sulfide Monitors 260

Chapter 12   PROCESS CONTROLLERS 265
Fundamentals of Controllers 266
Automatic Control Concept 266
Field I/O Signals 267
Systems and Applications 276
Control Program 282
Emergency Shutdown Systems 284
Human Machine Interfaces 285

Chapter 13   FUNDAMENTALS OF CONTROL 289
Fundamentals of Control 290
Control Applications 290
Control Modes 293
Advanced Control 298

Chapter 14   INSTALLATION OF CONTROL SYSTEMS 303
Systems and Applications 304

Contents

Chapter 15   PROCESS CONTROL LOOP CHECKING 321
Process Control Systems 322
Loop Checking 322
Example of Temperature Transmitter Loop Checking 326

Chapter 16   START-UP AND TUNING OF PROCESS CONTROL LOOPS 331
Plant Start-Up Process 332
Loop Tuning 336

Chapter 17   TROUBLESHOOTING PROCESS CONTROL LOOPS 341
Troubleshooting Procedures 342
Troubleshooting Devices 345

Chapter 18   DISTRIBUTED CONTROL SYSTEMS 353
Introduction to Distributed Control Systems 354
Definition of a DCS 354
Installation of a DCS 362
System Documentation 364

Chapter 19   PROJECT DOCUMENTATION AND MANAGEMENT ... 367
Managing Projects 368
Design Standards 368
Documentation Standards 369
Quality Standards 369
Instrument and Control Project Management 371

Contents

Appendix 379
Instrumentation Formula Sheet 380
Sample P&ID 381
Instrument Letter Chart 382
Industrial Electrical Symbols 384
Standard P&ID Symbols Legend | Industry Standardized P&ID Symbols 388
Ohm's Law 395
NATO Phonetic Alphabet 395
Resistor Color Code 396
ANSI and IEC Color Codes 397
Technical Data Section 398
Metric Prefixes 400
Decimal, Hexadecimal, Octal, and Binary Conversion Chart 401
4-20 mA Transmitter Output Chart 402
3-15 PSI Transmitter Output Chart 403
Logic Gate Chart 404
NFPA Hazardous Locations 405
Geometric Formulas 406

Glossary 407

Index 423

Introduction

Welcome to the first edition of *Applied Science of Instrumentation*. This text has been designed to provide a foundation of knowledge for the prospective Instrumentation technician enrolled in an apprenticeship program, vocational technical program, or a community college course.

Instrumentation is a vast field requiring the technician to become proficient in the physical principles that form the foundation of complex process control systems. The interactions in a multi-craft environment require a breadth of knowledge across converging disciplines including mechanical, electrical, and chemical systems.

The *electrical training ALLIANCE*, formerly the National Joint Apprenticeship and Training Committee (NJATC), has been the source for superior electrical training materials for more than 70 years. The *ALLIANCE* constantly works to improve curriculum and strives to provide the best support possible for apprentices, journeymen, and instructors in over 285 training programs nationwide.

About This Book

This book is the result of the *electrical training ALLIANCE*'s effort to blend the best available information and resources with practical examples, new technology, and multi-media enhancements. Combining the traditional textbook structure with technological advances accommodates multiple different learning styles and allows key concepts to be reinforced throughout the learning process in new and engaging ways.

This textbook has been written and designed to guide the reader progressively through the information, constantly building on the knowledge gained from previous chapters. Each chapter has been structured to provide the reader with an introduction of the material and clear learning objectives to identify key points at the beginning of the chapter. At the end, a brief summary and a selection of review questions serves to reinforce the information presented.

The use of diagrams alongside pictures of real-world applications and links to videos and web-based resources provide an engaging and highly visual learning experience. Through the use of QR code technology, information can be kept up to date with technological progress and advancements in the field.

Acknowledgments

Acknowledgments

Samantha M. Clancy
EdrawSoft
Endress+Hauser - Jeremy Farrow
Endress+Hauser - Jerry Spindler
Fluke, Inc.
Ken Haden
Ideal Industries
Industrial Scientific Corporation
Omega
Testo Inc.
T.P. Automation, LLC - Jim Berger

QR Codes

Altek Industries, Inc.
Emerson Process Management
Endress+Hauser
The Engineering Toolbox
Fluke, Inc.
Georgia State University
Hess
Omega Engineering, Inc.
Thermometrics Corporation
Transmation
U.S. Department of Labor Occupational Safety and Health Administration

NFPA 70E is a registered trademark of the National Fire Protection Association, Quincy, MA.

Workgroup

Contra Costa County Electrical JATC
Greg Arcidiacono
Kevin Ryan

Tampa Area Electrical JATC
Dave McCraw

Southwestern Idaho EJATC
Steven Nuxoll
Kelly Lamp

Cedar Rapids Electrical JATC
Jeb Novak

IBEW JATC Local 176
Michael Clemmons
John Warren

Lake County Electricians JATC
Ken Jania

Paducah JATC
Jarrod Shadowen

Ann Arbor Electrical JATC
John Salyer

West Michigan Electrical JATC
David Kitchen

Minneapolis Electrical JATC
Derrick Atkins
Greg Hayenga

IBEW LU 456 JATC
Warren Smith

Canton Electrical JATC
Tim McCort

NECA-IBEW Electrical JATC
John McCamish

Beaumont Electrical JATC
Billy Griffin

Northwest Wash. Electrical Industry JATC
Randy Ambuel
Frank Fagundes
Ed Loughney

Nashville Electrical JATC
Bobby Emery
Stephen Hall

Detroit JATC
Tom Bowes
Marty McLean
Jennifer Smith

Dakotas & Western MN Areawide JATC
Harvey Laabs

Parkersburg Joint App. & Training Committee
Jim Hickenbottom
Joel Thompson

IBEW LU 363 and Hudson Valley NECA JATC
Craig Jacobs

Indianapolis Electrical JATC
James Patterson

Kansas City JATC
John Candillo

About The Author

Paul J. Meyers is an instructor at the Electrical Training Institute apprenticeship program in Indianapolis, IN. He is a Journeyman Wireman in IBEW Local 481 and a graduate of Purdue University with a Bachelor's of Science degree. Meyers teaches courses for all five years of the apprenticeship curriculum, including AC theory, building automation, code calculations, instrumentation, and programmable logic controllers. He directs the instrumentation and building automation labs at the Electrical Training Institute and develops lab manuals for hands-on applications. Meyers' industry credentials include a Master Electrician license, ISA CCST Level III certification, IBEW-UA Instrumentation Test Administrator, LonMark Certified Professional, and BICSI Technician. Meyers and his wife Andra have one daughter, Norah. His interests include electronics, programming, motorcycles, and running.

Partnership

The International Brotherhood of Electrical Workers (IBEW) and the United Association of Journeymen and Apprentices of the Plumbing and Pipe Fitting Industry of the United States and Canada (UA), along with their employers, have operated a successful partnership to supply specialized technicians to industry for more than two decades. This program was developed because it is clear that there is and will continue to be a strong demand for highly skilled instrument technicians, with the development and construction of new power plants as well as all the other facilities that call for instrument technicians. These include pharmaceutical plants, semiconductor and other high-tech manufacturing facilities and sophisticated food processing plants, to name only a few.

This book is the result of a convergence of training information from both the *electrical training ALLIANCE* and the United Association Education and Training Department. Each of these entities actively train their respective members to specific guidelines as set forth within the agreement.

In order to supplement the combination of existing information there was a significant amount of new material which has also been added. Paul J. Meyers (IBEW Local 481) served as the primary author with support from Jason Lunardini (*electrical training ALLIANCE*) and John McFadden (IBEW Local 5) as well as technical review by Bill Boyd (UA Local 597), James Pavesic (UA Assistant Director of Education and Training), and Ken Ravel (UA Local 420).

The project team for this unique endeavor, in which two trade organizations embark on a joint educational project, would like to thank IBEW President Emeritus Edwin D. Hill, UA General President Emeritus William P. Hite, IBEW President Lonnie R. Stephenson, UA General President Mark McManus, and their partner contractor organizations as well as their respective training committees for their continued support of quality training development.

electrical training ALLIANCE
Todd Stafford, Executive Director
The mission of the *electrical training ALLIANCE* (formerly the NJATC) is to develop and standardize training to educate the members of the International Brotherhood of Electrical Workers and the National Electrical Contractors Association, ensuring they are providing the Electrical Construction Industry with the most highly trained and highly skilled workforce possible.

UA Education and Training Department
Chris Haslinger, UA Director of Education and Training
The mission of the UA Education and Training Department is to equip United Association local unions with educational resources for developing the skills of their apprentices and journeypersons. By facilitating the training needs of the membership, we maximize their employability and prepare them for changes in the industry. We are committed to making training opportunities available across North America, allowing members to acquire new skills and remain competitive in the industry regardless of geography. In this way, we are determined to meet the needs of the piping industry and enhance employment opportunities for our members, while remaining fiscally responsible to the beneficiaries of the International Training Fund, which is jointly administered by the UA and our contractor partners.

Instrument Contracting and Engineering Association (ICEA)
The Instrument Contracting and Engineering Association (ICEA) is the contractor association that negotiates and helps administer the "Joint National Industrial Agreement for Instrument and Control Systems Technicians" (the I & C Agreement). In this regard, the ICEA provides the management members of the Arbitration Board and Standing Committee referred to in the I & C Agreement. It also appoints the employer trustees to the Joint Labor / Management Corporation Trust.

Partnership

International Pipe Trades Joint Training Committee, Inc.

Christopher A. Haslinger, *President*, UA Director of Education and Training

Cornelius J. Cahill, *Vice President*, NFSA Representative

Robert Cross, *Vice President*, UA Local 68 Training Director

Tony E. Fanelli, *Vice President*, Canadian Mechanical & Industrial Contractors Representative

Joseph Labruzzo III, *Vice President*, MCAA Representative

Douglas Lea, *Vice President*, UAC/PHCC-NA Representative

W. Thomas McCune, *Vice President*, UA Local 25 Business Manager

Greg G. Mitchell, *Vice President*, UA Local 853 Business Manager

D. David Hardin, *Treasurer*, MCAA Representative

Joseph Shayler, *Secretary*, UA Local 170 Business Manager

Directors

Jason R. Amesbury, UA Local 516 Business Manager

Russell J. Borst, MCAA/MSCA Representative

David Dodd, UAC/PHCC-NA Representative

Julie Henderson, UA Local 177 Training Director

John W. Leen, UA Local 597 Training Director

Rory G. Schnurr, NFSA Representative

Partnership

Members of the IBEW-NECA *Electrical Training ALLIANCE*

Representing the National Electrical Contractors Association

Doug Hague, *Co-Chair,* Shelley Electric Inc.

Geary M. Higgins, *Vice President,* Labor Relations, NECA

John Amicucci, Oneida Electrical Contractors Inc

Luke R. Cunningham, West Side Hammer Electric

Kirk Davis, Bob Davis Electric Company Inc

Robert J. Turner, II, Turner Electric Inc

Dave McAllen, The Superior Group, Inc.

David J. Gill, Metro Electric Inc.

Ron Guarienti, Pueblo Electrics Inc.

Earl Restine Jr., Fuller Electric Corporation

Tom Parkes, O'Connell Electric Company

John M. Grau, *CEO,* NECA

Representing the International Brotherhood of Electrical Workers

Lonnie R. Stephenson, *International President,* IBEW

Jerry Westerholm, *Special Assistant to President, Construction & Maintenance & Business Development,* IBEW

David W. Fenton, L.U. 223, IBEW

Michael J. Gaiser, L.U. 41, IBEW

James R. Evans, L.U. 816, IBEW

David C. Svetlick, L.U. 728, IBEW

Tim Hutchins, L.U. 252, IBEW

Gregory Lucero, L.U. 66, IBEW

Charles Dockham, L.U. 322, IBEW

Marvin Kropke, L.U. 11, IBEW

Howard T. Hill, Jr., L.U. 379, IBEW

Patrick H. Wells, L.U. 347, IBEW

Features

Figures, including photographs and artwork, clearly illustrate concepts from the text.

Blue Headers and Subheaders organize information within the text.

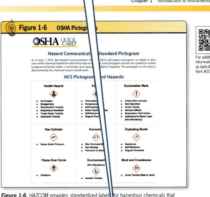

Code Excerpts are "ripped" from *NFPA 70E* or other sources. QR codes provide a link to the full *Code* online

TechTips and ThinkSafe boxes offer additional information related to Instrumentation.

For additional information related to QR Codes, visit qr.njatcdb.org Item #1079

Quick Response Codes (QR Codes) create a link between the textbook and the Internet. They can be scanned using Smartphone applications to obtain additional information online. (To access the information without using a Smartphone, visit qr.njatc.org and enter the referenced Item #.)

Features

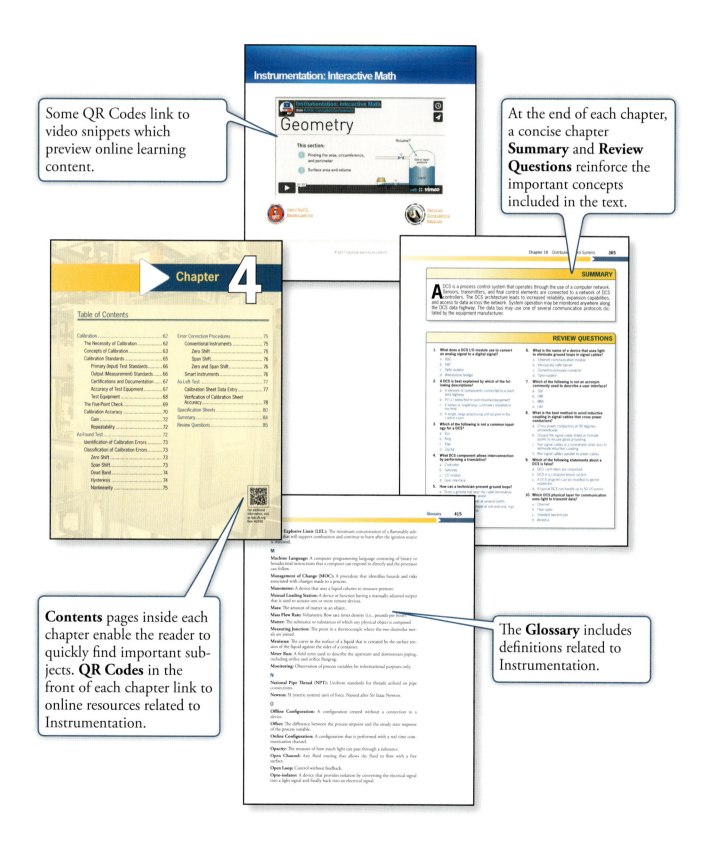

Some QR Codes link to video snippets which preview online learning content.

At the end of each chapter, a concise chapter **Summary** and **Review Questions** reinforce the important concepts included in the text.

Contents pages inside each chapter enable the reader to quickly find important subjects. **QR Codes** in the front of each chapter link to online resources related to Instrumentation.

The **Glossary** includes definitions related to Instrumentation.

Introduction to Instrumentation

The working environment of an instrument technician and the instrumentation systems should create a safer, more effective workforce. Anyone working in the process environment may encounter numerous safety concerns. General safety reminders are just the first step for all projects in all facilities. Technicians should review site-specific safety guidelines for detailed requirements for particular processes. It is also important to know some of the math, vocabulary, and electrical theory involved in instrument and control projects.

Objectives

» Explain in detail the importance of field instrumentation.
» Recall and implement the correct procedures for lockout/tagout requirements.
» Describe hazards associated with process line break procedures.
» Provide a list of requirements to prevent falls on the job.
» Explain the safety concerns associated with working in confined spaces.
» Demonstrate the proper techniques for safe use of ladders.
» Explain work practices and procedures when working near electrical circuits.
» Identify process hazards and explain methods to prevent exposure.
» Formulate the proper answer to math problems commonly used by technicians.
» Solve for electrical variables in direct current circuits.

Chapter 1

Table of Contents

- Purpose of Instrumentation 2
- Typical Working Environment 3
- Instrumentation and Control Safety 3
  - Lockout/Tagout 3
  - Process Line Break 4
  - Fall Protection 5
  - Confined Spaces 5
  - Ladder Safety 6
  - Electrical Safety 6
  - Process Hazards 8
- Math in Instrument and Control Applications 9
- Instrumentation Vocabulary 10
- DC Theory 12
- Summary 16
- Review Questions 17

For additional information, visit qr.njatcdb.org
Item #2690

PURPOSE OF INSTRUMENTATION

Instrumentation is the science of measurement and control to regulate a process. To *control* a process is to cause a system to act or function in a certain way. Instrumentation technologies are examples of attempts by humans to artificially control a process. The regulation of a process can be observed in living organisms in the control of body temperature. As body temperature increases, the body automatically produces sweat. The evaporation of the sweat causes evaporative cooling, reducing the body's temperature. This homeostatic phenomenon, or self-regulating process, is essential to all life forms. Manufacturers involved in industries such as petrochemical, pharmaceutical, or power generation require instrumentation systems to accurately measure and control systems to produce products. The quality of the final product, in addition to the safety of the process, relies on the ability to maintain rigid control of the elements of the process. Instrumentation technicians play an essential role in installing and maintaining the control systems that monitor and control the process.

Field technicians may be asked to troubleshoot, calibrate, loop check, and perform any number of adjustments to instrumentation system components. A large instrumentation project may contain several thousand instruments connected with miles of wires and tubing. An instrument technician must be capable of understanding the working principles of each instrument and performing tasks such as calibration, mounting, terminating, piping, and troubleshooting.

A *control system*, or a system in which deliberate guidance or manipulation is used to achieve a prescribed value of a variable, is shown on various types of drawings. These drawings display location, calibration, type, and other data. The connections made with controllers are called *input and output (I/O)*, defined as the interface between peripheral equipment and digital systems. The technician must understand instruments, control symbols, and identification standards to interpret how the devices are connected to the system and what data are to be sent to the controller. When field technicians understand how a control system functions and how a sensing device gathers information and transmits a signal to the

Figure 1-1. The typical working environment for an instrument technician varies greatly.

controller, they are considered proficient in the fundamentals of how instruments sense and transmit *process variables*, defined as any variable property of a process; the part of the process that changes and therefore needs to be controlled. This is the beginning of understanding the fundamentals of instrumentation.

TYPICAL WORKING ENVIRONMENT

The instrument technician is tasked with installing, replacing, testing, maintaining, and repairing sensitive electrical, mechanical, and pneumatic instruments. The field environment varies greatly. The working environment can range from working within a temperature-controlled, sterile room in a pharmaceutical facility to working outside on a natural gas control station for a thermal oxidizer. **See Figure 1-1.**

A typical day may have a technician begin by performing preventative maintenance on a pH analyzer and then respond to an urgent job order to address why a control valve is nonoperational. Many technicians enjoy the variety and challenging nature of the work. The working environment is often exposed to varying weather conditions, high noise, and hazards from electrical, mechanical, and chemical sources. These factors necessitate the instrument technician to be trained in the identification of hazards and the proper selection and use of personal protective equipment (PPE).

Successful instrument technicians must be able to read, interpret, and analyze a range of literature. The nature of instrumentation projects requires a range of skills. Perhaps the single greatest skill instrument technicians should develop is the ability to continuously learn and expand their knowledge of process control systems. In addition, communication is a critical component when working in cross-functional teams consisting of members with varying backgrounds and expertise. A typical project would involve process engineers, project managers, customer representatives, supervisory staff, and instrument technicians. Accuracy and timeliness of project completion rely heavily on excellent communication skills.

INSTRUMENTATION AND CONTROL SAFETY

Instrumentation safety is overseen by the Occupational Safety and Health Administration *(OSHA)* established by the federal Occupational Safety and Health Act of 1970 to protect the health and safety of U.S. workers. OSHA's mission statement declares that the act is intended "to assure safe and healthful working conditions for working men and women by setting and enforcing standards and by providing training, outreach, education and assistance."

Lockout/Tagout

Lockout/tagout is a system implemented by the technician to reduce the likelihood of injury from exposure to electrical, mechanical, thermal, or chemical energy. The National Fire Protection Association provides standards for such systems through its *Standard for Electrical Safety in the Workplace (NFPA 70E)*.

For additional information, visit qr.njatcdb.org
Item #2046

From *NFPA 70E*, 120.2(B)(1)
(B) Principles of Lockout/Tagout Execution.
(1) Employee Involvement. Each person who could be exposed directly...

For additional information, visit qr.njatcdb.org
Item #2105

From *NFPA 70E*, 120.2(E)(3)
(E) Equipment.
(3) Lockout Device.
(a) A lockout device shall include a lock
...

For additional information, visit qr.njatcdb.org
Item #2108

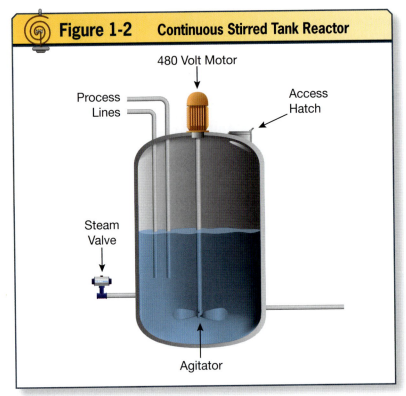

Figure 1-2. Multiple hazards are present in the process control environment. The motorized mixer, process lines, and steam valve of a continuous stirred tank reactor must have zero energy before work begins.

On complex or large projects, there may be multiple people working on the system. It is imperative that all technicians participate in the lockout/tagout procedure. Only the technician who installed and owns the lock may remove that particular lock. A multiemployer or multicraft project requires additional communication to ensure all workers are protected from process energy hazards.

Process Line Break

Process piping and equipment that contains hazardous material or may be under pressure must be opened under the guidance of the proper line break procedure.

From **OSHA 1910.119** *Process safety management of highly hazardous chemicals.*
1910.119(f) *Operating procedures.*
1910.119(f)(4) The employer shall develop and implement safe work practices...

For additional information, visit qr.njatcdb.org Item #2109

Instrumentation systems commonly present multiple sources of energy that could injure the technician. A common example of these potentially dangerous energy sources is seen in a continuous stirred tank reactor. **See Figure 1-2.**

In a continuous stirred tank reactor, steam under high pressure is used to heat a product to a desired temperature. The top of the tank has a motor to drive the mixer, which is powered by a 480-volt electrical supply. The steam valve that regulates the amount of steam allowed to pass through the valve is powered by a pneumatic actuator. Process lines containing hazardous chemicals are connected to the top of the tank. All three of these systems must be reduced to their zero energy state before work begins. The technician working in this environment must implement the proper lockout/tagout procedure before beginning work on this system.

From **OSHA 1910.146** *Permit-required confined spaces.*
1910.146(b) *Definitions.*
"Line breaking" means the intentional ...

For additional information, visit qr.njatcdb.org Item #2110

After the lockout/tagout procedure is implemented to prevent further hazards from entering the system, the technician may be required to remove equipment or open a line. Even after lockout/tagout is implemented, dangerous pressure or a hazardous material is often present in the process. Block

ThinkSafe!
All forms of energy must be isolated. Electrical power is the most common, but other areas, such as pneumatic, steam, and hydraulic sources, must also be considered.

and bleed valves are one means of relieving pressure in a process pipe, but the facility procedures must always be followed before releasing any substance to the atmosphere. When a process line must be opened, a safety data sheet (SDS; formerly known as a material safety data sheet, or MSDS) can be used as a source of hazardous material handling information. The proper PPE must be worn to prevent exposure.

Fall Protection

Falls accounted for 15.2% of all work-related deaths, according to the 2012 Census of Fatal Occupational Injuries, whose revised numbers were released by the U.S. Bureau of Labor and Statistics in 2014. Instrumentation technicians are often required to work on elevated surfaces or near openings to levels below the work area. A proper fall protection plan should include a prejob analysis of any fall risks, proper permitting to perform the work, proper selection of all PPE in case of a fall, and a rescue plan in case a fall occurs.

> From **OSHA 1926.501** **Duty to have fall protection.**
>
> **1926.501(b)(1)** *Unprotected sides and edges.* Each employee on a ...
>
> **1926.501(b)(4)** *Holes.* **(i)** Each employee ...
>
> **1926.501(b)(8)** *Dangerous equipment.* **(ii)** Each employee 6 feet (1.8 m) ...

For additional information, visit qr.njatcdb.org Item #2111

Technicians must inspect the fall arrest system for proper operation and ensure the equipment is in good working condition. Equipment that shows signs of damage or improper use must be tagged out of service and not used. Technicians should follow all manufacturers' instructions on the proper inspection and implementation of fall protection equipment. An *anchorage point*, a secure point of attachment for lifelines, lanyards, or deceleration devices, must be sufficient to withstand the force of a fall. Guardrails and process piping are often poor choices for attaching a fall protection system. A proper fall protection plan provides procedures to protect workers against injury from falls.

Confined Spaces

Instrument technicians may have to enter confined spaces to perform new construction or maintenance work in process environments.

> From **OSHA 1910.146** **Permit-required confined spaces.**
>
> **1910.146(b)** *Definitions.*
>
> "Confined space" means a space that:
>
> (1) Is large enough ...

For additional information, visit qr.njatcdb.org Item #2110

Storage tanks, boilers, manholes, enclosed spaces with process lines, and air handlers are just a few examples of spaces the technician may enter to perform work. A confined space presents numerous, serious hazards to technicians. These include trip or fall hazards, oxygen enrichment or deficiency, hazardous chemicals, entanglement and entrapment, exposure to mechanical or chemical hazards, and an explosive atmosphere. Before work begins, a thorough review of all possible hazards should be completed. This review should identify the proper method to mitigate the hazard, as well as emergency procedures in case of unforeseen hazards. A technician should never enter a confined space without an attendant outside the space. This attendant must remain in communication with the entrant and monitor the confined space hazards. In case of an emergency, it must be possible to retrieve the occupant without requiring entry

For additional information, visit qr.njatcdb.org Item #2048

ThinkSafe!
Continuous atmospheric monitoring must be used in all confined spaces.

> **ThinkSafe!**
> Never enter a confined space alone. An attendant should always be present outside the confined space.

by another person. Too often, an incident in a confined space causes the loss of multiple lives because additional, outside personnel enter the confined space and succumb to the hazardous environment. A lifeline attached to the occupant allows retrieval from outside the space.

Ladder Safety

The technician often uses ladders to access process control components that are not readily accessible. Both portable and fixed ladders are common in process environments. Fall protection may be required when working on ladders. Always survey the surroundings for overhead electrical or process lines before erecting a ladder.

For additional information, visit qr.njatcdb.org Item #2112

From **OSHA 1926.1053 Ladders.**
1926.1053(b) *Use.* The following requirements apply to the use of all ladders, including job-made ladders, except as otherwise indicated:
(1) When portable ladders are used ...
(5)(i) Non-self-supporting ladders ...
(5)(ii) Wood job-made ladders ...
(5)(iii) Fixed ladders shall be ...
(6) Ladders shall be used only ...
(9) The area around the top and ...
(13) The top or top step of ...
(15) Ladders shall be inspected ...

Use the following five-step procedure for ladder safety:
1. Select the proper ladder for the job. **See Figure 1-3.** Ladders that are too short or ladders made of conductive material (aluminum) should not be used.
2. Inspect the ladder for damage. Do not use the ladder if signs of excessive wear or damaged components are found.
3. Set up the ladder following the 4:1 ratio. **See Figure 1-4.** Make sure the floor surface is level and the ladder footing is not subject to sliding or moving. If the ladder reaches above a roof, a minimum of 3 feet must extend beyond the roof surface.
4. Climb and descend the ladder cautiously. Always face the ladder and maintain three points of contact. Never carry material up a ladder. Instead, use a hand line after climbing the ladder to hoist material.
5. Never stand on the top step of a stepladder. Do not overextend when working from a ladder.

Electrical Safety

The technician must be aware of electrical hazards present in process environments to prevent serious injury or death.

Figure 1-3   Ladder Capacity Ratings

Ladder Capacity Ratings	
Weight and Rating	**Application**
375-lb Type IAA duty rating	Special duty or professional use
300-lb Type IA duty rating	Extra heavy duty or professional use
250-lb Type I duty rating	Heavy duty or industrial use
225-lb Type II duty rating	Medium duty or commercial use
200-lb Type III duty rating	Light duty or household use

Figure 1-3. *Ladders must be used inside their design parameters for safety.*

From *NFPA 70E,* **110.1(F)**
(F) Electrical Safety Program Procedures. An electrical safety program ...

For additional information, visit qr.njatcdb.org Item #2113

From *NFPA 70E,* **120.1**
120.1 Verification of an Electrically Safe Work Condition. An electrically ...
(1) Determine all possible sources ...
(2) After properly interrupting ...
(3) Wherever possible. visually ...
(4) Apply lockout/tagout devices ...
(5) Use an adequately rated test ...
(6) Where the possibility of ...

For additional information, visit qr.njatcdb.org Item #2114

Electrical currents are the number of electrons that flow through a conductive material. An *ampere* is defined as 6.25×10^{18} electrons per second flowing through a conductor. An electrical current of 50 milliamperes direct current (DC) or 0.050 amperes DC (a cell phone charger typically draws more current) is sufficient to cause fibrillation of the heart that can lead to death. Electricity exposure can also lead to burns and nerve damage. The technician should always assume a circuit is energized and take appropriate precautions. When working around electricity, conductive material such as jewelry, watches, aluminum ladders, or metal tape measures can provide a path for electricity to enter the body.

When electricity enters the body, the three primary variables that determine the level of damage are as follows:

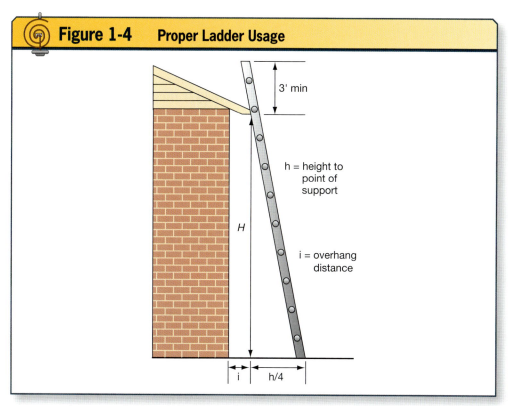

Figure 1-4. Installing a ladder at the proper slope and ensuring the ladder projects above the roof a minimum of 3 feet are important for ladder safety.

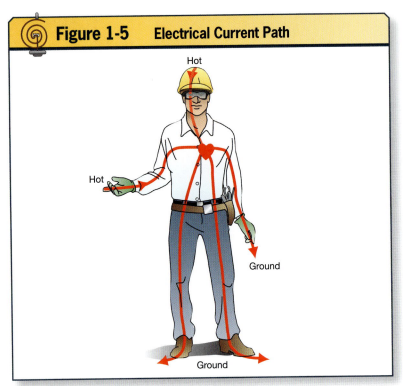

Figure 1-5. The current path through the body can cause severe damage.

ThinkSafe!
Never work on process pipes or storage vessels without understanding the dangers of the chemicals they contain.

special precautions to prevent alterations to the integrity of the system.

All circuits must be tested for the presence of electricity using proper test instruments before beginning work. The employer or facility should implement a procedure to assess the electrical hazards and protect workers against these hazards. These steps include verifying the circuit is deenergized, draining residual electrical energy, employing a lockout/tagout procedure, and verifying the lockout/tagout procedure has eliminated electrical energy. All technicians working on the system should participate in the procedure to reduce the risk of electrical hazards.

Process Hazards

The instrument technician's working environment can present a large number of hazards that vary depending on the materials, equipment, and processes that occur within the facility. Technicians often enter facilities with hundreds of hazardous chemicals in storage or undergoing process conversions. The employer is required by OSHA to perform a process hazard analysis. This information is used by technicians to develop safe working practices when they are required to work near the process.

1. Path—The route the current flows through the body. Current across the chest can damage the heart and bodily organs. **See Figure 1-5.**
2. Intensity—The amount of current flowing through the body. Higher currents typically lead to increased injury.
3. Duration—The amount of time the body is exposed to the current. Longer durations typically lead to increased damage.

Locations that contain or are likely to contain flammable or hazardous substances may employ a special, energy-limiting circuit design called intrinsically safe. *Intrinsically safe* refers to a circuit design that does not produce any spark or thermal effects under normal or abnormal conditions that may ignite a specified gas mixture. The design of the intrinsically safe circuit prevents the possibility of an arc or hot surface with sufficient energy igniting the hazardous atmosphere. Technicians must be aware they are working on an intrinsically safe circuit and observe

From **OSHA 1910.119** Process safety management of highly hazardous chemicals.
1910.119(e) *Process hazard analysis.*
(1) The employer shall perform ...
1910.119(h) *Contractors.*
(2) *Employer responsibilities.* **(ii)** The employer shall inform contract ...

For additional information, visit qr.njatcdb.org Item #2109

An SDS provides specific information concerning the substance identified in the process hazard analysis. Details included in the SDS include the

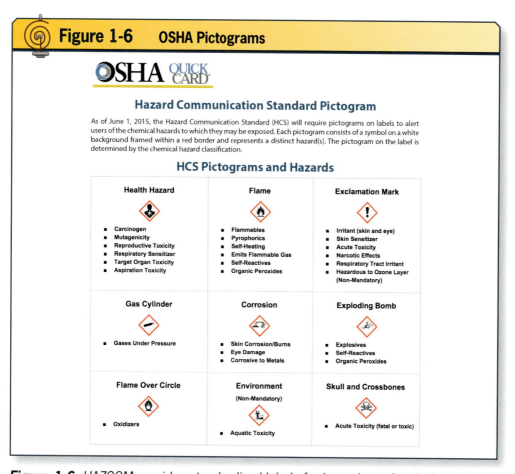

Figure 1-6. HAZCOM provides standardized labels for hazardous chemicals that contain both images and signal words.

For additional information, visit qr.njatcdb.org Item #2049

appropriate first aid for exposure, firefighting, and toxicology information.

In accordance with OSHA's hazard communication (HAZCOM) requirements, standardized labels must be used to identify hazardous chemicals. **See Figure 1-6.**

The labels must include signal words such as "Keep away from heat or sparks" and pictograms that identify the classification of hazards. The standardized labels are part of the United Nations Globally Harmonized System (GHS). These changes help to improve quality and maintain consistency in labeling of chemicals globally. These labels should be used to evaluate hazards and select the proper work procedure and PPE requirements for a specific job task.

All employees working near a process must fully understand the dangers of hazardous chemical exposure. Effective training programs and access to the proper chemical reference materials aid technicians in the safe execution of their work. Safe work practices, evacuation procedures, and evaluation of hazards before beginning work should all be addressed in training. Chemical release because of improper planning or failure to follow proper work procedures can have devastating impacts on human life, the environment, and plant operability.

MATH IN INSTRUMENT AND CONTROL APPLICATIONS

Instrument technicians are required to employ mathematical skills when working on process control systems. Measurement of process variables would be useless without numerical designations with proper units. Several formulas can be used to solve for the unknown variable.

For additional information, visit qr.njatcdb.org Item #2050

A *formula* is an equation used to explain a relationship among variables. For example:

$$E = I \times R$$

where
E = Voltage
I = Current
R = Resistance

By this formula, as R decreases, I must increase for E to remain constant. This formula is called Ohm's Law.

Geometry is the branch of mathematics used to investigate the properties of a shape. An instrument technician may need to find the volume of a cylindrical tank with a diameter of 5 feet and a height of 14 feet to determine how much product is in the tank. **See Figure 1-7.**

For example, consider the formula for the volume of a cylinder:

$$V = \pi R^2 H$$

where
V = Volume
π = 3.141 (a mathematical constant represented by the Greek letter pi)
R = Radius (half the diameter)
H = Height

$$V = \pi \times 2.5 \text{ ft}^2 \times 14 \text{ ft}$$

$$V = 274.84 \text{ ft}^3$$

Unit conversions are often required to present information in a more easily understood or usable manner. For example, a unit conversion table shows there are 7.4805 gallons in 1 cubic foot. Such a conversion can be used to present the answer to the previous problem in gallons instead of cubic feet:

1 cubic foot = 7.4805 gallons
Volume from the previous problem = 274.84 cubic feet

$$274.84 \text{ ft}^3 \times \frac{7.4805 \text{ gal.}}{1 \text{ ft}^3} = 2{,}055.94 \text{ gal.}$$

A solid understanding of algebra is necessary to perform calculations involving process variables. Geometry is also required for determining area and volume calculations. A fundamental understanding of calculus is useful when discussing process dynamics and controls.

INSTRUMENTATION VOCABULARY

Instrumentation technicians must learn the terminology of the systems in which they interact. Successful project management requires consistent communication among construction managers, project engineers, technicians, quality control, and customer representatives. Understanding specific terminology is essential for people in these various occupations to discuss installation and maintenance of the systems.

A *process* is the variable for which supply and demand must be balanced. It is a physical or chemical change of matter and/or conversion of energy. The process environment depends on *instruments*, a term used broadly to describe any device that performs a measuring or controlling function. The process variables measured by the instrument may be used for indication, recording, or control. Pressure, level, flow, temperature, and analytical measurements are the most common process variables.

For additional information, visit qr.njatcdb.org Item #2047

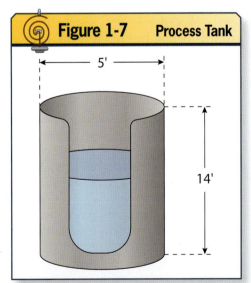

Figure 1-7. A formula can be used to determine the volume of a cylindrical tank with a known height and diameter.

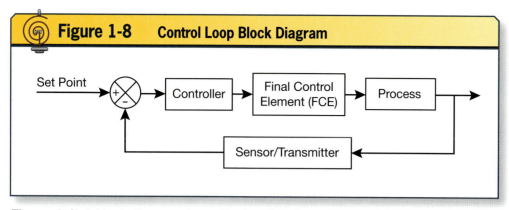

Figure 1-8. A block diagram shows the functional components of a control loop.

For any control system to function, accurate measurements are required that use a *primary element*, the part of a loop or instrument that first senses the value of a process variable and assumes a corresponding, predetermined, and intelligible state or output. A primary element is also known as a detector or sensor. For example, the fuel level in an automobile tank is sensed by the primary element, commonly a float. A transmitter must then send the information to the appropriate device—in this case, the dashboard-mounted fuel indicator. A *transmitter* is a device that senses a process variable through the medium of a sensor and converts the input signal to an output signal of another form. The output of the transmitter is a steady-state value that varies only as a predetermined function of the input (process variable). The sensor may or may not be integral with the transmitter.

The *signal* is information in the form of a pneumatic pressure, an electric current, or mechanical position that carries information from one control loop component to another. The most common signal is a 4- to 20-milliampere DC electrical or mechanical signal or a 3–15 pounds per square inch pneumatic signal. The primary element and transmitter cannot make a control decision based off of the information that has been measured. This decision-making role is played by the *controller*, a device having an output that changes to regulate a controlled variable in a specific manner. The controller is the device that processes the input signal and, based on the desired control parameters, generates an output. The *set point* is the desired value where a process should be maintained. The value may be manually set, automatically set, or programmed. It is expressed in the same units as the controlled variable. The set point is entered into the controller through an operator interface or directly through a physical keypad on the controller. The output signal is also commonly 4–20 milliamperes DC or 3–15 pounds per square inch and is sent to the *final control element*, the component of a control system that directly regulates the flow of energy or material to the process. It is the device that directly controls the value of the manipulated variable of a control loop. Often, the final control element is a valve.

This means that two or more devices make up the *control loop*, defined as a signal path in which two or more instruments or control functions are arranged so that signals pass from one to another for the purpose of measurement and/or control. **See Figure 1-8.** This process automatically repeats.

Fact

Some older instrumentation systems use a standard signal of 10-50 milliamperes DC.

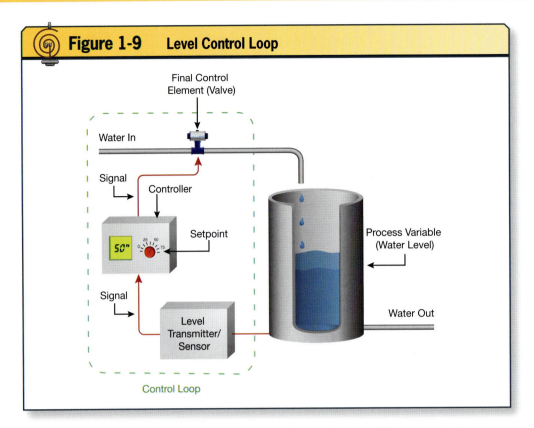

Figure 1-9. Components in a control loop work together to achieve a control objective.

These four devices are the primary element, the transmitter, the controller, and the final control element. **See Figure 1-9.**

DC THEORY

Electrical current is the flow of an *electron*, a negatively charged subatomic particle, through a conductive material. Electricity cannot be seen, but it can be used to spin a motor shaft or cause a lamp to produce usable light. Like charges (+ and +) repel, while unlike charges (− and +) attract. DC describes electricity that flows in a single direction. Alternating current describes electricity that flows in more than one direction.

An instrument technician must understand three variables that exist in all electrical circuits. These variables are voltage, current, and resistance. The *electromotive force (EMF)* is the electrical pressure that exists in a circuit; also called voltage. Put a different way, voltage is the difference in electrical potential between two points. Electrons in a circuit travel from areas of excess electrons to areas of fewer electrons. The voltage measures the difference in pressure between the two areas. The greater the pressure difference, the higher the level of voltage that is measured.

Current in an electrical circuit, given in amperes DC, is defined as the number of electrons that flow through a circuit in 1 second. Current cannot flow without a continuous path. In a DC circuit, current flows from areas of higher electron concentration (negative side of the battery or power supply) to areas of lower electron concentration (positive side of the battery or power supply).

Resistance is the opposition to current flow in an electrical circuit. All conductors have inherent resistance. As a conductor's diameter increases, its resistance decreases because of the increased cross-sectional area. Subsequently, as a

For additional information, visit qr.njatcdb.org Item #1735

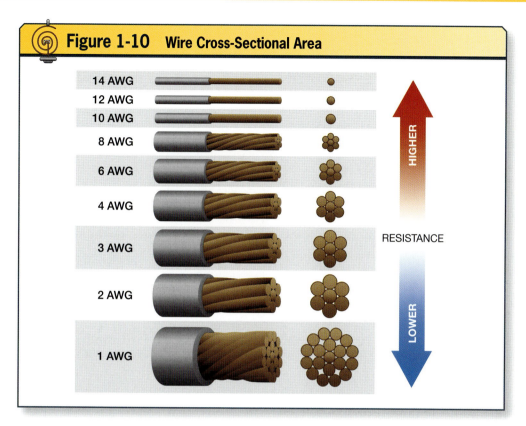

Figure 1-10. Wires with small cross-sectional areas have higher resistance than wires with large cross-sectional areas. Similarly, smaller-diameter pipes have higher resistance to fluid flow than larger-diameter pipes.

conductor's cross-sectional area decreases, the resistance increases. Resistance is expressed using the unit ohm, in tribute of Georg Ohm, who discovered the relationship among voltage, current, and resistance. Resistance through a wire can be compared to increasing or decreasing a pipe's diameter and examining the ability of fluid to flow through the pipe. **See Figure 1-10.**

Ohm's Law explains how these three variables interact in a circuit. When any two of the variables are known, the value of the third can be deduced mathematically. Take the following example:

$E = I \times R$

$I = 4 \text{ mADC } (0.004 \text{ ADC})$

$R = 250 \, \Omega$

$E = 0.004 \text{ ADC} \times 250 \, \Omega$

$E = 1 \text{ VDC}$

The technician would measure a voltage drop, or difference of potential, of 1 volt DC when using a voltmeter across the resistor. **See Figure 1-11.** Again, voltage is the difference between two points along an electrical circuit.

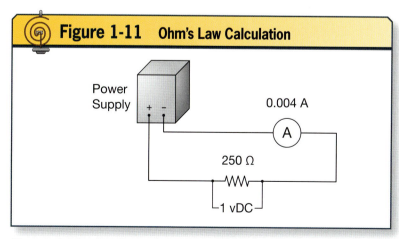

Figure 1-11. A voltage drop can be determined when the current and resistance are known.

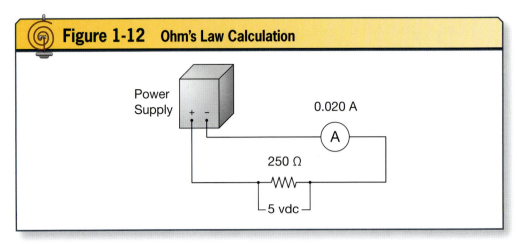

Figure 1-12. *Increasing the current through a resistor results in an increased voltage drop.*

The result changes when the current is increased from 4 to 20 milliamperes DC:

$$E = I \times R$$

$$I = 20 \text{ mADC } (0.020 \text{ ADC})$$

$$R = 250 \text{ }\Omega$$

$$E = 0.020 \text{ ADC} \times 250 \text{ }\Omega$$

$$E = 5 \text{ VDC}$$

In this case, the technician would measure a voltage drop of 5 volts DC across the resistor. **See Figure 1-12.**

Many devices, such as strip chart recorders or proportional–integral–derivative controllers, use a 250-ohm resistor to convert a 4- to 20-milliampere DC current into a voltage reading in the range of 1–5 volts DC. Multiple resistances in a circuit combine in different ways to affect the total circuit current. A series circuit has a single path for electrons to flow. In a series circuit, multiple resistances add together to form a total resistance, called R_T. **See Figure 1-13.**

Consider an example that uses the formula for resistance in a series:

$$R_T = R_1 + R_2 + R_3 + R_4 + \ldots R_N$$

$$R_T = 25 + 30 + 50 + 35$$

$$R_T = 140 \text{ }\Omega$$

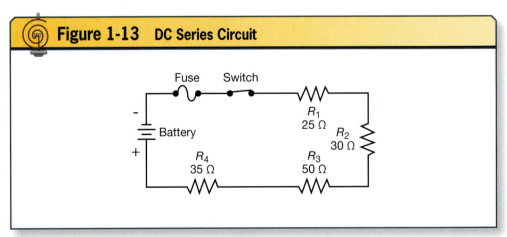

Figure 1-13. *To solve for total resistance, the resistance values for the four resistors in a series configuration are added together.*

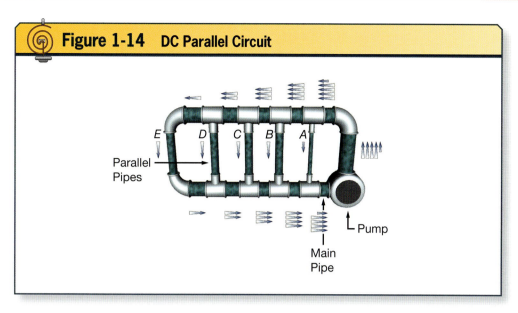

Figure 1-14. A water analogy helps in visualizing a parallel electrical circuit.

A parallel circuit provides more than one path for electrons to flow. **See Figure 1-14.** When provided with multiple paths, current flows on all paths. The adage "electricity follows the path of least resistance" is only partially true. Although more electrons flow on the least resistive branch, current will flow through any complete pathway back to the source. **See Figure 1-15.** In a parallel circuit, the reciprocal formula for parallel resistance must be used to solve for total resistance, again called R_T.

Consider an example that uses the formula for resistance in parallel:

$$R_T = \frac{1}{\left(\frac{1}{R_1} + \frac{1}{R_2} + \frac{1}{R_3} + \ldots \frac{1}{R_N}\right)}$$

$$R_T = \frac{1}{\left(\frac{1}{50} + \frac{1}{25} + \frac{1}{15}\right)}$$

$$R_T = 7.89 \; \Omega$$

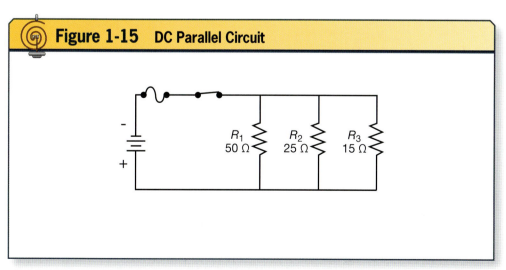

Figure 1-15. Different currents flow in each branch of a parallel circuit if the branch resistances are different.

SUMMARY

Instrumentation is the science of measurement and control. OSHA is tasked with enforcing safe and healthful working conditions in various environments, including those that involve instrumentation and control. Such safety precautions include lockout/tagout, which is used to reduce electrical, mechanical, thermal, or chemical energy to a zero state.

Falls are the number one cause of death to construction workers in the United States. Other hazards include exposure to electrical currents, which can cause death, and mechanical and chemical hazards. SDSs contain necessary information to determine the hazardous nature of a chemical.

Ohm's Law ($E = I \times R$) describes the relationship among voltage, current, and resistance. This and other formulas are useful when performing process variable calculations. Instrumentation vocabulary also must be understood to communicate and interpret information in a process environment, including the terms "series circuit" and "parallel circuit." Both circuits have distinct characteristics that determine the total resistance in the circuit.

REVIEW QUESTIONS

1. **What procedure should be enacted before working on an electrical or pneumatic system?**
   a. Confined space permit
   b. Lockout/tagout
   c. Loop check for continuity
   d. Set point change on the controller

2. **What is the total resistance of a parallel circuit with resistances of 25, 35, and 45 ohms?**
   a. 11 Ω
   b. 25 Ω
   c. 50 Ω
   d. 105 Ω

3. **Who is authorized to remove each lock and tag applied?**
   a. Any technician involved in the lockout/tagout procedure
   b. Only the technician who applied the lock
   c. The customer's representative
   d. The supervisor

4. **What should an instrument technician evaluate before opening a process line?**
   a. The final control element position
   b. The SDSs for the chemicals in the line
   c. The set point of the controller
   d. The setting on the digital multimeter

5. **What is the proper ratio of rise to run for an extension ladder?**
   a. 1:4
   b. 3' above the roofline
   c. 4:1
   d. 45°

6. **Find the volume of a tank with a diameter of 12 feet and a height of 37 feet.**
   a. 444 ft^3
   b. 1,332 ft^3
   c. 4,183.8 ft^3
   d. 16,738.4 ft^3

7. **Find the DC voltage drop across a 250-ohm resistor with a current of 16 milliamperes DC.**
   a. 0.064 VDC
   b. 4 VDC
   c. 15.6 VDC
   d. 4,000 VDC

8. **Convert 4,500 cubic feet to gallons.**
   a. 601.56 gallons
   b. 1,589.36 gallons
   c. 4,500 gallons
   d. 33,662.25 gallons

9. **Which of the following does not determine the severity of an electrical shock?**
   a. Current path
   b. Duration of shock
   c. Intensity of current
   d. Temperature

10. **This vocabulary term is defined as "a system in which deliberate guidance or manipulation is used to achieve a prescribed value of a variable."**
    a. Control system
    b. Controller
    c. Instrumentation
    d. Signal

Fundamentals of Process and Control Systems

All control systems can be broken down into specific devices and loops that measure and control the operating system. The function of a control system can be analyzed by first identifying its associated components, which include process loops, control devices, and measured variables. By identifying the basic control devices and loops, the operation of the system as a whole can be understood. Process control is a method for maintaining an efficient, automatic procedure of operation that produces a specific end result. Most people work with the basics of process control throughout their daily lives.

Certain information is needed to comprehend the devices used within a control system and describes its basic configuration. It is also important to understand the individual elements that are used together in process control to achieve the targeted objective.

Objectives

- » Describe the purpose of a control system.
- » Define automatic control.
- » List process variables that can be measured in a control system.
- » Describe common control signals.
- » Define analog, binary, and digital signals.
- » Identify an operator's control interfaces.
- » Recite and define the five elements of a control system.
- » Explain the operation of a transmitter.
- » Discuss the relationship of a process control loop to the process system.
- » Give an example of transient response.

Chapter 2

Table of Contents

- Historical Background 20
- Defining Control Systems 21
  - Manual Control 21
  - Automatic Control 22
- Defining Process Variables 23
- Control Signals 24
- Signal Types 27
  - Analog Signals 27
  - Binary Signals 27
  - Digital Signals 28
  - Wireless Signals 28
  - Units, Measurements, Standards, Signals, Conversions, and Tables 28
- Operator Interface 28
  - Graphic User Interface 28
  - Control Devices 29
  - Shared Control 30
- Components Of A Loop 30
  - Systems and Applications 32
  - Primary Element 32
  - Transmitter 33
  - Controller and Transducer 34
  - Final Control Element 36
  - Process Loops 37
- Summary 38
- Review Questions 39

For additional information, visit qr.njatcdb.org
Item #2690

HISTORICAL BACKGROUND

In the not so distant past, to fill a tank to a certain level, an operator would watch the rising liquid within the tank. After the desired level was achieved, the operator would manually halt the filling process. This was the first form of process control, which could be characterized as manual control.

In today's world, the intent of such process control is often to perform the filling process automatically, without human intervention. To achieve this, the control system consists of instruments and devices that initiate, record, and regulate the filling process of the tank. In other words, the instruments and devices in the process system are used to measure and regulate what is happening throughout the process. For example, one set of instruments could be measuring the liquid level in the tank, the flow of liquid into the tank, and the temperature of the liquid within the tank. These would be categorized as the primary elements, or sensing elements, of the control system. A corresponding signal would then be generated and transmitted as an input to the controller.

The controller, essentially a computer, would take the place of a human operator and make the decisions necessary to ensure an efficient and safe process. Output signals from the controller then would be sent to a separate set of instruments and devices, categorized as the final control elements. This set of devices exerts control over the many variables present within the process system. The measured variables, the controller, and the final control elements provide the necessary means to allow automatic control of a process.

To gain an understanding of control systems and their intricacies, the technician first must understand the structuring of the process control field, which can be defined by three topics undertaken by people in three roles. First, the design of the control system is a process usually undertaken by a controls engineer. The engineer's goal is to design a system that can successfully control a process automatically. **See Figure 2-1.**

The engineer is concerned with the overall response of the system, which includes variables such as linear response times, nonlinear response times, expansion coefficients of the process, dynamic responses, software development, and specific element specifications.

Second, a design technologist (sometimes the design engineer) undertakes the responsibility of the design of individual elements that make up the control system. This individual has a good understanding of the system components but may not fully grasp the inherent variables that are included in the control system engineer's responsibility. Third, a process control technician installs, maintains, and performs testing of the elements in a typical control loop. **See Figure 2-2.**

The technician must have a good working knowledge of the components of the control system. The technician is often the first to foresee a practical problem or the need for an enhancement to the system. In some areas, a person who performs this work is required to have a related 2-year associate

Figure 2-1. Complex arrangements of process systems and control devices necessitate proper engineering design and accurate installation requirements.

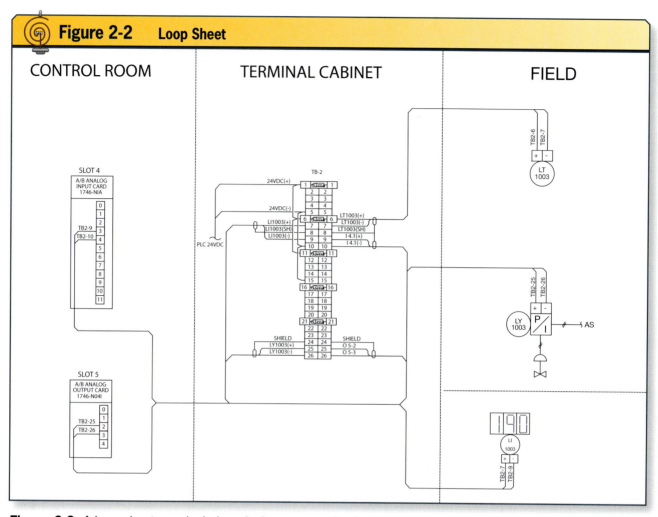

Figure 2-2. A loop sheet may include a device tag, terminations, grounding requirements, input loop, output loop, and local indicator.

degree, which provides extensive training in the technologic applications of instrumentation.

When approaching the subject of control systems, how the installation and maintenance personnel of the control systems can contribute to the working process and the design of the control system is often not covered in detail. However, it is important for the technician to understand how the elements in a system work together to achieve the final result. In this context, a *system* generally refers to all control components, including process, measurements, controller, operator, and valves, along with any other additional equipment that may contribute to its operation.

DEFINING CONTROL SYSTEMS

Control systems are used to regulate a process, monitor a process, or indicate when a process has reached a desired result (i.e., a set point). Control is obtained through any manual or automatic device used to regulate a machine and keep it at normal operation. There are two types of control: manual control and automatic control. With either type of control, the goal is to regulate a process at a predetermined value or set point.

Manual Control

Manual control functions, such as an operator manually filling a tank with

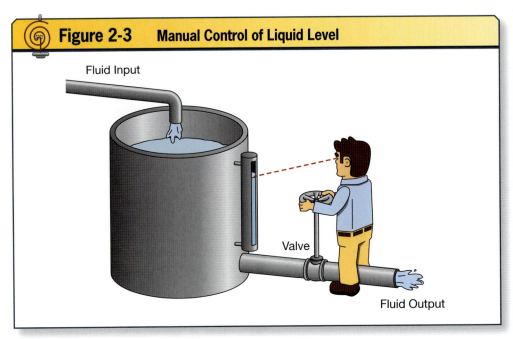

Figure 2-3. A sight glass and the operator's eye form a sensor measuring the liquid level in the tank. The operator bases the decision to manually control the valve on this observation.

liquid, are performed by individuals; in other words, a decision is made by a person. In manual control, an individual gathers information from a sensor, makes a decision, and adjusts the variables of the system to the desired values. **See Figure 2-3.**

Automatic Control

In automatic process control, the goal is to automatically regulate a process at a predetermined value. This is the primary purpose of a control loop. A control loop begins by measuring a process. A signal is then sent to a controller, which determines some form of control through the use of *software*, a collection of programs and routines associated with a computer. A signal then is sent to a final control element, which executes a control action.

The object of automatic process control is to perform the filling process automatically, without help from a human operator. To provide automatic control, the process system is outfitted with instruments and devices to perform the controlling process. Depending on the control variables that are defined, some or all process variables (e.g., flow rate, level, and temperature) may be needed for the controller to make the required decisions to accurately fill a tank to a proper level and maintain this level. The decisions are normally made by an *automatic controller*, a device or combination of devices that measures the value of the variable, quantity, or condition and operates to correct or limit deviation of this measured value from a selected reference. The automatic controller thus regulates the process. The most common method used allows control to be executed by a microprocessor-based controller. **See Figure 2-4.**

Microprocessor-based control systems can execute a user-defined *algorithm*, a detailed set of instructions executed by the central processing unit (CPU), which is the portion of a computer that decodes the instructions, performs the computations, and keeps order in the execution of programs. These instructions initiate and control the various processes. The control system is the collection of components needed to perform the

Figure 2-4. A typical control PLC provides the necessary control algorithms via software to achieve automatic control.

function of maintaining a safe and efficient process. It is defined as the *controlling means*, or the elements in a control system that contribute to the required corrective action, governed by the microprocessor. Depending on the system, the user may define various types of input and output (I/O) to be monitored or controlled by the microprocessor. I/O devices are the field instruments. The device for the *input*—or the incoming signal to a measuring instrument, control units, or system—gathers process data. The device for the *output*—or the signal provided by an instrument; for example, the signal that the controller delivers to the valve operator is the controller output—transmits the data from the controller in the form of electrical signals. The final control elements receive these signal instructions from the controller to modify the process parameters.

DEFINING PROCESS VARIABLES

A process is defined as any function or operation used in the treatment of a material. For example, the operation of adding heat, or thermal energy, to water is a process. Processes are taking place everywhere. Most technicians work with the basics of process and process control every day. Process variables include pressure and the following:
- *Level*—An expression of pressure in terms of the height of a fluid.
- *Flow rate*—The actual speed or velocity of fluid movement.
- *Temperature*—An inferred measurement based on the kinetic energy of molecules within a substance. As kinetic energy increases, temperature also increases. Temperature is expressed in units of Fahrenheit (F), Celsius (C), Rankine (R), and Kelvin (K).
- *Analytical measurement* — The measurement performed via an analyzer that monitors a process for one or more chemical compositions or physical properties.

As a mother drives a car, she is performing process control functions by controlling the speed of the car and its direction through steering. In a building, temperature and ventilation controls provide an adequate environment for comfort with minimal interaction from the workers.

Process control is the method by which a particular process is regulated. People perform process control when they vary the gas flow into an automobile engine and adjust a thermostat for

Figure 2-5 Process Control Components

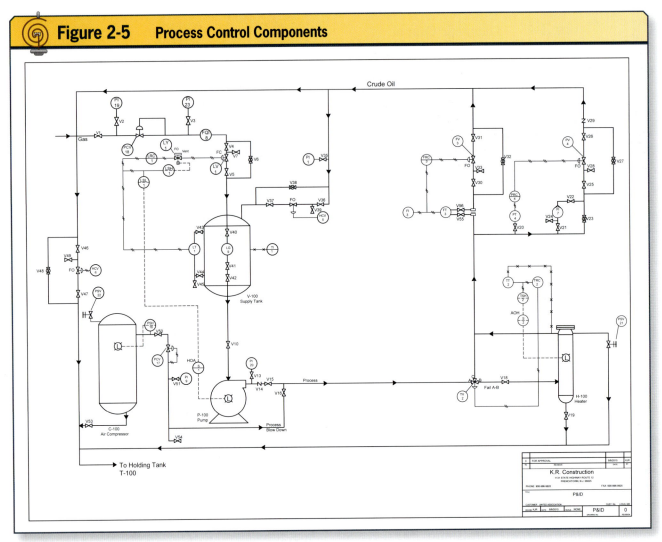

Figure 2-5. A working process can be controlled using various components, as illustrated by the process and instrumentation drawing.

environmental comfort. The concept of process control can be better understood by examining the components that enable someone to control a working process. **See Figure 2-5.**

CONTROL SIGNALS

An operating system depends on the accurate sensing, signaling, and transmission of data. These functions are performed by field devices. A common thread ties together all automatically controlled working processes from initiation to final control. Field devices provide process measurements to the controller, and final control elements receive signals from the controller to regulate the process.

After the I/O are defined and configured in the controller, they are capable of transferring information by a signal from the field devices to the controller and from the controller to the final control elements. The transmission of electronic signals to and from a controller is often accomplished through the use of twisted-pair wiring. Twisted-pair wiring is available in two configurations or types: shielded twisted pair (STP) and unshielded twisted pair (UTP). STP wiring is a physical construction of two current-carrying conductors and a bare ground. The wires are twisted around each other to negate capacitive and inductive effects on the current flowing in the conductors. The twisted pair is surrounded by a shield to further insulate and isolate the conductors. The sole difference between STP and UTP is the presence, or lack of, the shield. The use of either STP or UTP cable is dictated by job specifications. The wiring path creates a signal loop that allows the transfer of information from measuring devices to a controller to the final control element. **See Figure 2-6.**

After a controller performs its I/O processing, the transfer of information between the controller and the display used by operations to control the process may be executed internally or transmitted by means of a dedicated highway interface. Communication lines are often redundant (i.e., there is more than one) to allow the repair and maintenance of one line and to provide a path for checking the information as it is received by the controller. Depending on the level of control, software is used for diagnostic capabilities to ensure the process is performing as smoothly as the controller is directing.

The most important component of information gathered by the controller is the information received from field instruments that determine the appropriate response for the working process. This information is gathered by field-measuring devices that should accurately portray the working process and is transmitted to the controller as a signal. The signals can be in several forms, such as a current in milliamperes direct current (DC), a voltage in volts DC, or a pressure in pounds per square inch and, more recently, through the use of wireless technology. Field technicians should understand the working concepts of a system to know how their devices can interface with and change the operating algorithm of a controlled process. Control systems can contain a variety of devices and controllers that are used to control processes.

Installation, troubleshooting, maintenance, and documentation related to these devices are the responsibility of the field technician. So is *calibration*, the process of adjusting an instrument or compiling a deviation chart so that the instrument output can be correlated to a known standard input value. For field technicians to properly

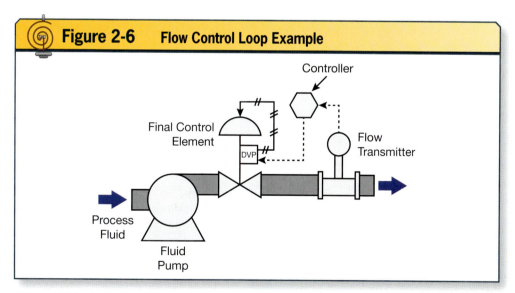

Figure 2-6. A measurement is received by the flow transmitter and converted to a signal, which is sent to the controller. The controller compares this input signal to the programmed set point and generates a corresponding output signal. The output signal is transmitted to the control valve (final control element) to achieve the desired result.

Figure 2-7. An accurate calibration process provides a thorough check of all control elements used during a typical control system project.

calibrate, troubleshoot, and loop-check various systems, they must have a working knowledge of the control systems used in various processes. Each field device used to transmit a variable to the controller must be accurate, underscoring the importance of understanding the steps required for installation and calibration of measurement devices. **See Figure 2-7.**

Inputs and outputs (I/O) must be connected to the controller by field wiring. The field wiring provides the path for the signal to the controller, but the termination point on the controller for each input or output must be located on the proper I/O card. Each signal type must land to the I/O card that is designated for that specific signal type.

A *programmable logic controller* (PLC) is a controller, usually with multiple inputs and outputs, that contains an alterable program. All PLCs have a chassis, power supply, slot, and point assignment to configure I/O, (see figure 2-4) and the termination point must be correctly assigned and terminated to ensure the proper input can be read when it is needed. Consider a single-chassis, three-slot PLC controller. **See Figure 2-8.**

Each slot is configured to perform a different function: digital input, analog input, and digital output. The proper interpretation of documentation ensures that the correct type of input is terminated in its correct location. A field loop sheet can be one source to provide the information needed to terminate I/O to a controller.

After the input is received by the controller, the inputs are converted to a usable form as a series of digital commands. Such a digital command is called a *bit*, short for binary digit, which is the smallest form of information indicated by a 0 or 1. Bits are

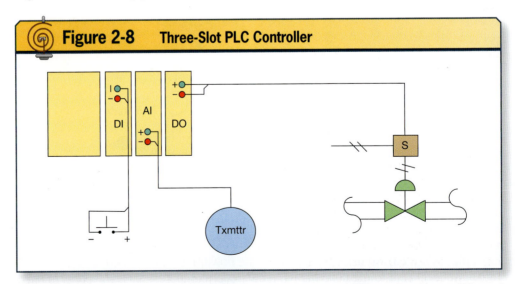

Figure 2-8. A three-slot PLC controller has discrete inputs (DI), analog inputs (AI), and discrete outputs (DO); a solenoid (S); and a transmitter (Txmttr).

packaged into groups as *bytes*, where a byte is an 8-bit representation of a binary character. These are then converted into words. In this context, a *word* is the number of bits treated as a single unit by the CPU. In an 8-bit machine, the word length is 8 bits; in a 16-bit machine, the word length is 16 bits. When the computer wants information it can read, it is in the form of words. A basic order of events needs to be established before work can be performed, and this task is accomplished internally by the CPU.

SIGNAL TYPES

Three typical signals used in process control are analog signals, binary or discrete signals, and digital signals. A microprocessor controller may transmit digital signals without the use of wires.

Analog Signals

An *analog signal* is a continuous operation signal. It is the output of a field transmitter that is continuously measuring a process variable, and it usually represents a range of values read by a *sensor*, the part of a loop or instrument that first senses the value of a process variable, also known as a detector or primary element. This analog signal is usually represented by a variable current (4–20 milliamperes DC), a variable voltage (1–5 volts DC), air pressure (3–15 pounds per square inch), or by other means. Analog signals can be used as inputs, recording a range of values, and can be used in the output signal by varying a position of a final control element. **See Figure 2-9.**

Before an analog signal can be used, it is converted by the controller to a digital format. When devices are calibrated correctly, the process measurement (i.e., input) is accurately represented by the 4- to 20-milliampere DC signal, and the output signal correctly positions an element for control. For example, when automatically filling a tank, the level transmitter is connected to the tank for the purpose of continuously monitoring the liquid level as the tank fills. The transmitter is sending a continuous analog signal to the controller in direct proportion to the liquid level in the tank.

Binary Signals

Another form of signal used is the binary, or discrete, signal. *Binary* is a term applied to a signal or device that has only two discrete positions or states (on/off), and *discrete signal* refers to a control signal that is on or off. A binary or discrete signal is easier to implement but provides less information than the analog signal. For example, with a level switch, the switch contact closes when a certain level in a tank is measured. This is called making, and it allows the signal to flow. The making of the switch contacts is the only indication of what level is in the tank.

A binary, or discrete, input is a signal received by the controller that is on or off. A controller determines whether an

Figure 2-9. A marshalling panel connects field wiring to controller termination points. Inputs and outputs are typically assigned to a rock, slot, box, or point assignment to terminate field wiring.

input signal is received by reading the voltage across the terminals for the discrete input point. If voltage is present, the signal is determined to be on, or high, and it is determined to be off, or low, if the voltage is absent. Such a discrete signal can be used to turn on or off a pump or other on/off devices to initiate control. An example of discrete control is the application of a sump pump with high and low switches that turn a pump on or off, depending on the level in the sump.

Digital Signals

The third type of signal that can be employed by a microprocessor controller is the digital signal. *Digital signal* is a term applied to a signal that uses binary digits to represent continuous values or discrete states. The digital signal can represent a range of values, or it can be a simple on/off reading. The digital signal is a packet of information that contains the process variable information as a series of discrete bits. This series of bits is transmitted to the controller using one of many standardized formats, such as RS-232, RS-422, RS-485, USB, or 1553B.

Wireless Signals

Advances in technology have allowed the ability of wireless signal transmission. Wireless communication provides the ability for the transfer of digital information between points that are not physically connected to each other by electrical conductors. As a result, one of the benefits of the use of wireless systems is ease of installation. In addition, wireless signals are able to travel rather long distances without suffering signal degradation. Because there are no physical electrical connections, the installation process is efficient and provides capabilities not easily accomplished by conventional wired methods. One such example would be temperature measurement within a rotating drum.

Still, wireless systems have downfalls. Because there are no physical electrical connections, wireless transmitters must use their own supply of power, typically a battery pack. Once the onboard batteries have been depleted, there is a loss of power to the transmitter, which means a loss of signal transmission as well.

Units, Measurements, Standards, Signals, Conversions, and Tables

The array of equipment used for industrial process control uses many forms of signals. A summary of these signals would include the following:

- *Electrical*—A voltage, current, or electronic signal. Common signal types are 1–5 volts DC and 4–20 milliamperes DC.
- *Mechanical*—Characterized by force or movement, as in an actuator imposing a force on a control valve assembly to initiate a change in position.
- *Pneumatic and hydraulic*—Characterized by pressure and flow. A common signal type is 3–15 pounds per square inch.
- *Optical*—High-speed digital signal transmission via fiber optics.
- *Radio or wireless*—Transmission via radio frequency (RF).
- *Ultraviolet*—Similar to radio or wireless but only useful over short distances.

OPERATOR INTERFACE

Although the overall intent of an automatic process control system is to operate independently and without human interaction, plant operations personnel must retain the ability to interact with and monitor the process system. This is accomplished through the use of various methods including graphic user interfaces, local and remote control and monitoring stations, and control devices.

Graphic User Interface

After field inputs have transferred information to the controller, operation control monitors allow visualization of the process. In industrial applications, operators use displays on control stations to gather information that allows them to see the process. **See Figure 2-10.**

These displays may be referred to by several names, including graphic user interface (GUI) and human–machine interface (HMI). A process can include many variables that may have the option of being monitored or used for control. *Automatic control systems* are operable arrangements of one or more automatic controllers along with their associated equipment connected in closed loops with one or more processes. With automatic control systems, the I/O can be viewed at various locations on displays by means of communications highways that allow an operator to view multiple processes.

Monitoring is the observation of process variables for informational purposes only. Several locations may have displays for monitoring I/O, or only one or two locations may be available for control functions. An entire process facility often can be viewed from a single machine with multiple views. **See Figure 2-11.**

Control Devices

The components monitored at a video screen are devices used to measure functions or devices that respond to control commands given by the controller. If a video display recorded the revolutions per minute of rotating machinery in a controlled process, the signal received by the controller would have been sent by a measuring device monitoring the rotating machinery. The device used to count the number of revolutions of the rotating machinery is the monitoring device. If the rotating speed of the machinery needed to increase, a signal would be sent to a device that controls the rotation speed. The device that controls the rotation speed is the final control element. These are simplified examples of field devices (or instruments) used in a process control system.

For all processes, the signals used to record, monitor, and control the process are generated by devices that provide an interface with a controller in the working process. This is another reason field technicians must understand how their devices reflect a working process.

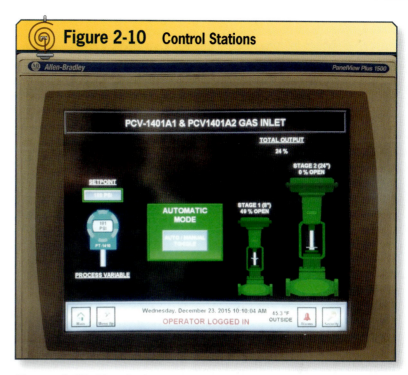

Figure 2-10. The displays on control stations enable easy viewing of the process through information gathered from field devices.

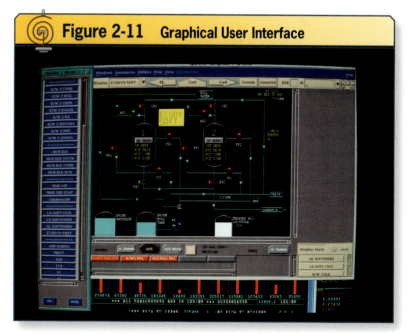

Figure 2-11. Graphical displays enable operating personnel to view and manipulate the process.

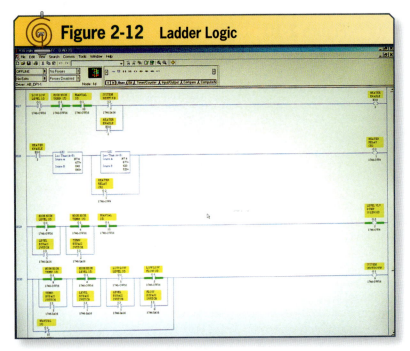

Figure 2-12. Ladder logic is one application of control software that may be used to provide control of a process or multiple processes.

Shared Control

As a necessity, a controller and its associated display sometimes perform the control functions for several processes. This is a controller containing preprogrammed algorithms that are usually accessible, configurable, and assignable, which permits a number of process variables to be controlled by a single device, and it is called a *shared controller*. It contains a user-defined algorithm that is changeable to permit flexibility for the user. The controller's algorithm is roughly a listing of instructions to be executed by the controller, and it often contains diagnostic features to verify information gathered by the system.

A process must be monitored continuously for accurate automatic control. The algorithm contained in the controller determines the frequency at which the various I/O processes are scanned. The scan of a program, or algorithm, is the repetitive reading of the I/O field to execute instructions contained in the program. Flow rates, levels, temperatures, pressures, and other characteristics are read by a controller, and there may be several hundred variables for each system. The outputs of a controller to valves, motor drives, solenoids, and heating elements are equally adjusted or maintained through each scan of the controller's program.

Reading of the process variables, output signal adjustments, diagnostic variables, and system informational statistics can be executed as quickly as a few milliseconds, and the resulting control actions are then taken. One reading and the resulting execution of the instructions entered into the controller are together called a scan. Since the scan time of a program can be fast and control actions are taken from the readings obtained from the scan, it is important to have an accurate sensing and signaling device sending information to a controller. **See Figure 2-12.**

COMPONENTS OF A LOOP

A loop is a combination of two or more instruments or control functions arranged so that signals pass from one

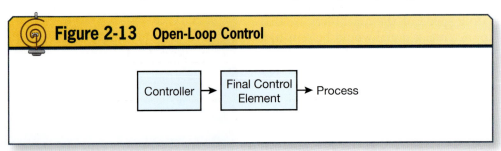

Figure 2-13. The block diagram shows a form of open-loop control. The process has a programmed set point in the controller and a variable is applied to the process, but the effects of the added variable are not sampled.

to another for the purpose of measurement, control, or both of a process variable. A simple representation of a control loop is shown by connecting a controller to a final control element. This creates an open control loop configuration. **See Figure 2-13.**

To provide automatic control, a process is equipped with sensing elements, signaling devices, automatic controllers, transducers, and final control elements. **See Figure 2-14.** A *transducer* is a device that converts information of one physical form to another physical type in its output (e.g., a thermocouple converts temperature to voltage).

The sensing element provides a method to measure the process. When information is gathered by a sensing element, a difference is found between the value of the set point and the value of the controlled variable. This difference is the *deviation*, defined as a departure from a desired value. To eliminate the deviation that exists between the controlled variable and the set point, action must be taken. This action involves a change or adjustment to the manipulated variable, which causes a change in the controlled variable and thus a return to the set point.

The signaling devices send the represented process measurements obtained by the sensing elements as pneumatic or electrical representations of the process to a controller. The sensing elements and the corresponding signaling devices can be thought of as the eyes and ears of the process. The controller, which could be thought of as the brain of the process, performs a calculated response to signals it receives and then initiates an output signal to control the process. The transducers are used when it is necessary to convert the present form of signal to a more readily usable form. For example, an I/P transducer performs the function of converting an electrical current signal (I) to a pneumatic pressure signal (P). In other words, an *I/P transducer* is a device that converts an electric current to a linear pneumatic pressure. The final control elements, which act as the hands of the process system, are used to adjust the process to a point at which the process is determined to be under control.

> **ThinkSafe!**
>
> Never attempt to communicate or perform work on a field device without verifying that the controller is in a safe state of operation (i.e., manual control). Failure to place the controller in the safe state can cause unexpected changes in the field device state, which can cause serious injury to the technician, irreparable damage to the process, or both.

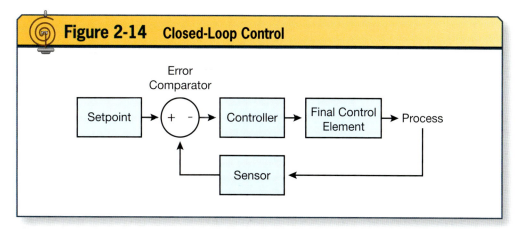

Figure 2-14. After the process variable is measured and compared with the set point, error corrections are adjusted out. This action is performed continuously while the process is in operation.

Systems and Applications

The control of a process system is achieved through the use and interconnection of multiple components. Accurate sensing, signaling, and control of the process depend on the primary element, the transmitter, the controller, the transducer, and the final control element. By gaining an understanding of how these components are related and interconnected via process loops, the technician achieves greater insight into the operation of automatic process control systems.

Primary Element

The primary element, or sensing element, of a process control system is the part of a loop or instrument that is in some form of contact with the process's manipulated variables and thus is the first to sense the value of a process variable. **See Figure 2-15.**

This primary element is contained in the body of the transmitter, but the

Figure 2-15. The primary element of any measurement device is the element that directly senses the desired measurement variable.

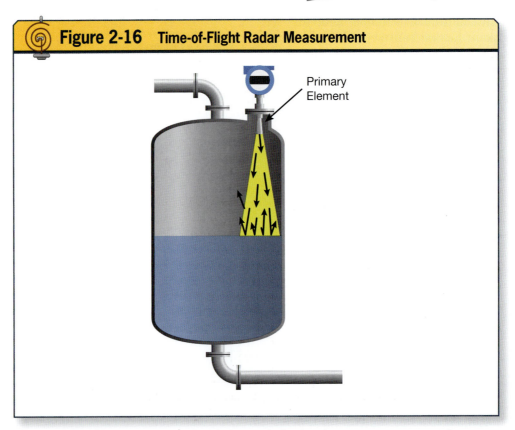

Figure 2-16. Measurements using radar to derive level measurement are unique applications.

sensor may or may not be physically located within the body of an instrument or transmitter. Some radar, infrared, sonic, nuclear, and other methods do not physically come into contact with the process, but examples of primary, or sensing, elements can also be used. **See Figure 2-16.**

Transmitter

The field name "transmitter" often is used as a reference to a device that transmits a signal to a controller, but a transmitter performs two functions. When a process is sensed, data are transmitted to a controller by the transmitter. The correct signal magnitude is determined by the calibration of the device, and an instrument must be calibrated accurately to send the correct signal. **See Figure 2-17.**

The transmitter is often thought of as the beginning of the process loop, but when the process is ongoing, the control loop has no beginning or ending point. The process is continually sensed, analyzed, and controlled. A block diagram demonstrates the process loop as it applies to an operating process. **See Figure 2-18.** A transmitter

Figure 2-17. A transmitter may include a digital display to indicate measured values.

TechTip!

Control systems using field I/O interfaces have many wiring connections. Wiring connection issues are a major cause of error readings on the controllers. Use special care for controller I/O connections to minimize connection errors.

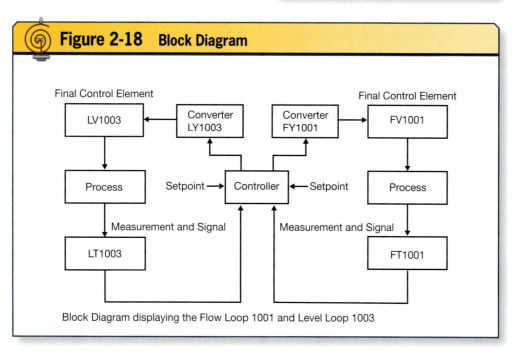

Figure 2-18. A block diagram shows the relationship of the individual instrumentation components.

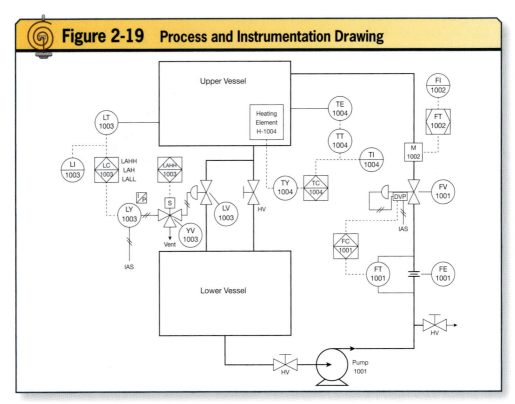

Figure 2-19. A process and instrumentation drawing shows the relationship between instrumentation components and the process equipment and piping.

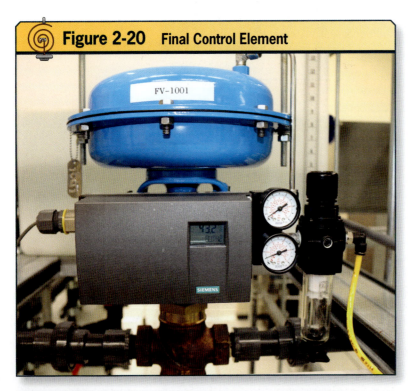

Figure 2-20. A final control element may contain an integral transducer, which converts an electronic signal into a proportional air pressure to operate the control valve.

could be labeled FT1001. **See Figure 2-19.** The transmitter senses a process flow rate through the use of a differential pressure measurement and then signals the controller about the magnitude of the flow rate. The only function the transmitter performs is signaling the process flow rate.

Controller and Transducer

The controller receives the signals from the transmitter. The controller performs predetermined instructions that are entered by the control system engineer. The software instructions are created and written in a manner that allows the controller to calculate and then respond to the signals provided by the transmitter. The controller is continually updating signals by adjusting its outputs to the transducers or final control elements. **See Figure 2-20.**

The controller does not measure the process. The controller can perform its calculations only on data that it gets from the sensing and signaling devices.

> **TechTip!**
>
> The technician should verify that the field device and the controller (e.g., distributed control system) are programmed in the same measurement units and ranges when performing maintenance on a field device. A different set of program parameters in the two devices is one of the most common causes of signal error.

It is therefore easy to conclude that the transmitter performs the critical function of delivering to the controller the information needed to establish control of the system.

The controller tries to establish a form of stable control of the process. The controller is continually comparing the signal received from the transmitter with the set point. This comparison is the result of the instructions written in the controller. The result of the calculations made is then used to determine an output signal that will position a final control element. The goal of the controller is to regulate the process to some value in a *steady state*, a situation in which process conditions have stabilized, or an unvarying condition in a physical process.

Achieving a steady state of process control implies that the process is controlled exactly to the point desired, the set point. The set point and process should be within some steady-state error limits. This error is usually expressed in units of the controlled variable (e.g., flow in gallons per minute).

The steady-state error is a result of the flow measurement, signaling, and controlling devices and includes the physical properties of the process, such as capacity, resistance, and response time linearity. The result is that the errors present in a controlled process may be the result of the process control system. The errors in the process loop can be the result of the individual components.

The process should be under control to the point that a process interruption or upset does not interfere with the automatic response of the system. When a sudden upset occurs that causes a variation in the process and set point beyond the normal steady-state error, the control system reacts by bringing the process back to the set point as instructed. Transient regulation describes the system response to a process deviation.

Transient response is another form of control system action, and it can be described as a process variable that depends on another process control variable and its associated control loop. A level loop is an entirely different loop. Studying the process and instrumentation drawings (P&ID) reveals that the level loop is controlling the liquid level in the storage tank supplied with a liquid through the use of flow loop 1001.

Suppose that a flow deviation occurred, causing a change of flow rate or even no *flow*, defined as the travel of liquids or gases in response to a force (gravity or pressure). The level loop must respond to the process deviation that will be sensed by the level transmitter LT1001. The time required for a deviated process variable to return to within steady-state error limits is known as *transient response*.

The transducer is a device that converts signals from one form to another. A common type is the I/P transducer. **See Figure 2-21.**

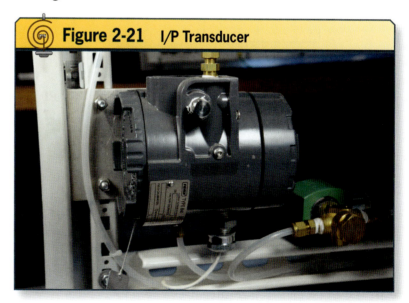

Figure 2-21. *An I/P transducer converts a current to the pneumatic form.*

Transducers provide a method for changing the usable form of signal for one element in the control loop to another form for a different element in the same loop.

Transducers are also susceptible to errors but must be accurate within the tolerances specified. Fortunately, it is easy to calibrate a device of this nature by attaching a gauge to the outlet pressure of the device and adjusting the transducer output as needed to provide the working range of the device.

For example, the current can be simulated at 4–20 milliamperes DC, and the outlet pressure should correspond to 3–15 pounds per square inch. Interpreting these data implies that at 4 milliamperes DC, the outlet pressure should be 3 pounds per square inch and that a 20-milliampere DC signal should correspond to an outlet pressure of 15 pounds per square inch. Different processes have different error tolerances, and those tolerances can be compared to actual readings of an I/P transducer under process operation.

Final Control Element

The final control element is often the control valve that regulates the process. Regulated processes must throttle the process between the minimum and the maximum values obtainable by the system. The control valve responds to a varying controller signal to position itself according to the value prescribed by the controller.

A final control element may be a simple on or off command (i.e., a fully open or a fully closed position) or other binary function. Final control operation may require steps to convert the controller's signal to a proportional valve (or other actuator) position. Several steps and variables can be observed and calculated between the functions of the controller signal and the final control position. **See Figure 2-22.**

Signal conversion is the first step of the final control valve positioning. The transducer receives a signal from a controller (usually a 4- to 20-milliampere DC signal). The transducer must then convert the signal to an equivalent pounds per square inch pressure.

The actuator also acts as a form of transducer, but it is a conversion of one signal type to another signal type. It is a conversion of a signal (usually in pounds per square inch) to mechanical force used to produce torque for valve positioning. The actuator may have many applications for control systems, but the most prevalent is the control valve.

The control valve is the only element of the process loop that can perform a direct action on the process. **See Figure 2-23.** The control valve must not prohibit the process from achieving maximum values, nor can it allow the process to force the valve position to change. Most engineers design the process to be controlled by starting with

Figure 2-22 Process Regulation

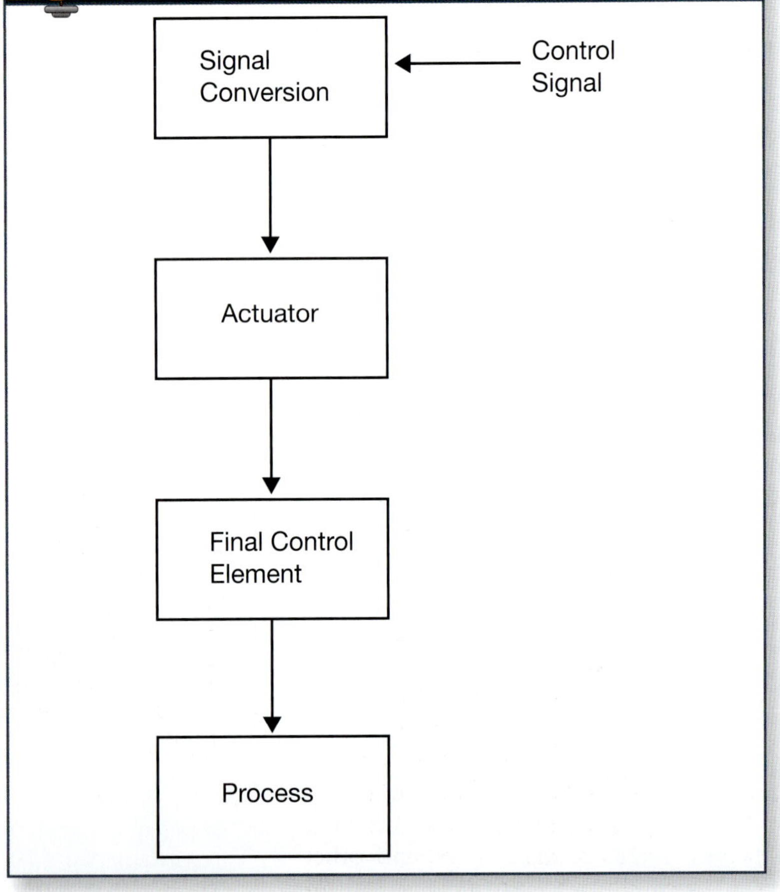

Figure 2-22. There are several steps that must be performed to regulate a process from a control signal.

the control valve and assemble the related components as they are required.

The petrochemical industries have many applications that require control of fluid processes (e.g., gases, liquids, and vapors). Other industries and their facilities, such as pharmaceuticals, chemical plants, and wastewater treatment facilities, also depend on the automatic control of processes for their operation.

Flow control is the primary objective for automatic control processes, and flow rate in process control is usually expressed as a volume per unit time (e.g., gallons per minute). The flow rate in a given process line is calculated to achieve minimum pressure loss, and it is easy to conclude that the control valve regulates flow by controlling the pressure loss in a process line, including the control valve. The relationship between flow rate and pressure differential provides a method for understanding control valve operation.

Process Loops

When each device in the process loop is understood, the concept of the process loop can be studied. The process loop is nothing more than the loop displayed by a block diagram. The two loops in a block diagram operate independently. This is a configuration of elements that is followed for all process control systems.

Loops always have a circular connection, from which the term "loop" is derived. Each loop follows the pattern of seemingly beginning at one element in the loop and proceeding to the other elements in the loop.

The elements in a process flow loop are the elements that need to be adjusted. The transmitter FT1001, the controller, the converter, and the control valve are all included in the loop. The same elements can be traced again to repeat the loop.

If a process deviation occurred in the process flow loop, it is the responsibility of one of the elements in this process loop only. In other words, the error present is not the result of one of the level loop devices. There are exceptions

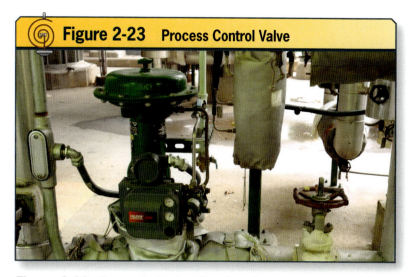

Figure 2-23. The control valve directly adjusts a process and is designed to be an integral part of the process.

to this rule; the controller is a device common to both loops. It is possible that the controller can be the cause of a process deviation or improper control response, but this is uncommon. The probable cause of a process upset is one of the devices in the flow loop. Likewise, a probable cause of a level loop deviation is one of the elements in the level loop. In cases of the level loop, the technician must understand how the process loops interact with each other.

The P&ID indicate that the level loop depends on the flow loop for supplying the process variable (e.g., liquid). If the response of the level loop is not as specified, the technician must be able to interpret the data to determine the cause of the deviation.

How well a process is performing is often a question posed by all three control system personnel: engineer, designer, and technician. Examining two areas can help them determine the answer.

First, compare the process under stable conditions and determine the steady-state error. This error should be within the tolerances specified by the engineer in charge of the design. This is one loop that is being evaluated, but this loop probably affects other loops and other processes downstream and upstream from the evaluated loop. Small errors here may not seem significant, but when

ThinkSafe!
Many controllers are supplied by 120-volt alternating current power, even though they output low-voltage DC power to the loops. Always wear the proper personal protective equipment to protect against electrical shock when working with controllers.

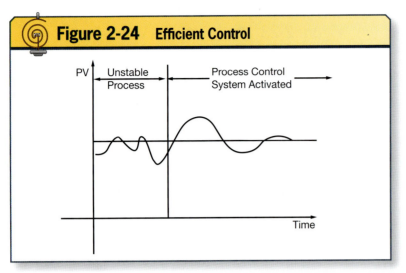

Figure 2-24. *A correctly designed and installed control system brings the system under control within a specified period and stabilizes it within a specified error tolerance. PV is the process variable.*

they are added to others and process action, they may cause an end product that does not meet specifications.

Second, evaluate the response of the loop in question when set point changes or other process upsets occur. Usually, many criteria determine whether the loop transients (regulated and response) provide the correct method and time to bring the loop under control as efficiently as possible. **See Figure 2-24.**

Loop tuning is the term used to describe the action needed to accomplish this goal. The controller responds to these variations in the process in a predetermined manner, and the transmitters provide the method for the controller to sense the process.

All workers in the instrumentation and controls field must understand how devices interact and work as a system of control. It is equally important to understand how each device performs its specified function to contribute to the proper working order of the loop.

SUMMARY

Instrumentation is the method by which a process variable is obtained in a usable form to position a final control element and thus control a process with minimum deviation. Analog, discrete, and digital signals are used in process control, and various control devices are employed to monitor systems. Identification of field devices is critical to the installation and maintenance requirements associated with field I/O.

Automatic process control equipment consists of sensors, signaling devices, automatic controllers, transducers, and final control elements. Sensors detect characteristics of the process media, such as flow rate. Signaling devices convert the process characteristics to a form that can be acted on by the controller. The controller reads electrical signals from signaling devices and performs logical operations based on those signals and on preprogrammed instructions. The controller then sends electrical signals directly to final control elements or transducers to control the process. The transducer (when required) changes electrical instructions received from the controller to a form of energy required to operate the final control element. The final control element converts the signal received directly from the controller or from a transducer into an action. A process system may require many separate control loops. Some of these control loops may be interdependent, whereas others are independent of one another.

By understanding the control system, technicians may begin to apply their knowledge to device checking, loop tuning, and adjusting controller operation parameters to achieve a stable, controlled, efficient process system.

REVIEW QUESTIONS

1. Where in the physical process would a primary element be located?
   a. In contact with the process
   b. Mounted in a marshalling cabinet
   c. Mounted in the main control room
   d. Mounted to an I/P transducer

2. What type of control signal is characterized as an on/off signal?
   a. Analog
   b. Digital
   c. Discrete
   d. Wireless

3. Automatic control systems relieve the operator of the responsibility to maintain a process.
   a. True
   b. False

4. Process variables in a system requiring control include which of the following?
   a. Density
   b. Flow
   c. Specific gravity
   d. Volume

5. How are inputs and outputs typically connected to the controller in a process system?
   a. Peer-to-peer links
   b. RS-232 wiring
   c. Serial cables
   d. Twisted-pair wiring

6. What type of signal is represented by a packet of information containing a series of discrete bits?
   a. Analog
   b. Binary
   c. Continuous
   d. Digital

7. What type of signal can be used as inputs to record a range of values and as outputs to position a final control element?
   a. Analog
   b. Digital
   c. Discrete
   d. Control

8. Which of the following is a process?
   a. The control actions exerted upon a loop
   b. The entire plant
   c. The final control elements
   d. The system being measured and controlled

9. A control valve is considered to be which of the following elements in a process?
   a. Final control element
   b. Primary element
   c. Transducer
   d. Transmitter

10. A control loop that uses float switches to turn a pump on and off is referred to as which type of control?
    a. Analog
    b. Discrete
    c. Manual
    d. Set point

Instrumentation Symbols and Diagrams

Effective methods of communication are necessary to achieve desirable and consistent results, whether from human to human, machine to machine, or reciprocity among humans and machines. This cannot be overly stressed in the realm of instrumentation and control systems. These systems range from small groupings of components performing clear, simple functions to hundreds or thousands of devices performing complex functions. Each device is carefully selected to perform a specific function, and two devices that appear to be the same often perform vastly different functions. The written language of instrumentation and control systems is used to communicate detailed information specific to the devices being calibrated. A thorough understanding of the various identifiers and documents unique to this work is essential. The information contained herein is based on widely recognized standards developed by various governing bodies, along with industry-accepted practices.

Objectives

- » Describe the typical drawings and documents used for instrumentation.
- » Identify the instrument and function symbols used on instrumentation drawings.
- » Use a device tag number to identify a device's function and loop number.
- » Interpret typical letter combinations for tag numbers.
- » Identify the line symbols shown on instrument drawings.
- » Identify symbols used to identify process equipment.

Chapter 3

Table of Contents

Instrument Symbols and Identifiers 42
    Lines and Instruments 43
    Instrument Identification 48
    Loop Numbers.. 50
    Equipment Symbols and Actions 52

Instrumentation Documents 53
    Process and Instrumentation Drawings 53
    Loop Sheets 54
    Wiring Diagrams 55
    Spec Sheets 56
    Instrument Index Sheets 56
Summary 59
Review Questions 59

For additional
information, visit
qr.njatcdb.org
Item #2690

INSTRUMENT SYMBOLS AND IDENTIFIERS

Instrumentation symbology is not unlike other types of symbols used within industrial documentation, where a graphical representation indicates a complex device. While the style and design of some symbols may be unique to a facility, a process, or even a designer's preference, they generally follow the guidelines of governing bodies such as the International Society of Automation (ISA), American National Standards Institute (ANSI), and International Organization for Standardization (ISO). The ANSI/ISA S5.1 standard is one such widely accepted guideline.

Symbols and their associated tags or identifiers allow devices to be tracked from drawings to specification (spec) sheets and beyond. First, a *process flow diagram* (PFD), a diagram commonly used to indicate the general flow of plant processes and equipment, is developed based on the intended process. Then, process and instrumentation drawings *(P&ID)* are developed to place the instrumentation components in relationship to other equipment. **See Figure 3-1.** To properly specify, procure, install, and maintain the proper types of instruments based on function, application, and environmental concerns, spec sheets, installation detail sheets (detail drawings), and instrument indexes are

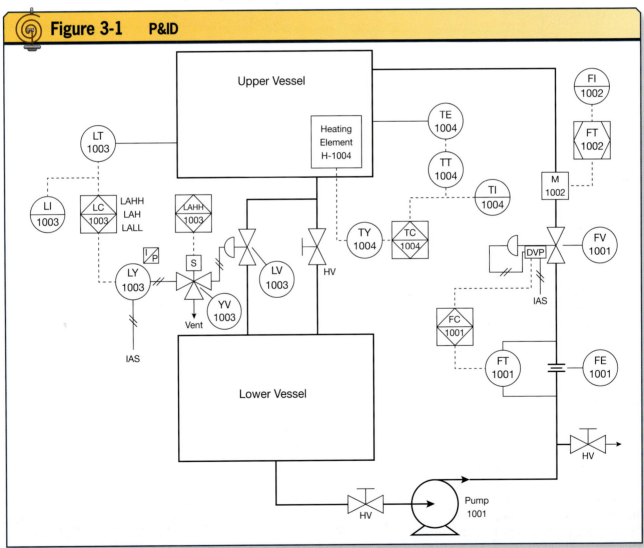

Figure 3-1. The P&ID shows all related components.

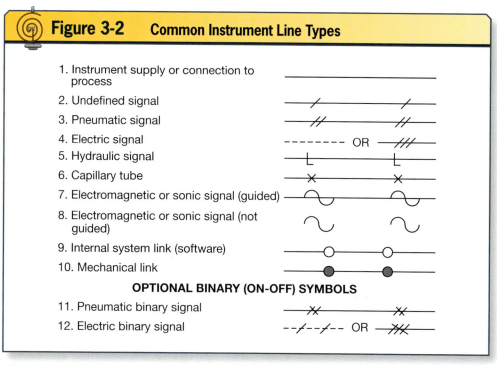

Figure 3-2. A variety of specialized markings identify the line function.

used to provide order to the array of information and options for these devices. Other diagrams, such as loop sheets and wiring diagrams, provide wiring interconnection details and signal paths.

Lines and Instruments

To understand how instrumentation processes work or how a system will function, it is necessary to interpret the lines that connect a process to an instrument and one instrument to another instrument. This network of symbols and lines can be likened to that of a road map. The major components of the process are similar to major cities, while the thick, heavy lines of a process that carry the process medium resemble the interconnecting roads between cities that carry the bulk of the traffic. The process instrumentation devices, similar to small cities and towns along the way, are indicated by various symbols. These lines are not designed to carry the bulk of the process medium; they are merely designed to be the on- and off-ramps of the process and therefore are shown with a thinner line weight. The line symbols not only indicate how instruments are connected to the process or to one another but also indicate what types of signals are being transmitted from one place to another. **See Figure 3-2.**

Electrical and pneumatic signals are commonplace within instrumentation cabinets.

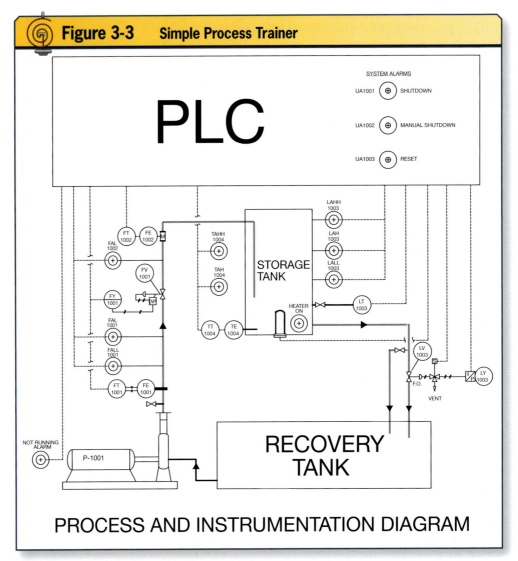

Figure 3-3. The process lines which interconnect the process components and carry the process medium are drawn in the bold line type.

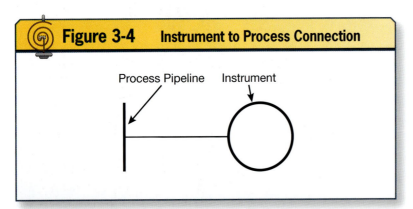

Figure 3-4. The instrument bubble is drawn with a thin line connecting it to the heavier, darkened line that is representative of the process line.

On a P&ID, process lines are shown with thick solid lines. **See Figure 3-3.** A thin solid line indicates an instrument-to-process connection. If a thin solid line extends from a process pipeline to an instrument, this is an indication that the instrument is connected directly to the process or the process piping. **See Figure 3-4.**

Symbols portraying process control devices such as primary elements, final control elements, termination points, and signal types are useful to the instrumentation technician. Symbols can also depict many devices and detailed information that is critical to the control of the process and is used by many people, in addition to the instrumentation technician.

On P&IDs, each instrument is represented by a symbol. The symbol is often a *bubble*, the circular symbol used to denote and identify the purpose of an instrument or function. It may contain a *tag number*, an alphanumeric sequence that identifies a device by unique code. However, in specialized cases, it can be another shape. Not only does a symbol explain what the device is, its function, and how it is connected to the process, it is also a quick way to know approximately where something is located.

The main control panel is generally the focal point of the control system. It may contain elements that are *panel mounted*, a term applied to an instrument that is mounted on a panel or console and is accessible for an operator's normal use. **See Figure 3-5.** Instruments may also be located *behind the panel*, a term applied to a location that is within an area that (1) contains the instrument panel, (2) contains its associated rack-mounted hardware, or (3) is enclosed within the panel. **See Figure 3-6.**

From this vantage point, anything that is not located at this panel is considered either auxiliary or local (field) mounted. An auxiliary location is not within the main panel. The term *local* refers to devices or conditions where a sensor or transducer is physically located, as opposed to a central monitoring or a processing station. For instance, a hot water heating vessel may be located a mile from the control room. This vessel contains a heater and a high-temperature cutout switch, which inhibit the heater if the heated fluid gets too hot. The switch is installed locally, because it is in the immediate vicinity of the heating device. Therefore, this location is remote from the control room, because it is out in the field.

Bubbles, or balloons, that indicate field locations do not have a line drawn through the center of the bubble. In contrast, bubbles that are drawn with a single, horizontal line through their middle are considered to be accessible

Figure 3-5. Panel-mounted temperature indicators give the operator an easy reference point for measurements.

to the operator at the main control panel. Lastly, bubbles shown with two horizontal lines are devices that are considered to still be accessible to the operator even though they are located at

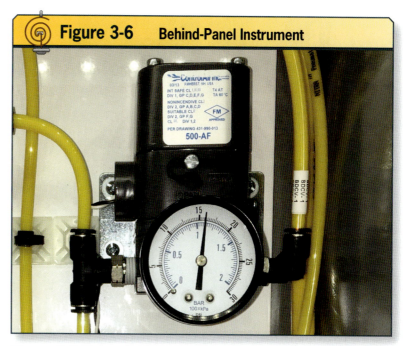

Figure 3-6. A panel-mounted current-to-pressure instrument may be hidden from the operator's view by an operator panel or other enclosure.

Figure 3-7. Location and Function Legend

	Primary location normally accessible to operator	Field mounted	Auxiliary location normally accessible to operator	Primary location hidden from operator
Discrete instruments	⊖	○	⊖	(dashed)
Shared display Shared control	⊟	▢	⊟	(dashed)
Computer function	⬡	⬡	⬡	(dashed)
Programmable logic control	◇ in ▢	◇ in ▢	◇ in ▢	(dashed)

Figure 3-7. The various shapes indicated make for quick identification of an instrument's location and function. If the horizontal lines shown within a symbol are dashed, then the instrument is hidden from the operator's view.

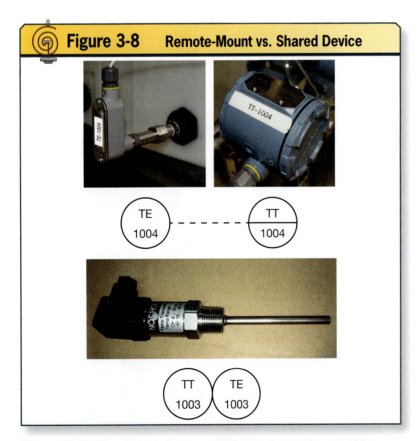

Figure 3-8 Remote-Mount vs. Shared Device

Figure 3-8. The shared device contains an integral transmitter, while the remote temperature sensor (RTD) requires an additional transmitter to transmit a 4- to 20-milliampere direct current signal.

an auxiliary location. **See Figure 3-7.**

Within instrumentation systems, typically a transmitter requires some type of sensing element. This sensing element may be either mounted remotely or integral to the transmitter housing. The temperature transmitter (TT) 1004 and its associated temperature element (TE) 1004 can be used to illustrate each scenario. **See Figure 3-8.**

Consider some examples of bubbles. First, the temperature sensor TE 1004 is drawn separately from TT 1004, and they are interconnected by a line type indicating an electrical signal. In addition, TE 1004 is shown to be inaccessible to the operator, while TT 1004 is accessible to the operator.

Next, TE 1003 and TT 1003 are drawn next to each other so that the bubbles touch. This is indicative of a shared housing, where both components are within the same assembly. Now, both bubbles are drawn as inaccessible to the operator because the temperature sensor placement is critical to the process. Instrument identification uses a variety of letters and symbols to identify the instrument function and location. **See Figure 3-9.**

Chapter 3 Instrumentation Symbols and Diagrams 47

Figure 3-9 Common Instruments and Their Symbols

Description	Image	Symbol
Conventional Pressure Gauge		PI
Indicating Gauge Pressure Transmitter		PIT
Indicating Differential Pressure Transmitter		DPIT
Thermometer	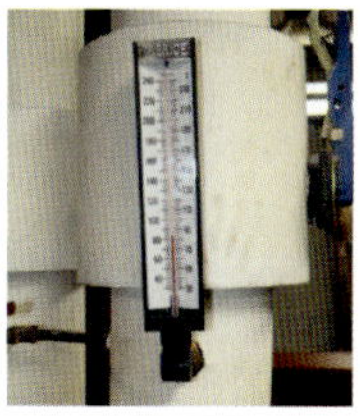	TI
Temperature Element		TE
Indicating Temperature Transmitter		TIT
Blind Temperature Transmitter and Temperature Element (Shared Housing)	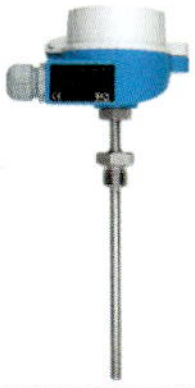	TT TE

Figure 3-9. *Common instruments are cross-referenced with their appropriate instrument symbol and description.*

Photos courtesy of Endress+Hauser

Figure 3-10 Tag Number

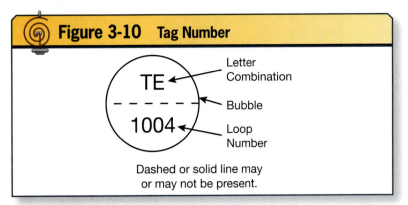

Figure 3-10. An instrument identification bubble contains a letter code, a loop number, and lines or additional shapes to indicate a specific function.

In addition to understanding how bubbles are used to represent instruments, it is necessary to understand some of the additional symbols used on the various drawings.

Instrument Identification

Within process systems, there are many types of instruments, valves, and other devices. To help locate and keep an accurate record of the different functions of these numerous devices, each instrument or function has an identifier label attached to it when it is shown on associated instrument drawings. The identifier, or tag, contains an *alphanumeric* string, with a character set that contains both letters and digits, and is determined by standards that are used for instrument identification. **See Figure 3-10.**

The unique letter code shown at the top of the bubble identifies the instrument function. Some common codes are:
PT = pressure transmitter
LT = level transmitter
PDT = differential pressure transmitter
IT = current transmitter
FT = flow transmitter
TT = temperature transmitter

Charts have been developed to aid in organizing the various letter codes based on nationally recognized standards like ANSI/ISA S5.1. The first two columns of this letter chart list a letter code for a specific variable to clearly identify what the succeeding letters are indicating, recording, or controlling. **See Figure 3-11.**

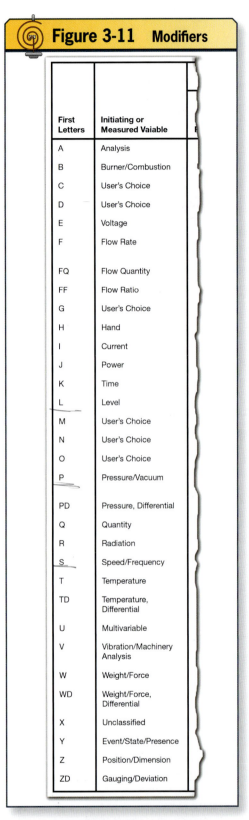

Figure 3-11. The first two columns of the letter chart are used to identify the type of device. Letters are used to identify various instruments.

The first column of the letter identifier indicates which variable is being examined within the process. For instance, when pressure needs to be monitored, the first letter of the tag begins with P. In some instances, there is a need for a first-letter modifier, such as D for differential. Care must be taken to notice that if differential pressure is desired, it is identified as pressure differential, or PD. (The description for D as a first letter is not differential; therefore, it cannot be used ahead of the P.)

Many first-letter identifiers appear to be nearly self-explanatory, such as T for temperature, L for level, and W for weight. However, some of the more obscure letters that have been used are Z for position, J for power, and I for current. Letters such as C, D, M, N, and O have been left blank for use at the design engineer's discretion or user's choice.

Typically the second letter, or third letter when a second letter modifier is present, indicates the device's function. There are a range of functions listed across the top of the chart, and in some cases, the function has been further broken down into specific categories. **See Figure 3-12.**

Figure 3-12 Instrument Letter Chart

First Letters	Initiating or Measured Variable	Controllers - Recording	Controllers - Indicating	Controllers - Blind	Self-Actuated Control Valves	Readout Devices - Recording	Readout Devices - Indicating	Switches and Alarm Devices* - High**	Switches and Alarm Devices* - Low	Switches and Alarm Devices* - Comb	Transmitters - Recording	Transmitters - Indicating	Transmitters - Blind	Solenoids, Relays, Computing Devices	Primary Element	Test Point	Well or Probe	Viewing Device, Glass	Safety Device	Final Element
A	Analysis	ARC	AIC	AC		AR	AI	ASH	ASL	ASHL	ART	AIT	AT	AY	AE	AP	AW			AV
B	Burner/Combustion	BRC	BIC	BC		BR	BI	BSH	BSL	BSHL	BRT	BIT	BT	BY	BE		BW	BG		BZ
C	User's Choice																			
D	User's Choice																			
E	Voltage	ERC	EIC	EC		ER	EI	ESH	ESL	ESHL	ERT	EIT	ET	EY	EE					EZ
F	Flow Rate	FRC	FIC	FC	FCV, FICV	FR	FI	FSH	FSL	FSHL	FRT	FIT	FT	FY	FE	FP		FG		FV
FQ	Flow Quantity	FQRC	FQIC			FQR	FQI	FQSH	FQSL			FQIT	FQT	FQY	FQE					FQV
FF	Flow Ratio	FFRC	FFIC	FFC		FFR	FFI	FFSH	FFSL						FE					FFV
G	User's Choice																			
H	Hand		HIC	HC						HS										HV
I	Current	IRC	IIC			IR	II	ISH	ISL	ISHL	IRT	IIT	IT	IY	IE					IZ
J	Power	JRC	JIC			JR	JI	JSH	JSL	JSHL	JRT	JIT	JT	JY	JE					JV
K	Time	KRC	KIC	KC	KCV	KR	KI	KSH	KSL	KSHL	KRT	KIT	KT	KY	KE					KV
L	Level	LRC	LIC	LC	LCV	LR	LI	LSH	LSL	LSHL	LRT	LIT	LT	LY	LE		LW	LG		LV
M	User's Choice																			
N	User's Choice																			
O	User's Choice																			
P	Pressure/Vacuum	PRC	PIC	PC	PCV	PR	PI	PSH	PSL	PSHL	PRT	PIT	PT	PY	PE	PP			PSV, PSE	PV
PD	Pressure, Differential	PDRC	PDIC	PDC	PDCV	PDR	PDI	PDSH	PDSL		PDRT	PDIT	PDT	PDY	PE	PP				PDV
Q	Quantity	QRC	QIC			QR	QI	QSH	QSL	QSHL	QRT	QIT	QT	QY	QE					QZ
R	Radiation	RRC	RIC	RC		RR	RI	RSH	RSL	RSHL	RRT	RIT	RT	RY	RE		RW			RZ
S	Speed/Frequency	SRC	SIC	SC	SCV	SR	SI	SSH	SSL	SSHL	SRT	SIT	ST	SY	SE					SV
T	Temperature	TRC	TIC	TC	TCV	TR	TI	TSH	TSL	TSHL	TRT	TIT	TT	TY	TE	TP	TW		TSE	TV
TD	Temperature, Differential	TDRC	TDIC	TDC	TDCV	TDR	TDI	TDSH	TDSL		TDRT	TDIT	TDT	TDY	TE	TP	TW			TDV
U	Multivariable					UR	UI							UY						UV
V	Vibration/Machinery Analysis					VR	VI	VSH	VSL	VSHL	VRT	VIT	VT	VY	VE					VZ
W	Weight/Force	WRC	WIC	WC	WCV	WR	WI	WSH	WSL	WSHL	WRT	WIT	WT	WY	WE					WZ
WD	Weight/Force, Differential	WDRC	WDIC	WDC	WDCV	WDR	WDI	WDSH	WDSL		WDRT	WDIT	WDT	WDY	WE					WDZ
X	Unclassified																			
Y	Event/State/Presence		YIC	YC		YR	YI	YSH	YSL			YT		YY	YE					YZ
Z	Position/Dimension	ZRC	ZIC	ZC	ZCV	ZR	ZI	ZSH	ZSL	ZSHL	ZRT	ZIT	ZT	ZY	ZE					ZV
ZD	Gauging/Deviation	ZDRC	ZDIC	ZDC	ZDCV	ZDR	ZDI	ZDSH	ZDSL		ZDRT	ZDIT	ZDT	ZDY	ZDE					ZDV

Note: This table is not all-inclusive.
*A, alarm, the annunciating device, may be used in the same fashion as S, switch, the actuating device.
**The letters H and L may be omitted in the undefined case.

Other Possible Combinations:
- FO (Restriction Orifice)
- FRK, HIK (Control Stations)
- FX (Accesories)
- (Scanning Recorder)
- LLH (Pilot Light)
- PFR (Ratio)
- KQI (Running Time Indicator)
- QQI (Indicating Counter)
- WKIC (Rate-of-Weight-Loss Controller)
- HMS (Hand Momentary Switch)

Figure 3-12. The use of letter codes simplifies identification of the device. See Appendix for a larger version of this chart.

Figure 3-13. (a) A non-indicating or blind transmitter does not have a display. (b) Indicating transmitters provide a quick reference to measured variables.

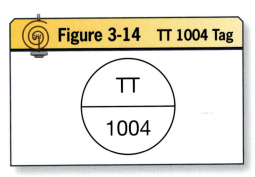

Figure 3-14. A temperature transmitter is installed within loop 1004.

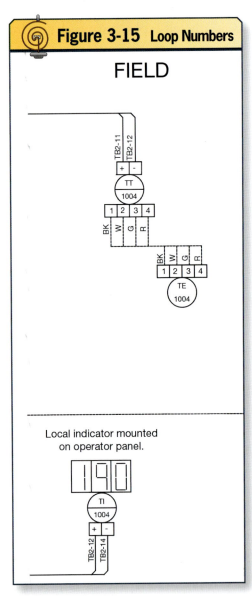

Figure 3-15. The loop number and letter description are highlighted here to indicate these devices are all part of the same control loop.

A non-indicating or blind transmitter does not utilize an integral display. Conversely, an indicating-type transmitter includes some type of integral display that reads out the current measured variables. Typically these displays can be configured to display a variety of information and units to indicate the measured variables, such as mADC output, range percentage, or even the actual measured value like PSI. **See Figure 3-13.**

Headers such as controller and recorder, also appear as sub headers under transmitters. While these devices are typically stand-alone devices, it is possible for a transmitter to include them as an onboard function. A recording transmitter is capable of storing the measured variable data while simultaneously transmitting an output signal. Additionally, some transmitters have the ability to act as a controller, circumventing the need for a separate loop controller.

Loop Numbers

The loop connection number is common to all instruments and connections to the loop. The loop number often has a prefix, a suffix, or both to complete the task of identifying an associated device. An example of a typical instrument identification or tag number is TT 1004. **See Figure 3-14.**

A letter combination chart indicates that the device is a temperature transmitter. The succeeding number is the loop identification number. A typical control loop requires a sensing device, a controller, and an output device; therefore, all devices within the control loop must use the same number, while the letter combination varies based on the particular device. **See Figure 3-15.**

This particular loop is numbered 1004. This does not necessarily mean there are 1003 prior loops; it is merely an identification tool. The particular numbering system used is determined by the design engineer, company policies, or other personnel and practices. In some instances, loop numbers that appear to be random are used to keep order to a particular process or design standard. For example, loop numbers beginning with 1000 may represent only temperature control loops to a particular system. Conversely, the first digit may indicate the first process within a plant that houses multiple processes, while succeeding digits may help to narrow down a specific location and transmitter. It is also possible that the previous numbers are reserved for future additions or specific options that have yet to be installed.

In a basic system consisting of only one process path, it is relatively easy to determine where a specific transmitter such as a TT 1004 exists. The additional numbers that are in square boxes beside the tag are used to indicate axillary connection points. TT 1004 is a transmitter-only device. **See Figure 3-16.** A sensing device is required to

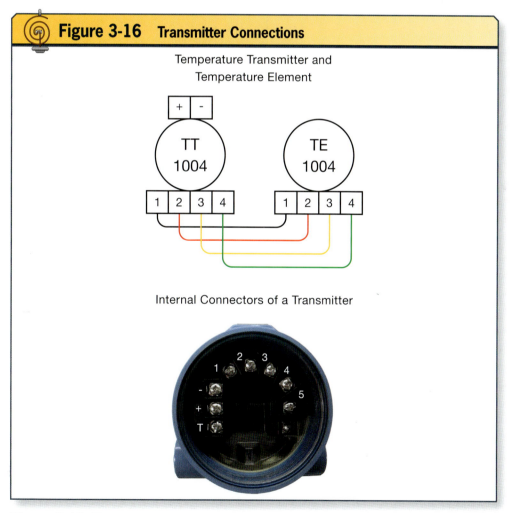

Figure 3-16. *Extreme care must be taken when interconnecting devices to avoid confusing the element terminals, numbered 1–4, with the loop terminals, indicated with the plus and minus signs.*

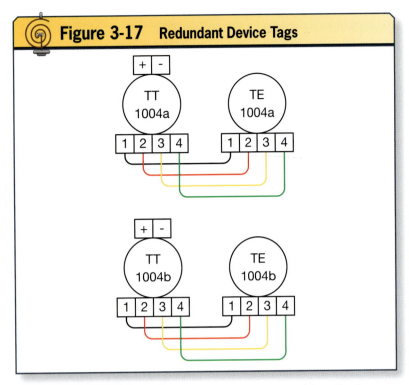

Figure 3-17. A succeeding letter has been entered into the tag number to identify that each device performs the same function under the same parameters.

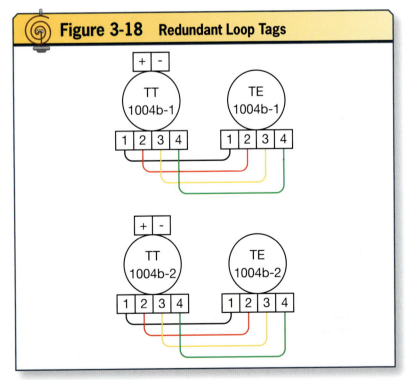

Figure 3-18. TT 1004b-1 performs a function identical to that of a similar device (TT 1004b-2) located in a redundant loop within this process segment.

transmit a temperature signal. The sensing device, TE 1004 (a four-wire resistance temperature detector), is connected by terminals 1–4.

If TT 1004 has a redundant backup, a suffix letter could be useful in identifying a device that has the same function, for example, TT 1004a and TT 1004b. **See Figure 3-17.**

In some systems, it is necessary to add an auxiliary process line. This additional line requires an additional control loop which operates identically to the first. Temperature transmitters TT 1004a and TT 1004b occur in both systems, so adding a dashed number could identify the particular system, such as TT 1004a-1 and TT 1004b-1. **See Figure 3-18.**

In terms of equipment numbering, the same convention should be used; however, the letter codes have different meanings. For example, P no longer indicates pressure; it may mean pump. In this example, P 1000 is the primary pump in loop 1000. Simple first-letter abbreviations may seem great until it is realized that vat, vessel, and valve all start with V. To overcome these scenarios, additional numbering systems and the use of hyphens allow numbered equipment groups.

While recognized standards exist to easily identify process components, it can be necessary for a designer or facility to implement variations in order to meet a particular need. What is most important is that the standard employed by any facility is understood, followed, and maintained.

Equipment Symbols and Actions

It is critical to be able to identify the various pieces of equipment that make up a process. Equipment symbols can be those of cooling towers, heat exchangers, pressure vessels, pumps, valves, and valve actuator. **See Figure 3-19.**

Valves and their associated actuators can present some special issues to the instrumentation technician. The action of a valve opening or closing can easily be changed by installing the wrong actuator or, if the correct actuator was installed, by controlling the actuator with the wrong pilot device. To understand how to recognize a valve symbol or an actuator symbol, an understanding of its related actions is necessary. When this valve is shown as an outline, the valve is shown to be open. If the valve is darkened, the valve is shown to be closed. In many cases, an arrow indicates position. **See Figure 3-20.**

This is the designated failure position such that upon loss of signal to the actuator, the valve body is positioned by the actuator to be either open or closed. In other circumstances, letter designators may be used such as follows:
FO = fail open
FI = fail indeterminate
FC = fail closed
FL = fail locked

INSTRUMENTATION DOCUMENTS

Just as specialized instrumentation drawing symbols are used to complement the conventional symbols used in both the electrical and mechanical system construction, there are an assortment of specialized drawings also. Familiarity with all types of drawings, specification books, and documents relative to a project are key. Additionally, it is imperative to keep the proper information updated on these documents to identify specific as-builds conditions.

In a recent trend, some project documentation is being made available electronically. These electronic versions use desktop, laptop, and tablet-style computers and even smartphones to view and alter the drawings. However, though the traditional paper-based method of drawing may be evolving, the information continues to be documented using the same standard documentation.

Process and Instrumentation Drawings

A P&ID is used to illustrate a process and how the related components are

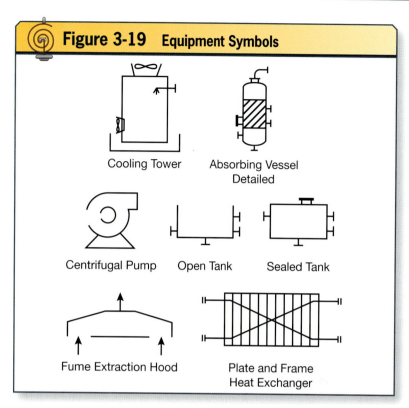

Figure 3-19. Process equipment is represented on drawings using a symbol library.

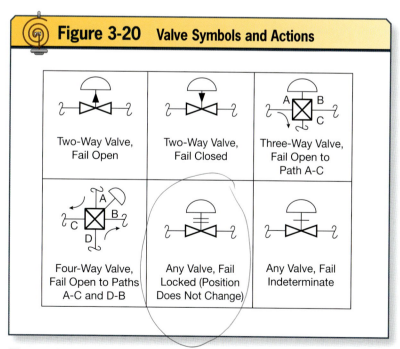

Figure 3-20. Failure mode indications vary for conventional, three-way, and four-way valves.

integrated. The P&ID is used to provide order to a great deal of information; however, it does not provide specific wiring paths or connections. To clearly understand a P&ID, a range of common symbols applies specifically to the equipment portion of the process. The loop numbering also should be maintained throughout the equipment portion of the P&ID. Simply using P-1 to identify the only pump with a particular process may be fine on a simple process; however, in some plants there may be 1,000 pumps, and numbering them sequentially could become confusing. As systems and processes grow in size and complexity, the use of equipment letter designations can become inadequate.

For additional information, visit qr.njatcdb.org Item #1736

Loop Sheets

Loop sheets provide a greater level of detail into the exact signal path that transmitter signals take and are arguably the documents that receive the most use by instrument technicians, and maintenance personnel. An instrument loop sheet depicts the schematic wiring diagram of the field wiring.

The current path on a loop sheet can be traced from beginning to end until it traverses a complete loop. Each device that functions within that wiring loop is depicted. The simplest loop sheet is one that shows only a transmitter and a controller. More complicated loop sheets have recorders, alarms, pilot lights, control valves, and other devices. The common point to a single loop

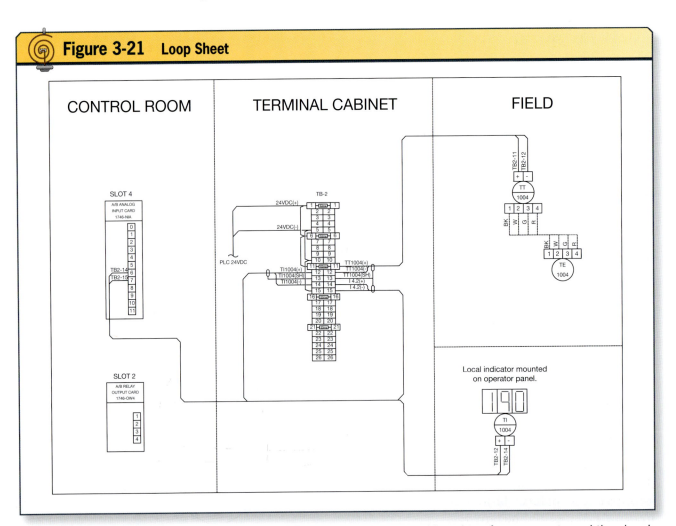

Figure 3-21. The loop sheet for control loop 1004 indicates the general location of components and the signal path from beginning to end.

sheet is that the devices on each sheet are generally categorized by instrument tag number. The tag number is used as an identifying method for records, accountants, material handlers, and the field technician. By having items with the same loop number shown on the same loop sheet, a simple method of data keeping is available to all involved. **See Figure 3-21.**

For the field worker, the loop sheet shows what devices are likely to be affected, where the device is located, the power requirements of that device, the controlling method for that device, and the intermediate wiring, junction boxes, terminal strips, and other data. Jobs that have pneumatic instrumentation involved also have a loop sheet. Instruments with tag numbers are shown, along with their hookup details.

The loop sheet is the instrument technician's link to the process. This sheet is useful during start-up and checkout procedures, because it is easy to follow the wiring path to verify the connections have been made properly. In addition, because it is used during the checkout period, it serves as an excellent as-built drawing. The key with all documentation is that as changes are made, the documents are kept up to date.

The example of a loop sheet with a transmitter and a controller can be checked by forcing a signal on the pair of wires that are connected to a transmitter and then monitoring the response of the controller to verify that the associated wiring is correctly installed, along with the controller's configuration limits. In other words, a signal of 4–20 milliamperes direct current is applied to the pair of wires to the controller, and then the current is adjusted. The signal is read in by the controller, and it should be compared with the sending current to verify proper working order.

A loop sheet should list calibration data such as calibration range and any associated set points. Field workers must be aware of the hierarchy of the drawing system at the site where they are working. If a loop sheet specifies a calibration range but the spec sheet does not agree, the field worker should call attention to this before proceeding.

There have been numerous accounts of industrial shutdowns because of incorrect data on loop sheets that were taken for granted to be correct when the hierarchy of the documentation system stated that the spec sheets had precedence over loop sheets. If the only drawings that are used for as-built drawings are the loop sheets, technicians must take extra care to ensure that the proper documentation updates are carried out.

Wiring Diagrams

An instrument loop sheet shows the field wiring schematically. In contrast, *wiring diagrams* illustrate the physical connection between electrical components in a machine or production system. When possible, wiring diagrams show the actual locations of components in the circuit. **See Figure 3-22.**

Manufacturers usually include wiring diagrams as part of the electrical equipment documentation. Although some wiring diagrams appear to be simple, many are complex. These diagrams are rarely used for troubleshooting purposes, because it is difficult to determine the circuit's intended operation. Typically, technicians use wiring diagrams in conjunction with *schematic diagrams*, which use graphical symbols

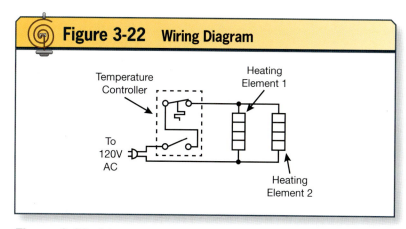

Figure 3-22. A temperature-controlled switch with an integrated isolation switch is shown within the dashed-line controller.

Figure 3-23 Schematic Diagram

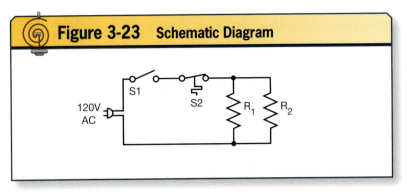

Figure 3-23. Schematic diagrams focus on identifying the path of current flow between components of a circuit.

to show the connections between components and the function of the circuit. Schematic diagrams are essential for troubleshooting a circuit, because they show the exact path of current flow. **See Figure 3-23.**

These diagrams exist for pneumatic instrumentation systems as well. Piping diagrams are the functional equivalent of wiring diagrams, and pneumatic schematics show the pneumatic path or logic of the circuit.

Spec Sheets

By the nature of control systems and their complexity, instruments are required to fit various scenarios, which often requires instruments to be available with a range of options based on range, sensed medium, body construction, fill fluids, outputs, manifolds, mounting options, and environmental concerns, among others. Instrument spec sheets are used to tell a variety of people a wealth of information about a device. Spec sheets are generally sorted by tag number and are usually located in a permanent records library.

Spec sheets serve as permanent records of devices that are in use or have been in use. The process conditions, calibration ranges, tag numbers, transmitter type, control functions, and other data are included on the sheet. Spec sheets are the primary source of instrument information when ordering, designing, or controlling a process.

Purchasing departments often refer to the spec sheets to determine the number of devices on hand compared with the average failure rate to get a rough idea about estimated repair costs. Spec sheets provide an easy path for interaction between the engineering staff and the front office personnel. **See Figure 3-24.**

Sometimes, the spec sheet is not completed until after installation to allow checking and reviewing a device. For such devices, the instrument index sheet is the guiding documentation until a spec sheet is completed.

Instrument Index Sheets

Instrument indexes provide a method for listing every instrument that will be used for the job. A summary that gives the present standing or location of the device usually accompanies each listing. An instrument index provides an accurate way to gauge the amount of work that must be done or remains to be completed on a job.

An index should be started at the beginning of the job, and as items are added or deleted, the list should be edited. Relevant shipping and receiving dates should be included. The changes that are made to the job and how they affect the instrument index should also be reflected. After a device is entered into the list, it should never be removed—just updated to show present status.

An instrument index should contain the devices that are to be included in the present job requirement. This is a minimum requirement; each job requires its own variations based on site-specific guidelines and procedures. **See Figure 3-25.**

For additional information, visit qr.njatcdb.org Item #2549

Figure 3-24 Spec Sheets

	INSTRUMENT DATA SHEET		NVS Cooling Tower Project
	Instrument Type	MAGNETIC FLOW METER	
	Data Sheet Number	862-1010	
	Issue Purpose	FOR BID	

#	GENERAL		#	TRANSMITTER	
1			41		
2	Instrument Number	0FIT-CW001	42	Accuracy	±0.25% OF RATE
3	Unit	0	43	Calibrated Range	0 - 2,000 GPM
4	Service Description	COOLING TOWER SUPPLY BLOWDOWN TO CATCH BASIN	44	Output Signal Type	4-20mA
5			45	Digital Communication	HART
6			46	Enclosure Type	NEMA 4X
7			47	Enclosure Material	POLYURETHANE COVERED ALUMINUM
8	**PROCESS DATA**		48	Certification / Approval Type	N/A
9	Fluid	WATER	49	Power Supply	120VAC
10	Design Pressure	60 PSIG	50		
11	Max Operating Pressure	35 PSIG	51		
12	Design Temperature	120 DEGF	52		
13	Max Operating Temperature	110 DEGF	53		
14	Piping Design Flow	1440 GPM	54		
15	Min Operating Flow	360 GPM	55		
16	Max Operating Flow	1440 GPM	56		
17			57		
18	**PROCESS CONNECTION**		58		
19	Flange Type	RF WELD NECK FLANGE	59		
20	Size	8"	60		
21	Rating	ASME B16.5 CLASS 150	61		
22	Material	CS	62		
23			63	**OPTIONS**	
24			64	Tag	SST WIRED TAGS
25	**SENSOR**		65	Local Indication	LCD
26	Electrode Type	2 MEASUREMENT ELECTRODES	66	Calibration Certificate	REQUIRED
27	Electrode Material	316L SST	67		
28	Lining Material	PTFE	68		
29	Upstream Straight Length Req.	5 PIPE DIAMETERS	69		
30	Downstream Straight Length Req.	2 PIPE DIAMETERS	70	**NOTES**	
31					
32					
33					
34					
35	**COMPONENT IDENTIFICATION**				
36	Manufacturer	ROSEMOUNT			
37	Sensor Model Number	8705TSA080D1W0NHB3Q4			
38	Transmitter Model Number	8732EST1A1NHMSQ4			
39					
40					

REV	DATE	REVISION DESCRIPTION	PREPARER	REVIEWER	APPROVER
A	3/13/2015	FOR BID SPEC S-7207	M. NUTELLA	T. PERKIS	N/A

Figure 3-24. Spec sheets contain the necessary information for field personnel to maintain and install the instrument.

Figure 3-25 Instrumentation Index

Instrumentation Index

Contractor:	Customer:		Revision No. 0
JPJ Instrument Services	Josh Manufacturing Plant		
A division of ELVIS Industries	Denali Assembly Line	Revision Comments: Issued for Procurement	
	Clearwater Florida Location		

PID Number	Tag Number	Description	Signal Type	Range	Signal	Operating Point	Immersion Length
BOR-PID-1101	PIT-0400	Pressure Indicating Transmitter	Analog	0 - 800 psig	4-20 mA	170 psig	N/A
BOR-PID-1101	PIT-1400	Pressure Indicating Transmitter	Analog	0 - 800 psig	4-20 mA	170 psig	N/A
BOR-PID-1102	PIT-1403	Pressure Indicating Transmitter	Analog	0 - 800 psig	4-20 mA	130 psig	N/A
BOR-PID-1102	LSL-1403A	Level Switch: float switch, 2" MNPT connection	Discrete	N/A	N/A	N/A	N/A
BOR-PID-1102	LSH-1403B	Level Switch: float switch, 2" MNPT connection	Discrete	N/A	N/A	N/A	N/A
BOR-PID-1102	LSL-1403B	Level Switch: float switch, 2" MNPT connection	Discrete	N/A	N/A	N/A	N/A
BOR-PID-1102	PIT-1404	Pressure Indicating Transmitter	Analog	0 - 800 psig	4-20 mA	70 psig	N/A
BOR-PID-1102	PIT-1404A	Pressure Indicating Transmitter	Analog	0 - 800 psig	4-20 mA	70 psig	N/A
BOR-PID-1202	PIT-2704	Pressure Indicating Transmitter	Analog	0 - 800 psig	4-20 mA	165 psig	N/A
BOR-PID-1202	AIT-2704A	Honeywell Apex Gas Detector, 4-20 mA Output	Analog	0 - 100%	4-20 mA	0%	N/A
BOR-PID-1202	AIT-2704B	Honeywell Apex Gas Detector, 4-20 mA Output	Analog	0 - 100%	4-20 mA	0%	N/A
BOR-PID-1202	BS-2704	Flame Eye	Digital	N/A	N/A	N/A	N/A
BOR-PID-1307	PIT-3500	Pressure Indicating Transmitter	Analog	0 - 800 psig	4-20 mA	40 psig	N/A
BOR-PID-1307	FE-3510	Flow Meter, Coriolis	Analog	0 - 250 GPM	4-20 mA	5000 bbl/day	N/A
BOR-PID-1307	TE-3702	Temperature Transmitter	Analog	-200 - 300°F	4-20 mA	60°F	2"
BOR-PID-1800	LSL-8400	Level Switch: Float switch, 2" MNPT Connection	Discrete	N/A	N/A	N/A	N/A
BOR-PID-1800	LSH-8400	Level Switch: Float switch, 2" MNPT Connection	Discrete	N/A	N/A	N/A	N/A

Figure 3-25. The devices required for a job are listed on the instrument index.

SUMMARY

Instrumentation diagrams show the relationship between instruments and industrial processes. Identification of field devices is critical to installation and maintenance requirements. Being able to use the tag numbers associated with instruments allows proper identification of devices and accurate interpretation of their functions. Proper identification also minimizes mistakes in the field.

REVIEW QUESTIONS

1. What do symbols portrayed on diagrams and drawings depict?
   a. Instruments
   b. Locations
   c. Ranges
   d. Wire type

2. What is the name of the unique identification code for devices shown on instrumentation drawings and diagrams?
   a. Address
   b. Locator ID
   c. Serial number
   d. Tag number

3. What is a device that is identified as TT 1004?
   a. Temperature element 1004
   b. Temperature transmitter 1004
   c. Thermocouple 1004
   d. Timed terminator 1004

4. What is a device that is identified as PSL 2345?
   a. Power switch 2345
   b. Pressure switch low 2345
   c. Process siren light 2345
   d. Pump speed light 2345

5. What does the tag number LAL 1003 represent?
   a. LASER alarm lamp
   b. Level alarm high
   c. Level alarm low
   d. Lowest allowable limit switch

6. What type of drawing shows all related (loop number) devices for controlling means and the wiring path, termination point, and device locations?
   a. Floor plan
   b. Loop sheet
   c. P&ID
   d. PID

7. What type of document indicates the specifics for which a device is ordered and the type of conditions under which it must perform?
   a. Calibration sheet
   b. Order sheet
   c. Product performance sheet
   d. Spec sheet

8. How are devices that are hidden from the operator's view indicated?
   a. Diamond
   b. Notation only; no symbol required
   c. Open circle
   d. Series of dashed lines within the tag number

9. A control valve that fails in position and does not move can be indicated with a graphic symbol or which letters?
   a. FC
   b. FL
   c. FL
   d. FO

10. A relatively quick way to determine the quantity of a specific type of device needed on a project would be to consult which documentation?
    a. Instrument index sheets
    b. Loop sheets
    c. P&ID
    d. Wiring diagram

Calibration Procedure and Documentation

Calibration is an essential step to ensure a process can be controlled accurately. The process of calibrating an instrument is necessary to compare the desired, calculated signal to the actual signal produced by the field device. It is essential that the calibration procedure identify calibration errors, properly record the errors identified, require accurate adjustments made in the correct order, and verify the device is within the accuracy tolerances specified when complete.

Objectives

- » Describe the importance of the calibration procedure.
- » List the calibration standards required for the calibration procedure.
- » Explain the purpose of the calibration procedure.
- » Determine the required test points for a five-point calibration check.
- » Identify and use the formula for device accuracy.
- » Identify and use the formula for gain.
- » Contrast repeatability and accuracy.
- » Identify five calibration errors.
- » Fill out a calibration sheet properly and completely when performing a calibration procedure.

Chapter 4

Table of Contents

Calibration 62
   The Necessity of Calibration 62
   Concepts of Calibration 63
   Calibration Standards 65
      Primary (Input) Test Standards 66
      Output (Measurement) Standards 66
      Certifications and Documentation 67
      Accuracy of Test Equipment 67
      Test Equipment 68
   The Five-Point Check 69
   Calibration Accuracy 70
      Gain 72
      Repeatability 72

As-Found Test 72
   Identification of Calibration Errors 73
   Classification of Calibration Errors 73
      Zero Shift 73
      Span Shift 73
      Dead Band 74
      Hysteresis 74
      Nonlinearity 75

Error Correction Procedures 75
   Conventional Instruments 75
      Zero Shift 75
      Span Shift 76
      Zero and Span Shift 76
   Smart Instruments 76

As-Left Test 77
   Calibration Sheet Data Entry 77
   Verification of Calibration Sheet
   Accuracy 78

Specification Sheets 80
Summary 84
Review Questions 85

For additional
information, visit
qr.njatcdb.org
Item #2690

CALIBRATION

The process of calibrating an instrument verifies the instrument is accurate within the requirements of the process. Calibration procedures vary according to the instrument and the requirements of the process. All calibration procedures are fundamentally based on the concept of comparing a known accurate signal with a signal of unknown *accuracy*, defined as conformity to an indicated, standard, or true value, usually expressed as a percentage deviation (of a span, reading, or upper-range value) from the indicated, standard, or true value.

All instruments fundamentally have two types of signals: input and output. The technician measures the output of the instrument being calibrated and records the output values. A properly calibrated instrument accurately and repeatedly produces the appropriate output signal when provided with an accurate input signal. The instrument may need adjustments to ensure the accuracy of an instrument over the measurement range.

The calibration procedure mandates that certain documentation be used to track the procedure. With instrumentation, records or a history of these calibrations also must be kept. Facilities may be required to maintain calibration records under regulatory controls by the U.S. Food and Drug Administration (FDA), the American Petroleum Institute (API), or the Nuclear Regulatory Commission (NRC). Every jobsite differs on how and what types of records are kept, but these sheets and the information they contain are similar.

The Necessity of Calibration

The operation of an automatic control process depends on the accuracy of each instrument in the loop. A correctly calibrated instrument ensures the safety and proper operation of the controlled process. To illustrate the significance of instrument calibration, consider the operation of a tank that must be maintained at a specified level. **See Figure 4-1.**

Consider a tank that has a process pipe coming from another part of the process that brings feedstock into the tank. The tank also has a level valve that maintains a specified set point of level in the tank. The feedstock must be maintained at a desired temperature to ensure good product quality. Two discrete level sensors are installed to alert operators if the tank becomes overfilled or underfilled. Finally, a pressure sensor allows the operators to monitor internal tank pressures and detect abnormal conditions.

Instrument calibration ensures the tank is maintained at a specified set point level.

The inputs of this system include the following:
- Level (analog)
- Pressure (analog)
- Temperature (analog)

The outputs of this system include the following:
- Level transmitter output (analog)
- Level alarms high and low (discrete)
- Pressure sensor (analog)
- Temperature sensor (analog)
- Level valve position (analog)

Now that the inputs and outputs have been identified, consider the following five calibration problems and their implications:

1. The level sensor erroneously reports the tank is 100% full when it is only 60% full. The process must be shut down until the problem can be fixed. This costs the company time and money.
2. The level alarm high does not work, resulting in tank overflow, a potentially catastrophic condition for both employees and the environment.
3. Because of improper calibration procedures, the pressure sensor tells the operator the pressure in the tank is normal. The pressure is actually dangerously high. A rupture disk releases excess pressure, venting hazardous gases into the atmosphere. The company faces fines for environmental damage.

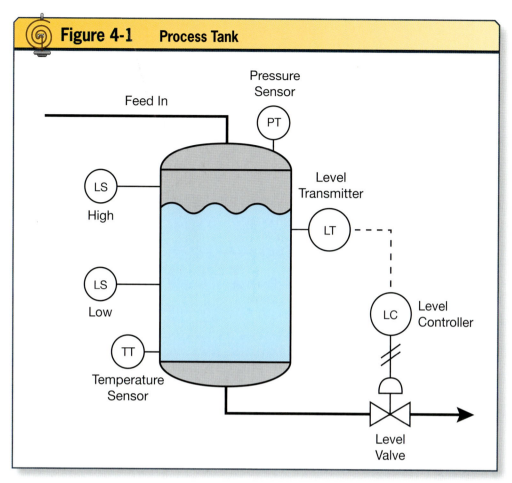

Figure 4-1. A tank may have several instruments that measure pressure, level, and temperature.

4. The temperature sensor reports the product is the proper temperature when in reality it is too low, causing the company to discard the entire batch. Money is lost and purchasers of the product must turn to another supplier.
5. The level valve should be open to 15%, but a faulty calibration allows it to open to 65%, allowing too much product to pass and ruining the product.

An *error*, or difference between the actual and the true values, often expressed as a percentage of the span or upper-range value, in any of these systems can lead to poor efficiency, lost time, damage to equipment, and harm to people and the environment. Proper calibration procedures are therefore necessary to ensure all process components are operating within the design specifications of the process.

Concepts of Calibration

To properly perform instrument calibration, it is necessary to understand the concepts of *range*, a set of values over which measurements can be made by a device without changing the devices sensitivity, and *span*, the difference between the upper and lower limits of a range which is expressed in the same units as the range. A typical pressure transmitter is used to demonstrate range and span. This example uses a pressure transmitter that can be calibrated with an input of pressure in pounds per square inch and an output of 4–20 milliamperes direct current (mADC).

ThinkSafe!

All connections made on the bench before calibration should be performed with the power disconnected from all supplies. This limits the risk of shock and limits the potential for damage to the instrumentation.

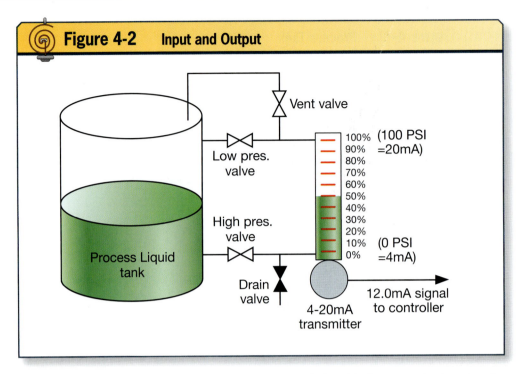

Figure 4-2. The process tank shows the correct relationship between the input and the output of the transmitter.

Suppose a transmitter is connected to a process tank with an input range of 0–100 pounds per square inch. **See Figure 4-2.**

The transmitter has a span of 100 pounds per square inch, which tells the instrument technician the difference between the upper and the lower limits. When the tank is empty, the pressure transmitter measures 0 pounds per square inch. This is called the lower-range value (LRV). When the tank is full, the pressure transmitter measures 100 pounds per square inch. This is called the upper-range value (URV). The transmitter has an output range of 4–20 mADC. The output span of the transmitter, the signal, is 16 mADC. The industry standard for a transmitter's output range is 4–20 mADC.

When an instrument is properly calibrated, its output is a direct representation of its input so that the following is true:

% Output span = % Input span

With an input of 0 pounds per square inch (0%), the output is 4 mADC (0%); with an input of 50 pounds per square inch (50%), the output is 12 mADC (50%); and with an input of 100 pounds per square inch (100%), the output is 20 mADC (100%). **See Figure 4-3.**

The calibration procedure requires the technician to solve for the span of the instrument. This measurement defines the instrument's ability to measure within a certain area of the process. To find the span of an instrument, the technician should use the following formula:

Span = Upper range value − Lower range value

TechTip!

Always verify that the calibration standard is within the certification date as defined by the standard records. If the certification date is past due on the calibration standard, do not use the device to perform calibrations.

For example, what is the span of a transmitter ranged for 0–200 inches of water?

200 − 0 = 200 "H$_2$O

In the preceding example, the LRV was zero. However, the LRV of a transmitter does not specifically start at a zero value. Both positive and negative values are encountered when performing calibrations. Consider another example: What is the span of a transmitter ranged for 35–165 inches of water?

165 − 35 = 130 "H$_2$O

Properly calibrated, the transmitter produces an output signal in mADC equivalent to the input signal in pounds per square inch. This mADC representation of the level in the tank is then transmitted via loop wiring to the appropriate device, such as a level controller or indicator. Every transmitter has limitations concerning accuracy and range. Every calibration performed must meet specifications for accuracy for the process to be considered calibrated.

Calibration Standards

It is essential to understand how to calibrate an instrument by using the appropriate input and output standards. Merely observing the instrument's output with respect to an actual process input does not constitute calibration. By correctly analyzing the input and output values of a transmitter using the correct standards, the technician can determine whether steps to adjust the transmitter are needed. This process simulates the process the instrument will measure and determines whether corrective action is needed. While many instruments are calibrated at a calibration bench, some instruments such as liquid level float switches and capacitance level probes must be calibrated in the actual working environment. Many companies produce written calibration procedures for technicians to use. These written documents include the necessary steps to ensure the calibration procedure is performed properly.

Figure 4-3   Pressure vs. mA Signal

Tank	Input	Output
0% full	0 PSI	4 mA
25% full	25 PSI	8 mA
50% full	50 PSI	12 mA
75% full	75 PSI	16 mA
100% full	100 PSI	20 mA

Figure 4-3. For a properly calibrated instrument, transmitter input and output across full range are directly related to each other.

At times, an instrument is calibrated after it has measured a process, such as for normal maintenance and repair, regularly scheduled calibrations, and adjustments because of process deviations. In these cases, care must be taken to follow established standards for the calibration of an instrument with respect to site procedures for decontamination, calibration, and other safety and site-specific requirements.

To determine that an output signal is correct for a given input signal, the technician must provide the input from a calibration standard, which ensures the input signal is correct. The technician must also measure the output using a calibration standard to ensure the measurement is accurate. A calibration standard is one that has been tested and certified as accurate. Most facilities send their calibration standards to a metrology laboratory to be certified. The equipment used by these metrology labs is typically larger and more precise than field-suitable calibration equipment. The level of accuracy achieved by the metrology equipment must be significantly higher than the field calibration standard. The intervals of recertification depend on the desired accuracy and facility requirements.

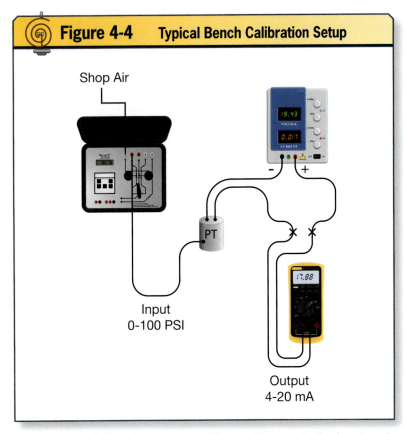

Figure 4-4. Calibration equipment is used to supply the input and verify the output of an instrument.

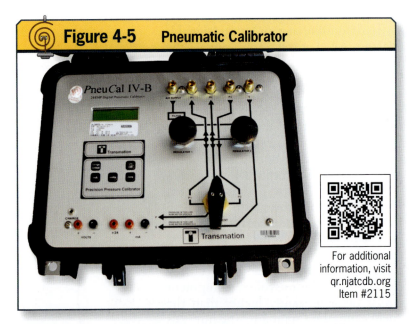

Figure 4-5. A pneumatic calibrator can be used to provide a precision pressure signal.

TechTip!
Before connecting air tubing to test equipment and instruments, the technician should blow clean and dry air through the lines to remove any contaminants.

Calibration standards can be used to measure both input and output signals of the pressure transmitter. **See Figure 4-4.**

A direct-current (DC) power supply is connected in series with the output loop to power the device. The input test standard provides a highly accurate pressure input signal. The output test standard measures the milliampere output signal.

Primary (Input) Test Standards

The instrument technician must determine whether an instrument is correctly calibrated within the specified accuracy requirements of the test procedure. The use of calibration standards ensures the accuracy of the instrument is acceptable. To provide an input signal to a transmitter, the technician must use an *input test standard*, defined as an item of test equipment used for the calibration of pressure, temperature, current, voltage, or other input measurements. For the transmitter in Figure 4-4, the input test standard must be capable of supplying pressure (in pounds per square inch) that covers the entire range of the instrument. An example of an input test standard capable of providing this signal is the pneumatic calibrator. **See Figure 4-5.**

Output (Measurement) Standards

The output signal must be verified using an *output test standard*, defined as an item of test equipment used for the calibration of pressure, temperature, current, voltage, or other output measurements. An example of an output test standard is a digital multimeter **See Figure 4-6.**

The output test standard measures the milliampere signal in the test loop.

Figure 4-6. A digital multimeter performs several electrical measurements.

For the output to be measured with sufficient accuracy, current must be measured with enough units of *resolution*, defined as smallest detectable increment of measurement.

Certifications and Documentation

On a typical jobsite, the documentation that is used or generated by workers doing calibrations consists of calibration sheets (calibrated data records), specification (spec) sheets, and instrument indexes. Calibration sheets, spec sheets, and instrument index sheets are needed to install, calibrate, and maintain the instruments used in industrial processes.

Calibration sheets are used to record calibration data. These sheets can be helpful when the instrument technician is attempting to determine an instrument's calibration errors. **See Figure 4-7.**

Accuracy of Test Equipment

All measurements exhibit some degree of inaccuracy. The acceptable level of inaccuracy is also called the tolerance,

Figure 4-7 Calibration Data Record

Manufacturer:

Instrument ID: | Model:

Calibration Range:

Input: | Output:

Test Equipment	Model	S/N

Accuracy of calibration is 0.5%

%	INPUT	DESIRED	AS FOUND	AS LEFT
0				
25				
50				
75				
100				
75				
50				
25				
0				

Performed by: _____

Date: _____

Calibration Procedure

- Select and connect input source as required.
- Select and connect for voltage supply and current readout.
- Input five cardinal test points, increasing and decreasing, while recording necessary information.
- Make adjustments, if required, to comply with accuracy limits.
- Record all necessary information.
- Complete all other information as per calibration sheet.

Figure 4-7. A calibration data record or calibration sheet contains information regarding the current calibration of equipment.

Figure 4-8   Calibration Sticker

Figure 4-8. A current certification sticker indicates that the equipment can be used as a calibration standard within its specifications.

TechTip!
When performing calibration procedures, it may be necessary to repeat a test to determine whether an instrument is functioning with the accuracy and repeatability required.

often expressed as a percentage deviation of span. These inaccuracies may be held to tight constraints, such as ±0.1°C, when the manufacturing process requires such accuracy. Other processes can tolerate a less accurate instrument without affecting the overall quality of the final product. Instrument manufacturers provide the acceptable ranges for the degree of accuracy of their devices, but most often the customer provides the requirements for the process concerning the level of acceptable accuracy of an instrument.

The degree of accuracy of the calibration equipment is equally important. To prove that a calibration is valid when the desired accuracy is obtained, certified test equipment should always be used to perform calibrate procedures. Certified equipment provides the only assurance that a desired value is recorded as accurate, rather than some incorrect value because of a faulty test equipment reading. **See Figure 4-8.**

A certified standard must have documentation to show it has been verified as accurate. This record of documentation provides *traceability*, an unbroken chain of measurements and associated uncertainties. For example, consider a company that manufactures automobiles. Assume that one factory produces the engine and a separate factory produces the transmission. For both parts to mate correctly, the assembly holes must be accurately spaced. Each factory must calibrate, or compare to a known accurate standard, their machines to precisely place the holes. The need for a standard that defines an inch, centimeter, or any other measurement of distance, mass, or temperature is apparent.

The National Institute of Standards and Technology (NIST) is a federal agency that maintains and develops standards for measurement in the United States. In addition, the International Organization for Standardization (ISO) works globally to develop many standards, including those for measurement and control. Technicians may see references to ISO 9001, which is a standard manufacturers use to maintain quality in manufacturing processes. Both NIST and ISO enable manufacturers to produce safe, high-quality products with consistent quality.

Test Equipment

Many instrumentation errors can be identified through the interpretation of input and output data. Calibrating an instrument means that it must be tested, recorded, and adjusted to perform as needed. Various pieces of test equipment are those used in the adjustment and calibration of process control sensors and transducers. **See Figure 4-9.**

The manufacturer of each piece of test equipment publishes user guides and spec sheets. These documents serve as references for the instrument technician and provide insight into the proper application and limitations of test equipment. Test equipment must be selected based on the ranges of the calibration procedure. For example, a hydraulic deadweight tester may be suitable for high pressure calibrations of pressure transmitters above 30,000 PSI. This same equipment would not appropriate for testing a pressure

Figure 4-9. Various styles of test equipment are available to perform instrument calibrations.

transmitter with a range of 0-5 "H_2O. For the second transmitter, test equipment such as an inclined manometer is more appropriate. Consult the test equipment manufacturer for information on the specifications and accuracy of test equipment.

Technicians should recognize influences on instrument calibration that may be outside their ability to control. These factors include temperature, humidity, atmospheric pressure, incoming alternating-current voltage, and variations in controller voltage, current, and frequency standards. Always refer to the manufacturer's documentation to ensure the equipment is used properly.

The Five-Point Check

Instrument calibration includes producing input signals that represent process values. Most calibrations have a five-point check. This five-point check is performed to verify that an instrument is properly calibrated over its full signal span. These five points establish baseline data needed for the evaluation of an instrument or process loop.

A simulated input is typically applied at 0%, 25%, 50%, 75%, and 100% of the input range of the instrument being calibrated or tested. These test points are also called checkpoints or cardinal points. The output is recorded for each input, and the measured output is compared with the expected output. Values other than 0%, 25%, 50%, 75%, and 100% may be used (e.g., 10%, 30%, 50%, 70%, and 90%).

Once the input process values are determined, output signals that represent these values are calculated. To see how these values are calculated, return to the example process transmitter with an output of 4–20 mADC, or a span of 16 mADC, and an input signal range of 0 to 100 pounds per square inch, or a span of 100 pounds per square inch. The next step determines that the test points are 0%, 25%, 50%, 75%, and 100%. The input signal for this transmitter is 0 pounds per square inch at 0%, 25 pounds per square inch at 25%, and so on. The output signal for this transmitter during the calibration is 4 mADC at 0% (0 pounds per square inch), 8 mADC at 25% (25 pounds per square inch), 12 mADC at 50%

Figure 4-10 Abbreviated Instrument Calibration Sheet

Input Values		Output Values	
Test Inputs	Inputs	Expected	Actual
0%	0 PSI	4 mA	
25%	25 PSI	8 mA	
50%	50 PSI	12 mA	
75%	75 PSI	16 mA	
100%	100 PSI	20 mA	

Figure 4-10. The calibration sheet shows the test inputs and expected output values.

(50 pounds per square inch), 16 mADC at 75% (75 pounds per square inch), and 20 mADC at 100% (100 pounds per square inch). **See Figure 4-10.**

A graph is often used to plot the ideal measured output signal values so that they can be compared with the actual output signal values. This visual approach can help in identifying errors. **See Figure 4-11.**

It is considered good practice to verify output signal readings on an upscale-and-downscale check. Checking upscale tests a transmitter's output signal starting at 0% and increasing to 100%. Checking a transmitter downscale measures output signals starting at 100% and decreasing to 0%. The values recorded for going upscale and downscale are expected to be identical.

The actual output signal values recorded may show a consistent shift higher than the expected values. A plot can be made of the points recorded during the initial test. **See Figure 4-12.**

If a consistent shift, known as a zero shift, can be recognized in the expected output signal, an instrument calibration error exists and should be corrected.

Calibration Accuracy

Accuracy describes the closeness between the measurement that is made and the actual value. All measurements have error. The calibration procedure ensures that the error is acceptably

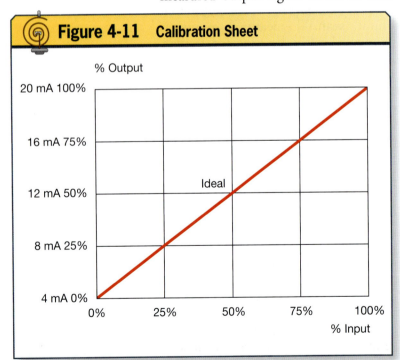

Figure 4-11. A calibration sheet includes the recorded output signal of the transmitter.

small so that it does not interfere with the process operation. Using the word accuracy frequently leads to some confusion. An instrument that measures 98 pounds per square inch when the actual value is 100 pounds per square inch is said to have 2% accuracy. This is not to say the instrument is 98% inaccurate but rather says the instrument is accurate within ±2%.

Consider the pressure transmitter in the earlier example. Applying a 50% input to the transmitter (50 pounds per square inch) should result in a 50% output signal (12.00 mADC). If an input value of 50% was applied and a measured output of 12.16 mADC was recorded, the transmitter would be considered accurate within 0.16 mADC. Comparing the difference of an instrument's input signal with its output signal calculates the instrument's deviation. Therefore, in this example, the transmitter is accurate within, or has a deviation of, 0.16 mADC.

This is one way of stating accuracy, but accuracy typically is expressed as a percentage of the transmitter span. This means that accuracy equals the difference between the true and the measured values (the deviation) divided by the instrument's span and then multiplied by 100:

$$\text{Accuracy} = \left(\frac{\text{Maximum deviation}}{\text{span}}\right) \times 100$$

The deviation and span can then be used to calculate the accuracy of the example transmitter, whose span is 16 mADC:

$$\text{Accuracy} = \left(\frac{0.16 \text{ mA}}{16 \text{ mA}}\right) \times 100$$

$$\text{Accuracy} = 0.01 \times 100$$

$$\text{Accuracy} = 1\%$$

This accuracy value can be used to determine whether an instrument must be recalibrated to meet specifications. The technician may be required to determine the maximum allowable

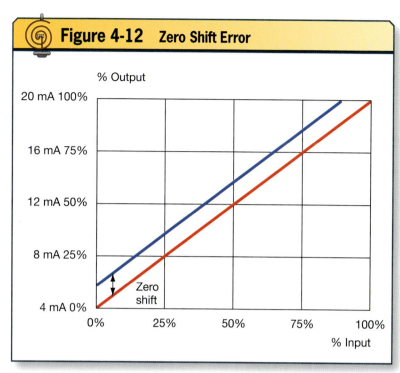

Figure 4-12. *The zero shifts for the transmitter are displayed by plotting the output signal.*

deviation based on information from the calibration sheet. For instance, assume the calibration sheet requires an accuracy of ±0.3% of the instrument output span. This requires manipulating the accuracy formula to solve for the deviation. The new formula would be as follows:

$$\text{Deviation} = \text{Span} \times \frac{\text{Accuracy}}{100}$$

$$\text{Deviation} = 16 \text{ mA} \times \frac{0.3}{100}$$

$$\text{Deviation} = 0.048 \text{ mA}$$

The instrument technician should ensure the output value of the transmitter at each test point is within 0.048 milliampere to satisfy the accuracy requirements for this calibration.

The resolution of a test standard reading in mADC affects the accuracy of a calibration. If an output standard was used and was only capable of measuring mADC in increments of 1.0 mADC, the calibration would lack the ability to measure an accuracy of greater than 6.25%. This is the best accuracy measurement that can be obtained:

$$\text{Accuracy} = \left(\frac{1.0 \text{ mA}}{16 \text{ mA}}\right) \times 100$$

$$\text{Accuracy} = 0.0625 \times 100$$

$$\text{Accuracy} = 6.25\%$$

Gain

The term *gain* describes the ratio of output divided by input of a control device. Gain can be thought of in terms of a mechanical example. The longer the handle of a pipe wrench, the more force that can be generated, much like an increased gain allows an instrument to generate a larger signal for a given input. Gain equals the output signal span divided by the input signal span:

$$\text{Gain} = \left(\frac{\text{Output signal span}}{\text{Input signal span}}\right)$$

Again, the example transmitter has an output signal span of 16 mADC and an input signal span of 100 pounds per square inch:

$$\text{Gain} = \left(\frac{16 \text{ mA}}{100 \text{ psi}}\right)$$

$$\text{Gain} = 0.16 \text{ mA/psi}$$

Thus, gain represents the magnitude of signal change for each pound per square inch of input. In this example, the transmitter signal increases or decreases to a respective input change of 1 pound per square inch by 0.16 mADC. If the input pressure was set at 0 pounds per square inch, the output would be 4.00 mADC. Increasing the input pressure to 1 pound per square inch would cause the output signal to increase to 4.16 mADC.

Repeatability

Emphasis has been placed on the accuracy of a transmitter, but repeatability is an important characteristic of measurement devices. *Repeatability* of an instrument describes the ability of a transmitter to reproduce output readings when exactly the same measured value is applied to it consecutively, under the same conditions, and approaching from the same direction. To put it simply, a repeatable instrument records an identical output signal (e.g., 7.93 mADC) each time the same input signal (e.g., 24.56 pounds per square inch) is applied. A relationship can be shown between accuracy and repeatability. **See Figure 4-13.**

In the target example, the group of holes that are shown tightly grouped are repeatable, but are not accurate in regards to the bullseye. In comparison, the group of holes that are near the bullseye are more accurate but less repeatable. Repeatability does not imply accuracy, and accuracy does not imply repeatability. The repeatability of an instrument is an inherent characteristic that cannot be modified or adjusted.

To clarify, accuracy is how closely an instrument reflects its input, and repeatability is a term given to an instrument that reflects the same output signal from a constant input signal, time after time.

AS-FOUND TEST

Calibration requires a linear approach, from beginning to end, to complete the procedure. It is important to document each step and record the data in the appropriate form. The current state of operation must be recorded for the instrument under test before changes are made. This *as-found test*, or calibration procedure during which the current, unaltered state of instrument operation is documented, can identify the change over time of an instrument. This information can be used for future maintenance scheduling and to ensure the process is operating at the optimum performance level. An instrument that exhibits a significant change, or drift, in accuracy over a 1-year calibration cycle may require more frequent calibrations intervals or may require replacement.

TechTip!

Mechanically based devices such as pressure switches are especially susceptible to repeatability errors. The technician should verify the repeatability by performing several tests to ensure the switch operates within the specified accuracy tolerance.

Identification of Calibration Errors

An instrument can be calibrated to within certain specifications of accuracy and must have sufficient repeatability to maintain accuracy. During the as-found test, the errors in the instrument must be identified. Instrument errors are discovered by comparing the measured output with the expected output for a specific input.

Classification of Calibration Errors

Calibration can adjust for several types of instrument errors. The five common instrument errors are zero shift, span shift, dead band, hysteresis, and nonlinearity.

Zero shift and span shift are the most common errors found in an instrument's calibration. Often, zero shift and span shift exist simultaneously.

Zero Shift

Zero shift, or a change resulting from an error that is the same throughout the scale, is the term used to describe an instrument error whose output is consistently higher or lower than the expected value. This shift is consistent throughout the entire output signal span. Another way to express a zero shift is to say the deviation is consistent throughout the signal span. Comparing the expected output for the transmitter and the actual output for the same transmitter may reveal a zero shift that elevates the output. An instrument with a zero shift graphed next to the ideal output exhibits a parallel set of lines. This indicates a zero shift. An elevated output suggests that the output signal started higher than expected. Zero shift also can start lower than expected, a condition called suppressed zero shift.

Span Shift

Span shifts are defined as an error identified as an output signal that does not reflect the desired span. These errors can be identified by performing the five-point check and recording the measured output signal values. With a span shift, the measured outputs vary from the expected values, but there is no equal shift value from the expected output. The technician can identify span shifts by examining the slope of the graphed line. The lines are no longer parallel but instead diverge, indicating the span shift. The measured outputs begin at the origin but do not follow the expected outputs. **See Figure 4-14.**

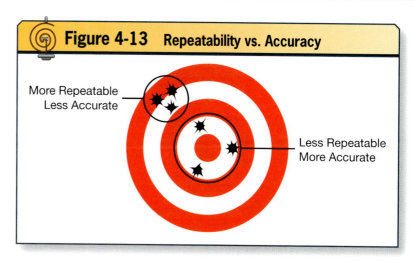

Figure 4-13. Repeatability and accuracy are two separate measures.

ThinkSafe!
The same level of safety awareness should be exercised in the shop and in the field. Always wear the same personal protective equipment when performing bench calibrations or field calibrations.

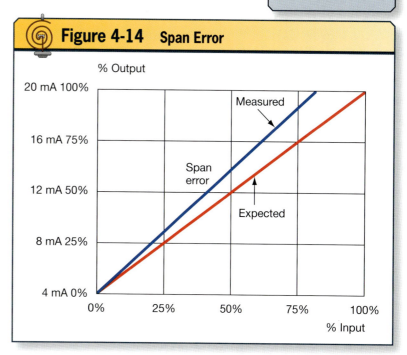

Figure 4-14. The span errors for the transmitter are displayed by plotting the output signal.

Figure 4-15 Hysteresis Error

Input Values		Output Values		
Test Points	Inputs	Expected	Upscale	Downscale
0%	0 PSI	4 mA	4.6 mA	3.4 mA
25%	25 PSI	8 mA	8.6 mA	7.4 mA
50%	50 PSI	12 mA	12.6 mA	11.4 mA
75%	75 PSI	16 mA	16.6 mA	15.4 mA
100%	100 PSI	20 mA	20.6 mA	19.4 mA

Figure 4-15. The results of hysteresis are noted in the different output value in the upscale and downscale columns.

Dead Band

Dead band is a change through which the input to an instrument can be varied without initiating an instrument response. Dead band can be a very small value within the specified tolerances. When the dead band becomes significant enough to stray beyond the specified tolerances, the instrument may be in need of repair. On certain instruments, such as pressure switches, dead band is commonly both desired and user adjustable. On other instruments, dead band may not be adjustable and the instrument must be replaced.

Hysteresis

Hysteresis is the difference between upscale and downscale results in instrument response when subjected to the same input approached from opposite directions. The five-point check should include both upscale and downscale test points. An instrument that produces different readings depending on whether the test is progressing upscale or downscale is said to have hysteresis. Only the direction from which the test point is approached causes the hysteresis. Hysteresis is not commonly found in electronic instruments; it usually occurs in mechanical measuring devices such as pneumatic transmitters. See **Figure 4-15.**

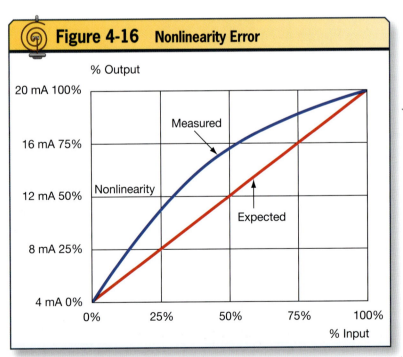

Figure 4-16. The nonlinearity errors for the transmitter are displayed by plotting the output signal.

Nonlinearity

An instrument's ideal output is a perfectly linear plot on a graph. *Linearity* is the proximity of a calibration curve to a specified straight line, which is expressed as the maximum deviation of any calibration point on a specified straight line during one calibration cycle. Instruments that are diagnosed as nonlinear produce an output that does not create a straight line when measured and plotted. Instead, the line follows a random path between the output LRV and output URV. A linearity adjustment is provided on some instruments but is less common than zero and span adjustments. A large linearity problem is not correctable, and the instrument must be repaired. In some cases, nonlinearity errors are small enough that their effect is scarcely noticeable, but these errors often increase with time. The technician can identify nonlinearity on an instrument calibration sheet when the output LRV matches the expected value and the output URV matches the expected value but the measured values between LRV and URV do not match the expected values. The five-point check tests for nonlinearity that could exist with an instrument that exhibits acceptable readings at 0% and 100%. **See Figure 4-16.**

ERROR CORRECTION PROCEDURES

The procedure to correct identified instrument errors varies by the type of instruments. General guidelines apply to most instruments. Conventional and smart instruments require different calibration procedures based on the operation of the device.

Conventional Instruments

Once the as-found test has been completed and the data are used to identify instrument error, the technician can begin the second step of calibration: error correction. Instruments that exhibit hysteresis, nonlinearity, and dead band are less commonly found.

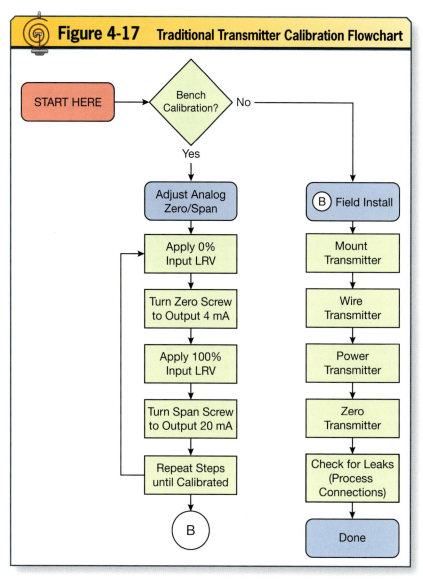

Figure 4-17. A calibration flowchart can be used for calibrating a conventional transmitter.

Consult the documentation provided by the manufacturer to determine whether an instrument with these errors can be calibrated. **See Figure 4-17.**

Zero Shift

An as-found test that shows a consistent zero shift requires an adjustment to the zero screw located on the transmitter. The calibration standards were already connected to the instrument during the as-found test procedure. The technician should set the input standard to provide the 0% value (LRV). While observing the output

For additional information, visit qr.njatcdb.org Item #2119

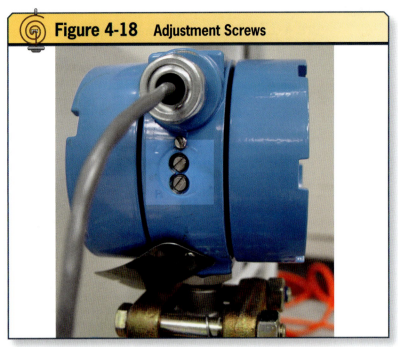

Figure 4-18. The zero and span adjustment screws of the pressure transmitter appear behind a transmitter access plate.

signal on the output calibration standard, the technician turns the zero screw until the output is within the specified tolerance.

Span Shift can fix

The procedure to correct a span shift starts by identifying the error from as-found data. The technician should set the input standard to provide the 100% value (URV). To correct a span shift in an instrument, the span adjustment screws can be turned to reduce the span shifts until they reside within the specified tolerance. **See Figure 4-18.**

Zero and Span Shift

An instrument that exhibits both zero and span shifts is commonly encountered during an as-found calibration test. When both errors are present, the zero output must be corrected and then the span is adjusted. When the span is adjusted, the zero setting should be checked several times to recalibrate. Many instruments have "interactive" zero and span adjustments (because of the construction of the transmitter). This means that when the zero screw is turned, the span is affected. Similarly, if the span is adjusted, the zero is affected. The adjustment process may need to be repeated several times to bring both zero and span shifts within tolerable levels. Before beginning an as-left test, check the output again to verify that it is correct.

Smart Instruments

The technician must apply a different set of procedures to calibrate a smart instrument. The operating principle of a smart instrument is similar to that of a conventional instrument, but instead of using analog electronic circuitry it relies on a microprocessor. The technician must distinguish between the concepts of setting a range (LRV to URV) for an instrument and actual calibration of the transmitter. When working on a conventional transmitter, both ranging and calibration operations are simultaneously performed. However, a smart instrument separates ranging and calibration into two distinct steps. Thus, a smart instrument can be

For additional information, visit qr.njatcdb.org
Item #2120

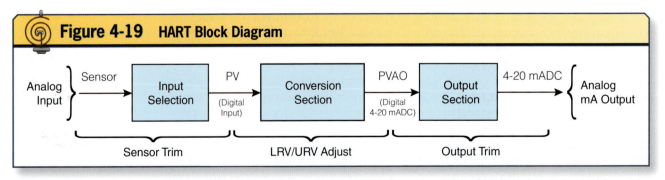

Figure 4-19. A block diagram can be used to show the analog input, process variable (PV), process variable analog output (PVAO), and the analog output.

reranged without affecting calibration. For example, a smart pressure transmitter can be reranged for 0–300 pounds per square inch to measure 100–500 pounds per square inch without a change in calibration. The technician may be required to fill out a new calibration data record if the device is reranged.

Because the smart instrument has a microprocessor, which receives and produces digital signals, and both the input and the output of the transmitter are analog values (i.e., pounds per square inch and mADC), there must be a conversion of the signals internally in the device. These conversions, performed by the analog-to-digital converter (ADC) and the digital-to-analog converter (DAC), are subject to inaccuracy and therefore must be calibrated. A relationship exists between the analog and the digital signaling that takes place within a smart instrument. **See Figure 4-19.**

A process called sensor trim is performed to calibrate the ADC and the DAC of a smart instrument. This trim process causes a change in how the instrument is characterized. There is a logical flow of the calibration procedure for smart instruments. **See Figure 4-20.**

AS-LEFT TEST

The next step in a calibration procedure is the as-left test. This step records the operation of the transmitter after the technician has identified and corrected all errors. The data contained in the as-left portion of the test are used to validate the proper operation of an instrument and provide a benchmark for comparison during the next scheduled calibration interval. The technician should provide the transmitter inputs at the required values and record the outputs the transmitter produces on the as-left data entry section. The final step is for the technician to verify that all test points are suitably accurate and that the transmitter exhibits no additional errors. Upon completion, the calibration sheet should be submitted to the proper contact person on the job and kept for future reference and regulatory compliance.

Calibration Sheet Data Entry

Calibration sheets are used when calibrating an instrument to record the actual values applied to and data response

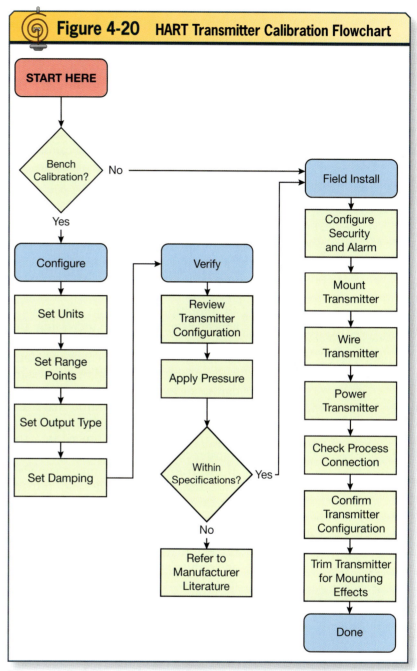

Figure 4-20. The steps to calibrate a smart conventional transmitter follow a logical procedure.

For additional information, visit qr.njatcdb.org Item #2121

of the device calibrated. A calibration sheet is the record used to determine whether a device is calibrated correctly, which equipment was used during calibration, and who performed the calibration.

Calibration sheets can vary from job to job, but in most cases the necessary information is universal. A calibration sheet not only shows that an instrument has been calibrated but also helps to create a permanent device history. The data listed on a calibration sheet provide information for maintenance personnel. In short, a calibration sheet shows how an instrument has performed in the past and what can be expected in the future.

Calibration sheets can either be a paper record or, with increasing frequency, a digital record that is uploaded to an instrument management database. Many newer calibration standards, such as the Fluke 754 documenting process calibrator, have the ability to record the data gathered during the calibration procedure and transfer this information electronically. The instrument management databases store the digital records and provide enhanced features, such as alert generation when an instrument is due for calibration, record access anywhere in the world with the proper credentials, and redundant backups so that data are more secure.

A calibration sheet can have different sections that are filled in. For example, a calibration sheet can be filled in to simulate a level transmitter 301 (LT 301). **See Figure 4-21.**
- Job—The jobsite or company where the calibration is being performed
- Instrument I.D.—The identification, or tag, number assigned to this instrument.
- Manufacturer—The manufacturer of the instrument
- Service—The type of process for which the instrument is used
- Model—The manufacturer's model number for the instrument
- Serial Number—The manufacturer's serial number for the instrument
- Input—The instrument's input or process signal
- Output—The instrument's output
- Calibration Data—The instrument's desired calibration range
- Remarks—Special notes or remarks on the instrument
- Performed by—Filled in by the worker according to jobsite procedures
- Date—The date of the calibration.
- Test Equipment—The equipment needed to perform the calibration. This shows that all calibration or test equipment used meets jobsite requirements. This also makes test equipment traceable, if needed at a future date. A Pneucal IV-B, Fluke 87, and a HART 475 communicator were used for this calibration. It also shows the model numbers and serial numbers of the three items used in this calibration.
- Input—The medium used to calibrate the instrument
- Output— The type of output from the calibration that is expected
- Calibration Test—The actual calibration of the instrument.

In this example, five test points have been selected, going both upscale and downscale. The calibration test section includes the following columns:
- Value—The actual value of the input signal at all test points
- Desired—The ideal output value at all test points when the instrument is calibrated perfectly
- As Found—The output values of the instrument after the initial calibration check
- As Left—The results of the calibration after all adjustments have been made

Verification of Calibration Sheet Accuracy

Data recorded in a calibration sheet must be recorded accurately and completely. A common reason for calibration sheet rejection is simple errors, such as listing the wrong units or forgetting a decimal point. The calibration sheet serves as a permanent document that ensures traceability for the instrument and therefore must be accurate.

The level of detail included in a calibration sheet is dictated by the facility. For example, the calibration sheet may require the milliampere values be listed to two decimal places (e.g., 4.00 mADC).

Figure 4-21 Calibration Data Record

CALIBRATED DATA RECORD

JOB: D&D DISTILLERY

INSTRUMENT I.D.: LT 301   **MANUFACTURER:** ROSEMOUNT

SERVICE: LEVEL

MODEL: 3051   **SERIAL NUMBER:** 125176

INPUT: 0-20 INCHES W.C.   **OUTPUT:** 4-20 mA DC

CALIBRATION DATA: 0-20 INCHES W.C.

REMARKS: LOCATED NEAR COLUMN DD-36

PERFORMED BY: D.RILEY, EPRI TECHNICIAN   **DATE:** CURRENT DATE

TEST EQUIPMENT	MODEL	S/N
Pneucal	IV-B	B0430G006
FLUKE	87	57680078
HART communicator	475	53158

INPUT: (INCH W.C.)   **OUTPUT:** mA DC

%	VALUE	DESIRED	AS FOUND	AS LEFT
0	0 IN. W.C.	4.00 mA	3.81 mA	4.00 mA
25	5 IN. W.C.	8.00 mA	8.25 mA	8.00 mA
50	10 IN. W.C.	12.00 mA	12.69 mA	12.00 mA
75	15 IN. W.C.	16.00 mA	17.12 mA	16.00 mA
100	20 IN. W.C.	20.00 mA	21.56 mA	20.00 mA
75	15 IN. W.C.	16.00 mA	17.12 mA	16.00 mA
50	10 IN. W.C.	12.00 mA	12.69 mA	12.00 mA
25	5 IN. W.C.	8.00 mA	8.25 mA	8.00 mA
0	0 IN. W.C.	4.00 mA	3.81 mA	4.00 mA

Calibration Sheet

Figure 4-21. A calibration data record has many different sections that can be filled in.

A calibration sheet that is filled out to only one decimal place (for example, 4.0 mADC) does not sufficiently describe the accuracy of the reading and, therefore, is considered inadequate by the facility. The engineering units must be entered into the calibration sheet on each numerical entry. A data entry listing only 4.00, without the mADC unit, does not comply with the requirements of the calibration sheet. In all cases, attention to detail and strict adherence to the calibration procedure are mandatory to complete proper calibration on an instrument.

SPECIFICATION SHEETS

Instrument specification sheets (spec sheets) are used to list the capabilities and applications of an instrument. They show how a certain instrument should perform and how it should be installed. Spec sheets are used to specify what instrument should be purchased when building a new instrumentation system. They are also used to determine which instrument should be purchased to replace a failing instrument on an existing loop. Spec sheets are generally stored in a plant database, in either paper or electronic form, and are often sorted by tag number.

The main purpose of spec sheets is to assemble, in one place, all information related to the instrument hardware. Spec sheets serve as a permanent record of devices that are in use. Typical data that spec sheet would contain include pressure range, temperature range, process conditions, calibration ranges, tag numbers, transmitter type, control functions, and materials of construction of the instrument. Spec sheets are the primary source of instrument information when designing and ordering instruments for a process. Spec sheets contain the necessary information for field personnel to maintain and install the instrument.

Consider spec sheet examples of a control valve, pressure transmitter, and a resistance temperature detector. **See Figures 4-22 through 4-24.**

TechTip!

Pipe flanges are rated according to the ASME 16.5 standard. Class designations include 150, 300, 400, 600, 900, and 1,500. These designations are used to specify the proper pressure and temperature ratings of a pipe flange and gasket assembly. The technician must verify that the proper rating class is used to prevent leakage or rupture of the process piping connections.

Figure 4-22 Specification Sheet for a Control Valve

			CONTROL VALVES SANITARY, ON/OFF				DS-SYV-006	
			No	By	Date	Revision	Spec. No. SYV	Rev. B
Project:			A	VG		IFA	Contract	Date
Customer:			B	VG		IFA		
Job No.:							Req. P.O.	
Plant:								
Location:							By	Chk'd	Appr.
File: \cvalv30a.doc							VG	KG	
Tag: YV-83102	Service:	63mm-EPU-SST4-83004 to Room # Z 7				PID:	PID-947-831	

			Units	Max Flow	Norm Flow	Min Flow	Shut-Off
1	Fluid	Water			Crit. Press PC		
2	C	Flow Rate:	LPM (GPM)	100			
3	O	Inlet Pressure:	BAR (PSIG)	6.15(90)			6.15(90)
4	N	Outlet Pressure:	PSIG				
5	D	Inlet Temperature:	DEG C	22			
6	I	Spec Wt/Spec.Grav/Mol Wt.:	SG				
7	T	Viscosity/Spec Heats Ratio:	CP				
8	I	Vapor Pressure PV:	PSIA				
9	O	*Required CV:	----				
10	N	*Travel:	%				
11	S	Allowable/*Predicted SPL:	dBA	<85			
12							

13	L	Pipe Line Size & Schedule	In: 63 mm (2-1/2")		53		*Type:	PNEUMATIC, REVERSE
14	I		Out: 25 mm (1")		54		*Mfr & Model:	ITT, ADVANTAGE
15	N	Pipe Line Insulation:			55		*Size: A 2086	Eff. Area*:
16	E	*Type:	SANITARY, DIAPHRAGM, PURO-FLO		56	A	*On/Off: YES	Modulating:
17		*Size: DN 25 (1")	ANSI Class*:		57	C	Spring Action Open/Close:	CLOSE
18	V	Max Press / Temp*:			58	T	*Max Allowable Pressure:	120 PSIG
19	A	*Mfr & Model No.:	ITT / DN25/9876-R2-A2086		59	U	*Min Required Pressure:	80 PSIG
20	L	*Body/Bonnet Matl:	316 L SS, FORGED, Zero Static		60	A	Available Air Supply Pressure:	
21	V	Liner Material / ID:			61	T	Max: 100 PSIG	Min: 60 PSIG
22	E	End Connection:	In: 63 X 63" U-tubing w/buttweld ends		62	O	Actuator Orientation:	
23	/		Out: DN 25 (1") Tri-Clamp		63	R	Handwheel Type:	NONE
24	B	Flg Face Finish:			64		Solenoid Valve Tag No: SV-83102, 220 VAC, 50 Hz, manual maint.	
25	O	End Ext. / Mtl:			65		operation, NEMA 4X, 1/4" NPT ports, Class F Coil. 3/32" Orifice	
26	N	*Flow Direction:			66		*Mfr & Model:	ASCO 8320-G184
27	N	*Type of Bonnet			67		Positioner – YES/NO:	NONE
28	E	Lube & Iso Valve:	Lube:		68	P	Input Signal:	
29	T	*Packing Material			69	O	*Type:	
30		*Packing Type			70	S	*Mfr & Model:	
31					71	I	*On Inc Signal Output Incr. / Decr	
32		*Type:			72	T	Gages:	By-pass:
33		*Size:	Rated Travel: 0.25"		73		*Cam Characteristic:	
34		*Characteristic:			74		Type: N/A	Quantity:
35		*Balanced/Unbalanced:			75	S	*Mfr & Model:	
36	T	*Rated CV: 18.6	FL: XT:		76	W	Contacts/Rating:	
37	R	Diaphragm Assembly: Grade R-2, PTFE w/backing cushion			77		Actuation Points:	
38	I	*Seat Material:			78			
39	M	*Cage / Guide Material:			79		*Mfr & Model:	
40		*Stem Material:			80	A	*Set Pressure:	
41		*Stem Diameter:			81	I	Filter:	Gauge:
42					82	R		
43		NEC Class:	Group: Div:			S		
44						E		
45		NOTE: 1. Internal polish to be 180 grit.			83	T	*Hydro Pressure:	
46	O				84	T	ANSI/FCI Leakage Class:	IV
47	T				85	E		
48	H				86	S		
49	E					T		
50	R					S		
51								
52								

*Information supplied by manufacturer unless already specified

Figure 4-22. Information contained in specification sheets varies from product to product, and will usually be provided by the manufacturer.

Figure 4-23 Specification Sheet for a Pressure Transmitter

Forms for Process Measurement and Control Instruments, Primary Elements, and Control Values

		PRESSURE TRANSMITTER (SANITARY)				DS-PT-001	
		No	By	Date	Revision	Spec. No. PT 1	Rev. A
Project:		A	VG		prelim	Contract	Date
Customer:							
Job No.:						Req.	P.O.
Plant:							
Location:						By VG	Chk'd KG	Appr.
File: \prsntr1a.doc							

G E N E R.	1	Tag No.:	PT-83001	PT-84001
	2	Service:		
	3	P&ID:	PID-947-830	PID-947-840
	4	Line Identification Number:		
	5	Line Size/Sch.No.:	63 mm (2.5")	50 mm (2")
D A T A	6	Fluid:	Water	Water
	7	Operating Pressure:		40 PSIG
	8	Max. Pressure:	103 PSIG (7 BAR)	103 PSIG (7 BAR)
	9	Operating Temperature:	30 deg C/80 deg C	30 deg C/80 deg C
T R A N S M I T T E R	10	Function:	Pressure Transmitter	Pressure Transmitter
	11	Body Material:	316 SS	316 SS
	12	Pressure Rating*:		
	13	Element Type*:		
	14	Element Material*:		
	15	Process Connection:	1-1/2" Tri-Clamp Sanitary fitting	1-1/2" Tri-Clamp Sanitary fitting
	16	Adjustable Range:	0-15 to 0-150 PSI	0-15 to 0-150 PSI
	17	Calibrated Range:	TBD	TBD
	18	Accuracy:	+/- 0.25% of span	+/- 0.25% to span
	19	Indicating Meter & Range	N/A	N/A
	20	Output Signal:	4 – 20mA DC., 2 wire	4 – 20mA DC., 2 wire
	21	Power Requirements:	External 24 V DC	External 24 V DC
	22	External Load Resistance:	750 ohms at 24 V DC	750 ohms at 24 V DC
	23	Electrical Class*:		
	24	Mounting:	Direct	Direct
	25	Conduit Connection:	½" NPT, Conduit housing	½" NPT, Conduit housing
	26	Manufacturer:	Rosemount Inc.	Rosemount Inc.
	27	Model No.	2090 FG 2 A 2D E 1	2090 FG 2 A 2D E 1
A C C E S S O R I E S	28	Diaphragm Material:	316 L SS	316 L SS
		Type:		
		Manufacturer:		
		Wet Part Matrl.:	316 L SS	316 L SS
		Fill Fluid:		
		Process Connection:		
		Transmitter Connection:		
	29	Capillary:		
		Connection/Length		
		Material:		
		Other:		

Notes: 1. Vendor to provide engraved SS tag with SS wire.

*: Vendor to specify.

Data Sheet

Figure 4-23. Spec sheets contain all information needed to install and maintain each piece of equipment in a process.

Figure 4-24 Specification Sheet for an RTD Sensor

Forms for Process Measurement and Control Instruments, Primary Elements, and Control Values

		Resistance Temperature Sensors (sanitary)				DS-TE2-001	
		No	By	Date	Rev.	Spec. No.	Rev.
						TE-2	A
		A	VG		prelim	Contract	Date
Project:							
Customer:						Req.	P.O.
Job No.:							
Plant:						By	Chk'd	Appr.
Location:						VG	KG	
File: \RTDSAN1A.DOC								

ELEMENT

1. RTD Type: 3-wire, sanitary
2. Ice Point Resistance: 100 ohm
3. DIN Coefficient: 0.00385 ohms/ohm/deg C
4. Temp Range: -45 to 176 deg C
5. Accuracy: +/- 0.03% of range
6. Single or Dual: dual
7. Sheath Material: 316L SS
8. Sheath OD: 0.25"
9. Construction: 1-1/2" Tri-Clamp Fitting
10. Housing: Conduit Housing, 304 SS
11. Cable length: N/A
12. Manufacturer & Model No.: Anderson & SW32004- 0370000

Well or Tube

13. Type: N/A
14. Manufacturer & Model No.:

RTD Transmitter

15. Mounting: Remote (Control Panel CP-3)
16. Output:
17. Power Supply:
18. Accuracy:
19. Integral Meter:
20. Manufacturer:
21. Model No.:

Tag No.	Rev. No	Well Dim "T"	Well Dim "U"	Element Length	Calib. Range deg C	Line Id. No.	Oper. Temp.deg C	P & ID	Notes.
TE-83002	A			32 mm		63 mm-EPU-SST4-83004	22/80	PID-947-830	
TE-83003	A			32 mm		63 mm-EPU-SST4-83004	22/80	PID-947-830	
TE-84002	A			32 mm		50 mm-EPU-SST4-83004	22/80	PID-947-830	
TE-84003	A			32 mm		50 mm-ETU-SST4-83004	22/80	PID-947-830	

Notes: 1. Vendor to provide engraved SS tag with SS wire.
2. Surface finish (Wettable parts): Ra max=0.2 microns

*: Vendor to specify.

Data Sheet

Figure 4-24. Spec sheets serve as a permanent record of all information related to the instrument hardware, including function and materials.

SUMMARY

The calibration procedure is essential to ensure the proper operation of an instrumentation process. The adherence to a set procedure is mandatory for consistency and proper operation of all instruments. Technicians must have knowledge of each piece of equipment to be used during the calibration so that it is used properly. Following the procedures described for calibrating devices should allow the technician to become familiar with calibration equipment and to perform the procedure correctly. Properly identifying and correcting calibration errors are fundamental tasks for an instrument technician.

Documentation is used by technicians to help them install and maintain a control system. Technicians must use the proper documentation for the job and know how to apply it in the field. Every project involving control systems requires time for reviewing documentation to ensure the safe installation and maintenance of the equipment.

REVIEW QUESTIONS

1. What is the accuracy of a transmitter that has a maximum deviation of 0.18 milliampere and an output span of 16 mADC?
   a. 1.80%
   b. 1.13%
   c. 0.11%
   d. 2.3%

2. What is the maximum allowed deviation (in mADC) for a device that has an accuracy requirement of 0.5% and an output span of 16 mADC?
   a. 0.08 mADC
   b. 0.5 mADC
   c. 0.8 mADC
   d. 5 mADC

3. What is the span of an instrument with a LRV of 37 inches of water and an URV of 210 inches of water?
   a. 37–210 inches of water
   b. 210–37 inches of water
   c. 173 inches of water
   d. Insufficient information to solve

4. Determine the five input test points (0%, 25%, 50%, 75%, and 100%) used for a temperature transmitter with calibration range of 32°F to 212°F and an output of 4–20 mADC.
   a. 32°, 77°, 122°, 167°, 212°
   b. 32°, 45°, 90°, 135°, 212°
   c. 4, 8, 12, 16, and 20 mADC
   d. 32, 77, 122, 167, and 212 mADC

5. What is the gain for a transmitter with an input span of 20 inches of water column and an output signal of 4–20 mADC?
   a. 1.25 mADC per inch of water
   b. 1.25 mADC per pound per square inch
   c. 0.8 inches of water per mADC
   d. 0.8 mADC per inch of water

6. What calibration error is present when a device records a consistent error of 0.2 mADC at each of the five test points?
   a. Hysteresis
   b. Nonlinearity
   c. Span shift
   d. Zero shift

7. When the as-found calibration shows that the output varies depending on the direction of test points (upscale or downscale), the instrument has this type of error.
   a. Dead band
   b. Hysteresis
   c. Nonlinearity
   d. Span shift

8. What should the output signal be at 73 inches of water on an instrument with an input range of 0–250 inches of water and an output range of 4–20 mADC?
   a. 7.32 mADC
   b. 4.67 mADC
   c. 8.67 mADC
   d. 15.68 mADC

9. A smart instrument differs from a conventional instrument because of the presence of this feature.
   a. Digital readout
   b. Microprocessor
   c. Pressure manifold
   d. Wireless signaling

10. Which of the following describes the use of an instrument spec sheet?
    a. Defines the calibration procedure
    b. Documents the technician who performed the calibration
    c. Referenced for purchasing and maintenance
    d. Shows a map of loop components

Principles of Pressure

Pressure is a physical property that plays an important role in the existence of life. To understand the concepts of the different pressure measurements encountered while performing instrumentation calibration work, the fundamentals of pressure are examined first. With an understanding of what pressure is, further study into how pressure is observed, measured, and indicated occurs. The following are some examples situations in which pressure measurements are valuable:

- Determining whether sufficient pneumatic pressure exists for a machine to operate properly
- Checking the quality of an air filtration system as a filter becomes clogged
- Monitoring oil pressure within an engine
- Monitoring the fluid level within a tank by measuring the fluid's pressure
- Measuring the velocity of a fluid, such as compressed gas or a liquid

Four types of pressure measurement are studied: differential, absolute, gauge, and vacuum. Given the nature of pressure characteristics, it is important to understand the operation of pressure sensors and their applications to properly handle, install, configure, and calibrate. Once a usable pressure signal has been determined, this value can be interpreted as a measure of fluid flow, level, and even temperature.

Objectives

» Identify the concepts of pressure that are found in everyday life.
» List the types of elements used to measure pressure.
» Describe the functions of pressure elements.
» Describe pressure measurement applications.
» Perform unit conversions between common pressure scales
» Examine the relationships that are present with pressure and temperature.

Chapter 5

Table of Contents

Pressure Fundamentals 88
Pressure Scales 91
    Differential Pressure 94
    Absolute Pressure 94
    Gauge Pressure 95
    Vacuum Pressure 96

Pressure Devices and Their Functions 96
    Manometer 97
    Bourdon Tube–Type Detectors 98
    Bellows Pressure-Sensing Element 98
    Pressure Switch 99
    Pressure Transmitter 101
        Capacitance-Based Transducers 102
        Piezoresistive Devices 102
        Piezoelectric-Based Transducers 103
        Linear Variable Displacement Transformer 103
Pressure Units of Measure 105
Multivariable Relationships 106
Summary 109
Review Questions 109

For additional information, visit qr.njatcdb.org Item #2690

PRESSURE FUNDAMENTALS

Pressure is defined as force over area:

$$P = \frac{F}{A}$$

where

P = Pressure

F = Force

A = Area

To fully understand the concept of pressure, several other concepts must first be examined. The *mass* of an object is the amount of matter in an object. *Matter* is the substance or substances of which any physical object is composed. There is often confusion when the terms "mass" and "weight" are used interchangeably. In actuality, though they seem the same, there are specific differences. Mass is the amount of matter in an object, and weight takes another factor, known as gravity, into consideration. *Gravity* is the force that attracts a body of mass toward Earth's center or to another body of mass.

Pressure is the force exerted by a fluid per unit area; pressure units include pounds per square inch (psi; English system) and pascals or newtons per square meter (Pa or N\m²; SI units). If a 10-pound box of fluid that measures 1 foot by 1 foot square regardless of height (1 square foot) is sitting on a table, the pressure exerted on that table is 10 pounds per square foot. If a second box of identical mass and size is placed directly on top of the first, the force is effectively doubled and the pressure exerted is 20 pounds per square foot. If the second box is instead placed beside the first, the force is doubled but so is the area, because the combined surface area of the two boxes against the table is 2 square feet. Therefore, the force remains the same, and the pressure remains 10 pounds per square foot. If half of the fluid in each box is removed, the force is reduced by half. Hence, the pressure exerted by each box is now 5 pounds per square foot. **See Figure 5-1.**

Pressure plays a vital role in allowing life to exist. An examination of the atmosphere here on Earth can help to define what pressure truly is. It can be easy to forget that the surrounding air is matter. Because it is matter, there is a mass quantity. This mass quantity, when combined with Earth's gravitational pull, generates a force. This force is called atmospheric pressure. In outer space, there is no air and therefore no matter. Without any type of matter, regardless of the minimal gravitational pull, no pressure can be generated. This lack of pressure is known as absolute zero. **See Figure 5-2.**

At sea level, atmospheric pressure is measured at 14.7 pounds per square inch absolute, or *PSIA*, the pressure referenced to an ideal vacuum, because the pressure exerted by the force of the atmosphere on Earth is 14.7 pounds per square inch greater than that in outer space. The suffix letter A is used to indicate that this measurement is taken from absolute zero. As elevation from sea level increases, the amount of atmospheric pressure decreases because less atmospheric mass is present. By measuring atmospheric pressure,

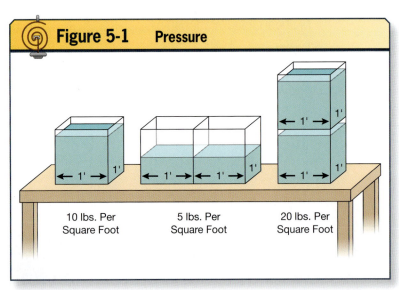

Figure 5-1. The force exerted by the fluid in the boxes varies depending on position of and amount of fluid in the boxes: (a) one full box, (b) two half-full side-by-side boxes, and (c) two full stacked boxes.

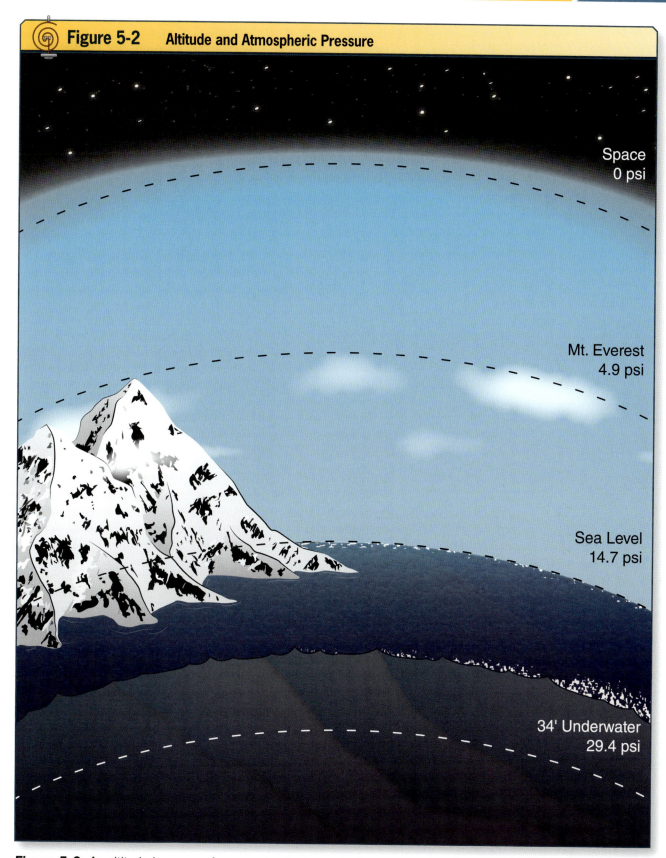

Figure 5-2. As altitude increases from sea level, atmospheric pressure decreases because there is less weight associated with the air.

elevation change can be determined with great precision. **See Figure 5-3.**

Without pressure, specifically atmospheric pressure, it is impossible for the human body to sustain life. When a breath of air is taken, the diaphragm expands the lungs, causing the volume of space in the lungs to increase. As this happens, the amount of matter in the lungs remains the same; therefore, the pressure decreases in the lungs. This creates a brief pressure differential. The atmospheric pressure counteracts this and forces air into the lungs to bring everything back into *equilibrium*, the condition of a system when all inputs and outputs (supply and demand) are in balance. **See Figure 5-4.**

When the volume of air enters the lungs, a number of oxygen molecules are brought with it. When the atmospheric pressure is reduced, a smaller volume of air carrying oxygen molecules is introduced. Therefore, reduced numbers of oxygen molecules are carried into the human body, and supplemental oxygen is required. This is why aircraft are pressurized: to allow relatively normal breathing. As altitude increases, the relative atmospheric pressure decreases, because there is less mass of air and therefore less air pressure.

Figure 5-3 Atmospheric Pressure Chart

Pressure vs Altitude

Altitude (feet)	Absolute Pressure		
	in of Hg	mm of Hg	psia
0	29.921	760.0	14.70
500	29.38	746.4	14.43
1,000	28.86	732.9	14.18
1,500	28.33	719.7	13.90
2,000	27.82	706.6	13.67
2,500	27.31	693.8	13.41
3,000	26.81	681.1	13.19
3,500	26.32	668.6	12.92
4,000	25.84	656.3	12.70
4,500	25.36	644.2	12.45
5,000	24.89	632.3	12.23
10,000	20.58	522.6	10.10
15,000	16.88	428.8	8.28
20,000	13.75	349.1	6.75
30,000	8.88	225.6	4.36
40,000	5.54	140.7	2.72

Figure 5-3. Atmospheric pressure changes proportionally with a change in altitude. Pressures taken at sea level are highlighted in yellow.

Figure 5-4 Human Breathing

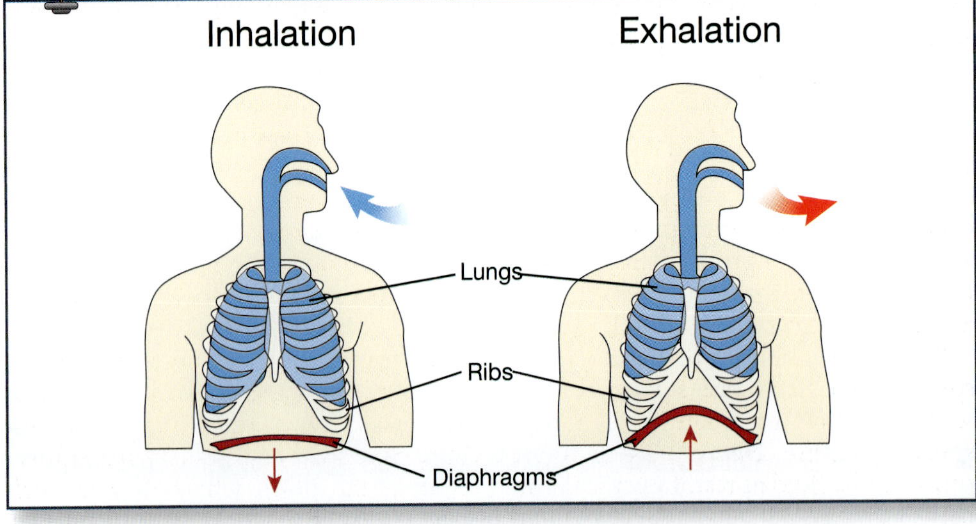

Inhalation | Exhalation

Lungs
Ribs
Diaphragms

Figure 5-4. Inhalation and exhalation of air from the lungs create equilibrium.

Conversely, when descending into a body of water, not only is atmospheric pressure exerting force on the surface of the water, but the water is exerting additional force on the human body exterior. For example, with a deflated balloon on a table and a book placed on top of the balloon, significant force is required to overcome the additional force being applied to the deflated balloon. Underwater scuba gear is designed to compensate for the additional ambient pressure surrounding the body by raising the pressure of the air that is breathed into the lungs to maintain equilibrium. Without this increase, air would not flow into the lungs. This has to happen automatically to compensate for any changes in depth, and care must be taken because the force generated by changes in pressure and surrounding conditions can result in rupture of the lungs.

PRESSURE SCALES

Pressure measurement requires a comparison between force and area to accurately measure pressure. Sir Isaac Newton discovered that for every force, an equal and opposite force is generated. For example, a table supporting a 10-pound box is applying a force opposite gravity to support the box. If the table cannot withstand this force, it breaks. Likewise, when a balloon is inflated, the pressure is pushing out on all sides of the balloon approximately equally. **See Figure 5-5.**

At the same time atmosphere is pushing back equally, so the balloon inflates relatively symmetrically. Once the structural integrity of the balloon is overcome by the internal pressure of the air, the balloon ruptures. This can also happen with pressure-rated vessels, such as steel tanks, that are used in industry.

Within the confines of Earth's atmosphere, all objects are subjected to a static pressure force known as atmospheric pressure. *Static pressure* is the pressure of a fluid at rest. In general, humans do not perceive the change in pressure, because the pressure within the

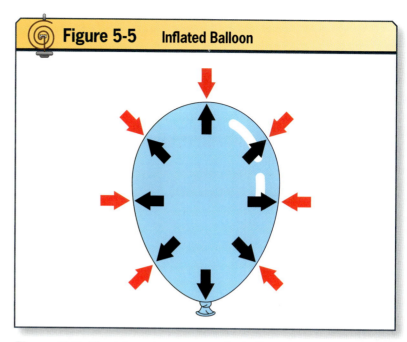

Figure 5-5. *The internal pressures of a balloon push equally against the atmospheric pressure on the exterior of the balloon.*

body is equal to the pressure outside of the body. If the body is submerged in water and the inner and outer pressures do not acclimate properly, there can be small pockets of trapped low pressure, especially within the inner ear canal. As the body is submerged, the pressure on the outside of the body is increased; however, the trapped pressure cannot equalize, so pain is felt. Conversely, as the human body is moved to a higher altitude, trapped pressure within the body generally is higher than the pressure on the outside of the body. This is the phenomena of the ears "popping." In either case, the sensation or pain felt is a result of the difference between the two pressures.

To determine a pressure value, the sought or measured pressure must be related to some type of a reference pressure. For instance, by comparing a pressure to a column of pure water at standard temperature and pressure (STP), the applied pressure can be related to the mass of the water. The applied pressure supports a column of water equivalent at the point where they are balanced. By measuring the change in height of the water, a relationship can

TechTip!

Even if the process fluid is not water, pressures are commonly read in inches of water. This is because water is a common reference standard across the world.

be drawn that pressure is equivalent to an x-inch water column.

Another way to relate pressure to a physical quantity is to determine the force in pounds per square inch exerted by pure water. **See Figure 5-6.**

One cubic foot of water weighs 62.43 pounds. If the cubic foot of water is placed in a box with a bottom side surface area of 1 square foot, the water exerts a force of 62.43 pounds per square foot. By dividing 62.43 pounds per square foot by 144 (the number of square inches in a square foot), it is determined that this water is exerting a force of 0.434 pounds per square inch.

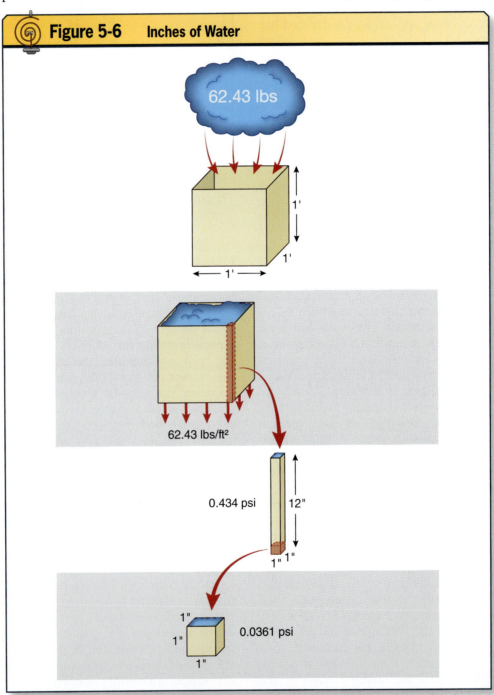

Figure 5-6. A series of calculations can be used to derive the inch of water ("H_2O) unit.

However, because this calculation is based on the 1-square-foot (12-square-inch) surface area of the box, the resulting value is for a column of water 12 inches high. By dividing by 12 (the number of inches in a foot), it is determined that a column of water 1 inch high is equivalent to 0.0361 pounds per square inch. Therefore, air pressure that is supporting a column of water 1 inch high, 1 inch wide, and 1 inch deep is exerting a pressure of 0.0361 pounds per square inch:

$$\frac{62.43 \text{ lbs}}{1 \text{ ft}^3} \times \frac{1 \text{ ft}^2}{144 \text{ in}^2} \times \frac{1 \text{ ft}}{12 \text{ in}}$$

$$= \frac{62.43 \text{ lbs.}}{1728 \text{ in}^3} = 0.0361 \text{ lbs/in}^3$$

The use of the inch-of-water-column measurement is effective for measuring small pressures; however, it can be impractical for measuring large values of pressure because of the numbers. For instance, 100 pounds per square inch exerts an equivalent pressure of 2,770 inches of water. **See Figure 5-7.**

Conversely, accurately measuring a value between 0.25 and 0.5 pound per square inch is considerably easier when using the inches of water column method. **See Figure 5-8.**

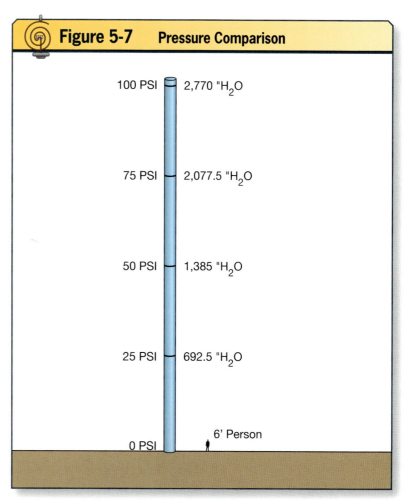

Figure 5-7. This graphic indicates the actual height of an equivalent water column at 100 pounds per square inch.

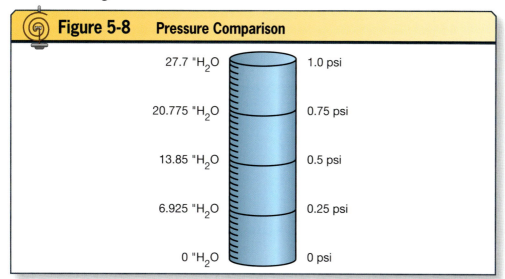

Figure 5-8. The fractional inch markings allow for accurate measurements of pressure less than 1 pound per square inch, without specialized digital pressure indicators.

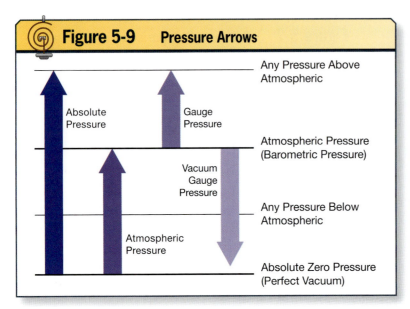

Figure 5-9. This graph shows the relationships among pressure scales.

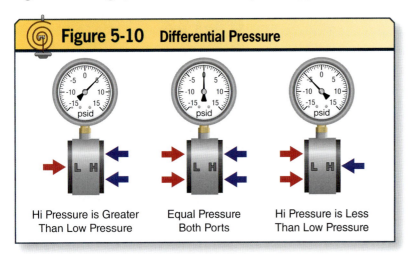

Figure 5-10. The indicator movement is represented by the effects of different differential pressures.

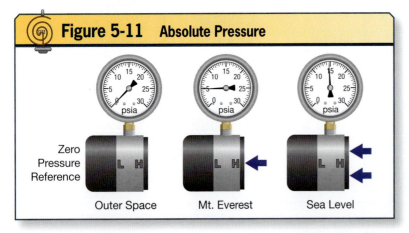

Figure 5-11. The indicator movement is represented by the effects of different absolute pressures.

Four common pressure measurements can be made: differential, absolute, gauge, and vacuum. **See Figure 5-9.**

Differential Pressure

Differential pressure is the difference in static pressure between two identical pressure taps. In differential pressure, both taps, also called ports, become active measuring ports. They are labeled high (HI) and low (LO). Depending on the calibration, a higher pressure on the HI port than on the LO port is indicated by a positive reading, whereas a lower pressure on the HI port than on the LO port is indicated by a negative reading. By measuring differential pressure, many other properties can be measured, such as velocity of fluids, filter condition, tank levels of both open and sealed vessels, and altitude. However, depending on how these devices are calibrated, it is easy to get an inversely proportional (high pressure, low signal) output as opposed to a directly proportional (high pressure, high signal) output. These devices are classified as direct acting or indirect (or reverse) acting. **See Figure 5-10.**

Absolute Pressure

Absolute pressure is the sum of gauge pressure plus atmospheric pressure. In absolute pressure measurements, the differential is between the measuring port (HI side) and a sealed vacuum reference chamber (LO side). These measurements use the unit of pounds per square inch absolute.

Absolute pressure is measured from the point of zero atmospheric pressure, also known as a perfect vacuum. Therefore, an absolute pressure measurement indicates the amount of pressure present at any point within the confines of the atmosphere. Outside of Earth's atmosphere, outer space is a point of zero atmospheric pressure. Pressure measurements have to have a reference. When making absolute pressure measurements, the reference must be to a point of zero pressure. Because it is impractical to stretch a tube into outer space to be a zero pressure reference, an absolute zero reference vessel must be used. To make this vessel, mechanical means are used to extract the air molecules from the sealed vessel, thereby creating a perfect vacuum condition. The vessel structure still has to withstand great pressure, because there is nothing internal of the vessel to counteract the external atmospheric pressure being placed on it.

Without any opposing pressure from the reference chamber, the measuring chamber is free to move to the full zero position. Therefore, the atmosphere exerts a positive pressure on the measuring port, thereby indicating an increase in pressure. For instance, an absolute pressure gauge lying at sea level would indicate 14.7 pounds per square inch absolute. This value is equivalent to the pressure exerted by the atmosphere. **See Figure 5-11.**

Because the pressure is referencing absolute zero, the unit of measure is succeeded by the letter A to show it is an absolute value. Because absolute pressure gauges require a near-perfect vacuum reference, they are not only extremely fragile but also expensive to manufacture.

Gauge Pressure

Measurement of *gauge pressure*, absolute pressure minus local atmospheric pressure, is common. In gauge pressure measurements, the differential is between the measuring port (HI side) and the vent to atmosphere or ambient pressure (LO side). These measurements use the unit of pounds per square inch gauge.

The advantage of gauge pressure measurements is that because they are referenced to the surrounding atmosphere, those effects are automatically compensated. If a gauge pressure gauge is sitting on a table at sea level, it would read zero. The effects of atmospheric pressure are still prevalent, but they are balanced because the force is both measured and reference. **See Figure 5-12.**

When a gauge pressure reading is taken of a tire, the pressure read is the pressure exerted on the inside of the tire. This pressure also has the value of atmospheric pressure as an added force; however, the interior value of atmospheric pressure is negated by the exterior atmospheric pressure. A gauge pressure device automatically compensates for this, because the pressure gauge is vented to the same atmosphere that is acting upon the tire. **See Figure 5-13.**

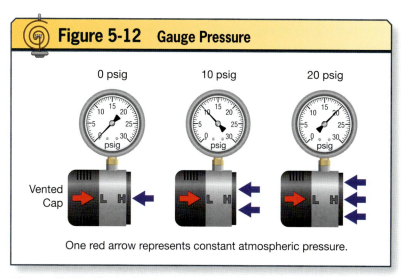

Figure 5-12. *The indicator movement is represented by the effects of different gauge pressures.*

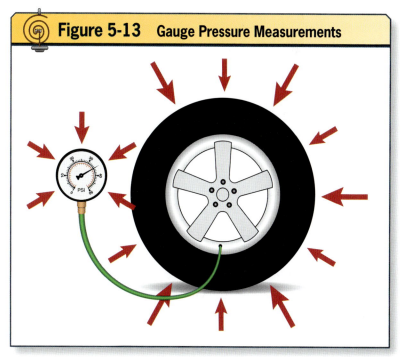

Figure 5-13. *Atmospheric pressure equally affects both the tire and the pressure gauge.*

Vacuum Pressure

A *vacuum* is a space that is devoid of matter. *Vacuum pressure* is any pressure less than atmospheric pressure as measured using a gauge pressure measuring device. When a vacuum pressure is generated, it is measured as a negative departure from zero on a gauge pressure device and is generally indicated by negative sign or the suffix VAC. **See Figure 5-14.**

When a vacuum is developed, it is a point of pressure lower than the surrounding pressure. The airflow that is felt is the surrounding pressure pushing or rushing toward the lower pressure to equalize the pressures.

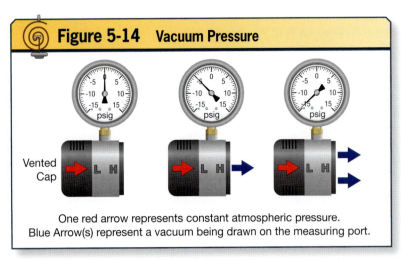

Figure 5-14. The indicator movement is represented by the effects of different vacuum pressures.

For additional information, visit qr.njatcdb.org Item #2489

When gauge pressure measurements are taken, the suffix is G, resulting in measurements in the unit of pounds per square inch gauge. However, it is common practice to give gauge pressure readings with only the acronym "psi" as the unit of measure. In such cases, pounds per square inch gauge is normally assumed.

PRESSURE DEVICES AND THEIR FUNCTIONS

Pressure devices are instruments used to sense, measure, or indicate pressure. These devices can take many forms; the most common ones are visual indication, discrete signaling, or continuous monitoring of specific values.

When selecting these devices, many factors must be anticipated regarding the medium being sensed, conditions of the sensing, mounting methods, accuracy, and sensing range. There are a variety of options available from manufacturers for most scenarios, with new and improved technology in constant development.

One of the most important concerns in dealing with any type of pressure-related devices is the possibility of over pressurizing. Over pressurizing a device can result in damage to the device and harm to personnel in the vicinity. In addition, inaccurate measurements can occur when the device is not properly configured for the variable being monitored. When selecting a pressure range, care should be taken to use a range that is centered on the measured value. Inaccuracies can occur in at the extreme limits of the device's range, typically within 10% of its minimum value and maximum value. For instance, using a pressure gauge that measures 0–1,000 pounds per square inch gauge to monitor a steady pressure supply of 100

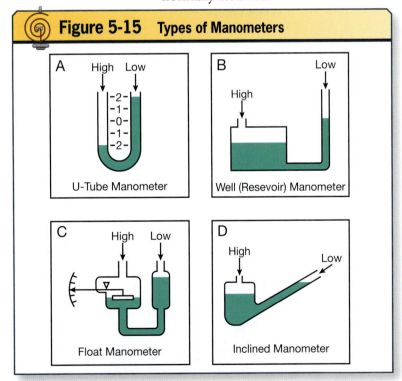

Figure 5-15. Styles of manometers include (a) U-tube, (b) well or reservoir, (c) float, and (d) inclined.

ThinkSafe!
Always verify that the rated pressure of the device is not exceeded by the supply pressure. Overpressure conditions can cause serious injury to personnel and equipment.

pounds per square inch gauge is impractical in terms of cost and accuracy. Most likely, a gauge that measured 0–200 pounds per square inch gauge would be better. The type of the product the pressure device is measuring is also a concern, because some chemicals are not compatible with standard pressure transmitter materials and require devices that are made from compatible materials.

Manometer

The *manometer* is a device that uses a liquid column to measure pressure. One of the most common types of manometers is the U-tube manometer. However, other styles are available, such as well, float, and inclined. **See Figure 5-15.**

The U-tube manometer is a piece of glass or plastic tube bent into a U shape and attached to a stand. Both ends of the tubes are open, and there is a measuring device between the tubes. The tubes are filled with a reference liquid, such as pure water, to the point that the measurement begins. **See Figure 5-16.**

For gauge pressure measurements, one tube is connected to what is being measured (whether positive or vacuum), and the other tube is open to atmosphere. When a positive pressure in relation to the atmosphere is placed on this tube, the liquid is forced down and back up the opposing tube. If a pressure lower than the atmosphere, or the vacuum, is placed on the measuring port, atmospheric pressure then pushes the fluid down and up toward the measuring port. For differential measurements, one tube is connected to the HI pressure, and the other tube is connected to the LO pressure. In either case, a volume of water is supported by the pressure change. This total difference is called the column height and is typically expressed in millimeters or inches. In some cases the surface tension of the

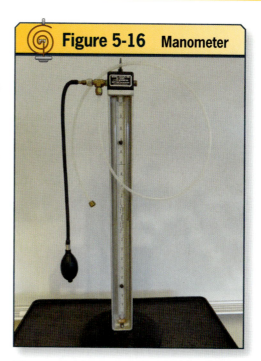

Figure 5-16. A U-Tube manometer is shown with a tube connected to make gauge pressure measurements in inches of water column.

fluid present within the tube may form a curve. This curve is known as the meniscus. Measurements should be taken below the meniscus. **See Figure 5-17.** This measurement is representative of

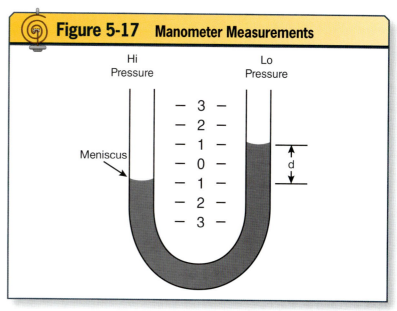

Figure 5-17. The total pressure is equivalent to the distance (d) between each column of water level.

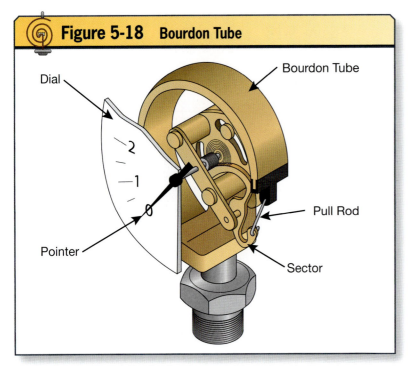

Figure 5-18. As pressure is increased inside the Bourdon tube, it expands outward, causing the needle to move.

For additional information, visit qr.njatcdb.org Item #2490

Bourdon Tube–Type Detectors

The *Bourdon tube* is a pressure sensor that converts pressure to displacement, a coiled, flattened tube that is straightened when pressure is applied. It is one of the oldest pressure-sensing instruments still in use. It consists of a thin-walled tube that is flattened diametrically on opposite sides to produce an elliptical cross-sectional area that has two long, flat sides and two short, round sides. **See Figure 5-18.** The tube is bent lengthwise into an arc. Pressure applied to the inside of the tube causes distention of the flat sections and tends to restore its original, round cross-section. This change in the cross-section causes the tube to straighten slightly. Because the tube is permanently fastened at one end, the tip of the tube traces a curve that is the result of the change in angular position with respect to the center. Within limits, the movement of the tip of the tube can be used to position a pointer or to develop an equivalent electrical signal to indicate the value of the applied internal pressure.

the pressure and is expressed in terms of both the unit of distance measure and the column fluid type, such as inches of water or millimeters of mercury. For vacuum measurements, the units must include the suffix VAC to indicate it is a vacuum pressure.

Bellows Pressure-Sensing Element

A *bellows* pressure-sensing element is a pressure sensor that converts pressure to linear displacement. Bellows-style pressure-sensing devices work on the premise that the bellows expands and contracts to cause a mechanical movement.

The bellows is a one-piece, collapsible, seamless, metallic unit that has deep folds formed in thin-walled tubing. System pressure is applied to the internal volume of the bellows. As the inlet pressure to the instrument varies, the bellows expands or contracts. The moving end of the bellows is connected to a mechanical linkage assembly. As the bellows expands or contracts, the associated linkages move an indicator needle or the device that alters an output signal. **See Figure 5-19.**

Pressure gauges are simple visual indicators of pressure conditions within a process. The most common method for

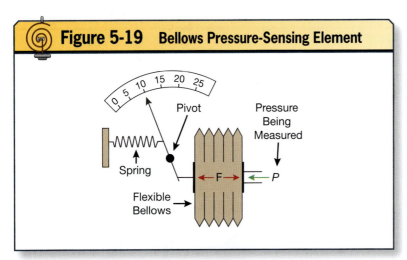

Figure 5-19. As the measured pressure (P) increases or decreases, the resulting force (F) within the bellows causes it to expand or contract. The bellows movement moves the needle in proportion with the change in pressure.

indicating pressure uses a Bourdon tube-style arrangement, which is connected through a linkage to a movable needle. In some instances, the needle can be adjusted or calibrated to compensate for changes in this linkage or needle assembly.

Pressure gauges can be installed in virtually any scenario; however, they must be specified based on process conditions. Basic gauges are available in a variety of styles. **See Figure 5-20.** In some cases, pressure gauges can also have a set of axillary contacts to interface with other devices electrically.

When looking at a typical process and instrumentation drawing (P&ID), pressure gauges are identified by utilizing the letter combination of PI, as in "pressure indicator."

Pressure Switch

A pressure switch is a single contact or group of contacts that changes state based on a change in pressure. Pressure switches can be configured to perform many functions. Many common pressure switches are basic and operate at fixed pressures, fixed dead band, and low tolerance, whereas others offer adjustable pressure and dead band ranges. By using adjustable pressure and dead band ranges, a switch can be tuned to operate reliably for a variety of applications. **See Figure 5-21.** Some pressure switches can also provide pressure indication and are designated as PISL or PISH on a P&ID.

Pressure switches are classified as point measuring devices, whereas they only operate within a set point and reset range. Once the set point has been reached and the contact changes position, it does not change further until the reset point has been reached. It is possible to perform some rudimentary continuous monitoring by using several devices; however, this is

> **TechTip!**
> Many pressure devices come with dual pressure measurement units (e.g., pounds per square inch, millimeters of mercury, and bar). Always verify whether a device has more than one scale indication, and be certain the scale reference being viewed is the proper scale for the situation.

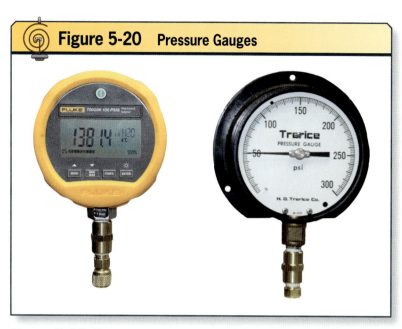

Figure 5-20. Pressure gauges are available in a variety of types in both conventional and electronic formats.

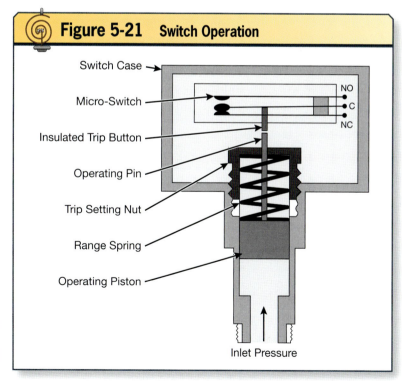

Figure 5-21. The internal components of a pressure switch.

generally reserved for a continuous monitoring device, such as a pressure transmitter.

Low pressure switches change state upon a decrease of pressure below a set point. These are indicated on a P&ID by the basic letter combination PSL for "pressure switch low." These switches are configured so that as the pressure drops below a set point, the switch contacts change state. The contacts do not revert to their original state until the reset point has been reached with an increasing pressure. An example is a low pressure switch on a facility's compressed air system. If the system pressure is 150 pounds per square inch gauge, a pressure switch may set off an alarm if the pressure drops below 150 pounds per square inch gauge, thereby alerting personnel to an issue such as a compressor malfunction. Once the issue is corrected and sufficient air pressure is available, possibly above 170 pounds per square inch gauge, the switch resets to a normal contact state. The reset point is determined by the dead band of the switch, an interval during which no action occurs, which can be either fixed or adjustable.

High pressure switches do the opposite: they change state upon an increase of pressure above a set point. These are indicated on a P&ID by the basic letter combination PSH for "pressure switch high." These switch contacts change state once the pressure increases above a set point. **See Figure 5-22.**

An example is a high pressure switch, again on a facility's compressed air system. If the system pressure is 170 pounds per square inch gauge, a pressure switch may sound an alarm if the pressure rises above 170 pounds per square inch gauge; in this case, the issue may be a failed air pressure regulator. Once the pressure falls below the deadband range, a switch with an automatic reset feature will reset, and the contacts will be repositioned to a normal state. If the switch is a manual reset type, the reset button must be pressed following the pressure returning to normal conditions in order for the contacts to be repositioned.

As with all pressure devices, there has to be a measured port and some type of reference. The opposite side of the pressure component is often open to the atmosphere, not unlike how a pressure gauge operates. As long as the atmosphere where the switch is located is the same as the atmosphere of the monitored vessel, accurate operation should be expected.

Differential pressure switches are indicated on a P&ID as PSDH or PSDL (for "pressure switch differential high" and "pressure switch differential low," respectively), because these devices activate upon a change in differential pressure. This is accomplished by

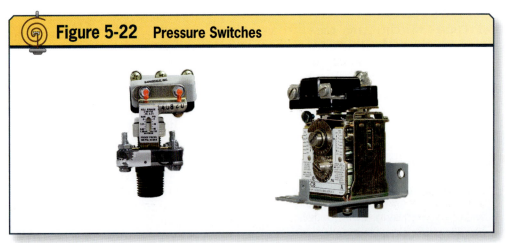

Figure 5-22. Enclosures have been removed from these pressure switches. The left image shows the fixed dead band. The right image shows the adjustable dead band.

connecting the reference or LO port into the system, not relying on atmospheric pressure to be the reference.

Devices such as pumps, fans, filters, and valves can create a change in pressure within a system. A differential pressure switch that has a low reading across the inlet and outlet of a pump indicates fluid flow through the pump. A differential pressure switch that has a high reading across the inlet and outlet of a strainer by the pump indicates a clogged strainer or other problem is reducing flow through the pump. **See Figure 5-23.**

Typically, pressure switch contacts are shown in their nonactuated states, meaning that if the switch is designed to trip on an increase in pressure to 100 pounds per square inch gauge, normally open contacts do not close until this set point is reached and then re-open once the pressure falls below reset. This becomes a little more complicated with low pressure switches. If a contact is desired to open on a falling pressure, the normally open contact would still be selected. At set point, these contacts open, but at reset point, these contacts close. Care must be taken because drawings and documentation may show a contact that is open but the drawing notes may state that the switch should be closed under normal operating conditions.

Pressure Transmitter

The pressure transmitter produces a signal type that is constantly transmitting data in direct relationship to the pressure that is being measured. Though it is commonly assumed that pressure transmitters are purely electric devices, they can be purely mechanical devices.

ThinkSafe!

Be aware that some pressure devices contain mercury. Always use caution and wear the proper personal protective equipment to protect against exposure from possible device failures.

Figure 5-23. *This differential pressure switch is shown with the enclosure removed.*

For now, consider pressure transmitters to be devices that combine a method of pressure sensing with electronic circuitry, which transmits the electrical signals to other pieces of equipment. A common industrial pressure transmitter outputs a signal of 4–20 milliamperes direct current (DC) that is directly proportional to the measured pressure or the measured differential of two pressures. The signal output of the transmitter is great for long-distance transmitting, easily converted to a signal of 1–5 volts DC, and reliable overall. However, it cannot be created by the pressure being measured. Various other methods can be employed to compare a relative change in pressure or force to another material that has electrical property changes based on the applied forces. This describes the function of a sensor or transducer. These materials can change capacitance, resistance, and even generate voltage; however, it is generally in small amounts and delicate. Therefore, additional electronic circuitry, the transmitter, is necessary to manipulate a pressure sensor- or transducer-generated signal into a usable and reliable control signal.

To measure a pressure value, the relationship between what is being sought and a reference must be examined. This

For additional information, visit qr.njatcdb.org Item #2491

relationship involves a force—and therefore resulting movement that can be measured. This movement is imparted on a sensor or transducer, which converts the movement into an electrical property using capacitance, piezoresistance, piezoelectricity, or linear variable displacement.

Capacitance-Based Transducers

Capacitance-based transducers use the principles of a capacitor. The dielectric and the area of plates are impractical to change. However, the distance between these plates can be more easily altered. As the distance between the plates changes, the transfer of electrons between those plates is altered. **See Figure 5-24.**

Capacitance can be determined using the following equation:

$$C = 0.225 \times K \frac{Area}{Distance}$$

where

C = Capacitance (in picofarads)

A = Area of one capacitor plate (in square inches)

k = Dielectric constant of the material between the plates (silicone oil is 2.7–2.8)

d = Distance between the plates (in inches)

Change in capacitance can be related to change in pressure. This is a commonly used relationship, and it is found throughout industry as the basis for devices such as pressure transmitters. **See Figure 5-25.**

Capacitive elements are rarely suitable to place in direct contact with process fluids. Instead, metallic diaphragms are used to isolate the capacitive element from the process fluid. **See Figure 5-26.**

Piezoresistive Devices

Piezoresistive devices, commonly referred to as strain gauges, are another method of measuring a movement. These devices change resistance as they are pulled, compressed, or flexed. This is known as the *piezoresistive effect*. However, the actual strain gauge is generally delicate and impractical to be directly affected by the force of the pressure. By combining the strain element and a flexible substrate that is exposed

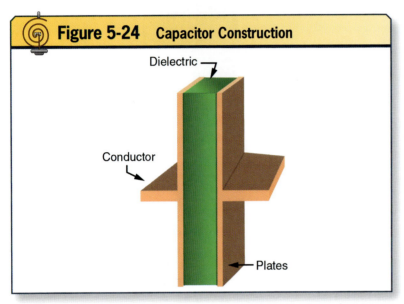

Figure 5-24. *The basic capacitor construction consists of two or more plates separated individually by a dielectric material.*

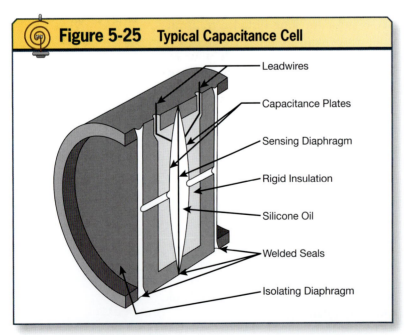

Figure 5-25. *A typical capacitance-based pressure transducer is shown, including the internal components.*

Figure 5-26. The transducer portion of a transmitter is shown with the cells exposed.

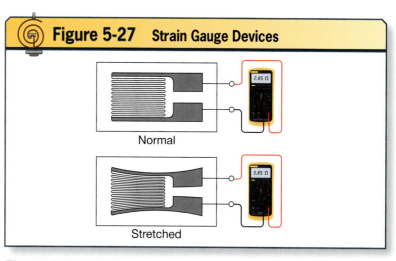

Figure 5-27. An increase in pressure deforms the strain gauge, causing a resistance change.

to the measured and reference pressure, the strain gauge changes resistance based upon the amount of deflection that happens between the pressures. This resistance change can then be measured, and a relationship identified by a net change in pressure is equivalent to a value of resistance. **See Figure 5-27.**

Piezoelectric-Based Transducers

Piezoelectric-based transducers use the principles of the *piezoelectric effect*. The application of pressure to a piezoelectric material results in a voltage being generated. This phenomenon is best used within pressure sensors that are used in dynamic or ever-changing conditions. This style is not good for static applications, because only a finite amount of energy is generated for a given amount of pressure. **See Figure 5-28.**

Linear Variable Displacement Transformer

A linear variable displacement transformer (LVDT) consists of three windings and a movable core. The movable ferrite core provides the link between the primary and the secondary windings, similar to that of a conventional transformer. This transformer device uses dual secondary windings, one on each side of the core's travel. When excitation voltage is applied to the primary winding, and the coil is in the center position, equal amounts of voltage are induced into each of the secondary windings. Because of the series

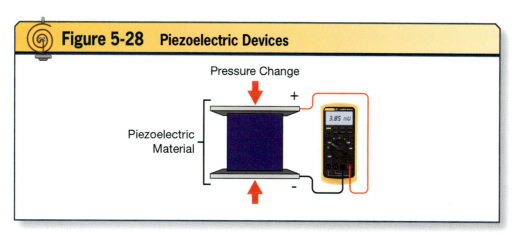

Figure 5-28. As the pressure changes, the material generates a millivoltage DC.

For additional information, visit qr.njatcdb.org Item #2492

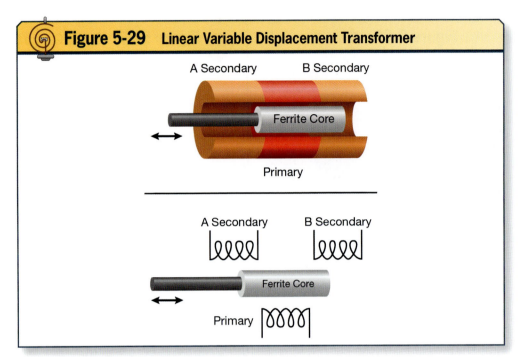

Figure 5-29. This type of transformer can be found not only in pressure transmitters but also in many places where accurate interpretation of movement is required.

connection within the secondary windings, the induced voltage is flowing two directions, which equates to a nearly 0 voltage DC output. As the core is moved from center, the voltage increases proportionally with the distance of travel from center. How the windings are set up determines whether the value of the induced voltage is in phase or 180° out of phase with the primary voltage. The change in phasing is used to determine the direction the core has traveled, and the value of the induced voltage equates to the distance the core has traveled. **See Figure 5-29.**

Transmitter devices commonly output a signal that is 4–20 milliamperes DC. In contrast, the sensor in capacitance-based, piezoresistive, piezoelectric-based, and linear variable displacement devices does not directly create a signal of 4–20 milliamperes DC. Each of these devices requires some type of amplifier or additional circuitry to convert the capacitance, resistance, inductance, or voltage into a usable signal type. This conversion typically takes place inside the transmitter component of a control loop. In some cases, transmitters may be mounted separately from sensors; in other cases, they are integrated into each other's enclosure.

Figure 5-30. This typical calibration setup indicates the transmitter output is below 4.00 milliamperes, which is not proportional to the input pressure.

The calibration technician must ensure that the pressures being input to the pressure sensor are in direct proportion with the output from the transmitter. For example, a transmitter for 0–100 pounds per square inch gauge is supposed to output 4–20 milliamperes DC, where 0 pounds per square inch gauge is equivalent to 4 milliamperes DC and 100 pounds per square inch gauge is equivalent to 20 milliamperes DC. This is accomplished by placing an input pressure standard on the sensor and adjusting the transmitter's zero and span. **See Figure 5-30.**

The input reference standard to the pressure transmitter is created by the small test pump and verified by the test gauge. The pressure present on the sensor is 0 pounds per square inch gauge; however, the output of the transmitter, read by the digital multimeter, is 3.85 milliamperes DC. **See Figure 5-31.**

Though current-based signals (in milliamperes DC) are common, there can be other signal types, such as voltage-based or digital-based outputs. Voltage-based signals are simple to transmit; however, they are not the most accurate over long distances because cable resistance greatly affects their performance. Digital-based outputs offer an array of information that can be transmitted over two wires. Some examples of digital signal types are highway addressable remote transducer (HART), Foundation Fieldbus, and even serial communication such as Modbus.

PRESSURE UNITS OF MEASURE

Using this concept as the basis for pressure measurement, several common units can be compared. **See Figure 5-32.**

While many pressure units exist for a variety of scenarios, this figure shows some of the more common units: pounds per square inch (first gauge, on the left), inches of mercury (second gauge), inches of water (third gauge), and the kilopascal SI unit of pressure

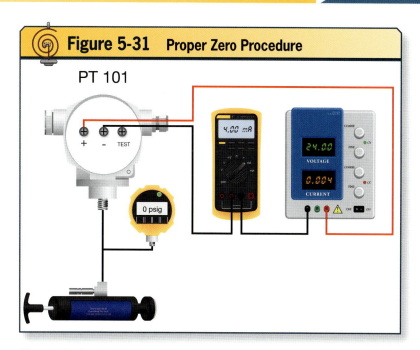

Figure 5-31. *The input reference standard to the pressure transmitter is created by the small test pump and verified by the test gauge. The pressure present on the sensor is 0 pounds per square inch gauge; however, the output of the transmitter, read by the digital multimeter, is now 4.00 milliamperes DC after a proper zero procedure.*

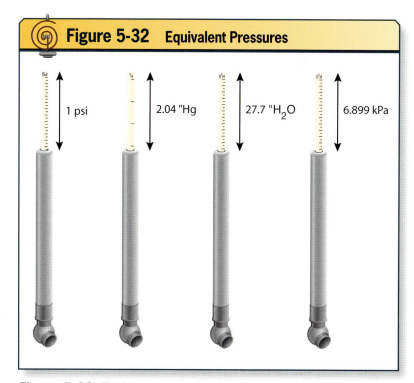

Figure 5-32. *Each gauge represents the equivalent scale for the specified unit, but all indicate the same pressure.*

(fourth gauge, on the right). Mercury (Hg) is 13.6 times the density of pure water (H₂O). Therefore:

$$1 \text{ PSI} = 27.7 \text{ "H}_2\text{O}$$

and

$$1 \text{ PSI} = 2.04 \text{ "Hg}$$

So

$$27.7 \text{ "H}_2\text{O} = 2.04 \text{ "Hg}$$

which is a 13.6:1 ratio.

The SI system of measurement uses a unit known as the Pascal (Pa), named after the French physicist Blaise Pascal. There are 249.089 pascals per inch of water; however, since this is such a small unit the kilo (1×10^3) and mega (1×10^6) are added to the Pascal for simplicity.

An instrumentation technician may desire to convert from PSI to Pa, utilizing a conversion factor such as 1 "H₂O = 249.089 Pa. In the example below, 1 PSI has been converted to kilo-Pascals (kPa):

$$1 \text{ PSI} = 27.7 \text{ "H}_2\text{O}$$

And

$$1 \text{ PSI} = 27.7 \times 249.089$$

So

$$1 \text{ PSI} = 6,899.7653 \text{ kPa}$$

These and other values can also be determined using a pressure conversion table. **See Figure 5-33.**

MULTIVARIABLE RELATIONSHIPS

A basic example of a multivariable relationship can be demonstrated by a tire on the car. Assuming that there are no air leaks in the tire, mounting, or valve mechanism, this example considers what happens after driving for some distance. As the tire is rotated along pavement, friction creates heat, which transfers to the air within the tire. Though the tire will warm and most likely expand, the

Figure 5-33 Pressure Conversion Table

	PSI	PASCAL	BAR	MILLI-BAR	IN. Hg	IN. H₂O at 4°C	mm HG	mm H₂O	ATM	kg/cm²
PSI	1	0.000	14.504	0.015	0.491	0.036	0.019	0.001	14.696	14.223
PASCAL	6984.6	1	100000	100	3386.5	249.089	133.32	9.807	101.20	98067
BAR	0.069	0.000	1	0.001	0.034	0.002	0.001	9.8068 E-05	1.013	0.981
MILLI-BAR	68.946	0.01	1000	1	33.865	2.491	1.333	0.098	1031.2	980.68
IN. Hg	2.036	0.000	29.529	0.030	1	0.074	0.039	0.003	29.92	28.959
IN. H₂O at 4°C	27.68	0.004	401.47	0.401	13.596	1	0.535	0.039	406.78	393.72
mm HG	51.714	0.008	756.06	0.750	25.401	1.868	1	0.074	760	735.59
mm H₂O	703.05	0.102	10197	10.197	345.32	25.399	13.595	1	10332	10000
ATM	0.068	9.8692 E-06	0.987	0.001	0.033	0.002	0.001	9.6788 E-05	1	0.968
kg/cm²	0.070	1.0197 E-05	1.110	0.001	0.035	0.003	0.001	0.000	1.033	1

Figure 5-33. A pressure conversion table is a quick way to find equivalent values for pressure units of measure.

air warms quicker and therefore expands quicker than the physical constraints of the tire. This increase in temperature creates an increase in pressure within the tire. Conversely, on a cold day when a tire is at rest and has low air temperature, the pressure is lower inside the tire, even though there are no apparent leaks. Volumetric changes to a tire are generally minimal because of the tire's construction and ratings.

Another relationship shows that boiling points of liquids are affected by changes in atmospheric pressure. Recipes are often written with additional instructions for preparation based on the altitude above sea level at which the recipe is being prepared. The change in elevation can alter the lengths of cooking times, as the reaction of the ingredients will differ based on the change in atmospheric pressure.

Discoveries regarding the relationships among a volume of gas, the number of molecules of the gas, the pressure of the gas, and the temperature of the gas have been made to show distinct relationships among these properties. Scholars such as Jacques Charles, Robert Boyle, Joseph Louis Gay-Lussac, and Amedeo Avogadro made discoveries regarding how gases behave based on these factors.

The ideal gas law was developed by Emilie Clapeyron in the mid-1830s and is a combination of individual gas laws. This law describes how a theoretical ideal gas will behave based on a number of factors, such as a change in temperature or change in pressure, in relation to one another. While there is not a known true ideal gas, the theories related to this gas can be applied to many known gasses. There are limitations, but consider them excluded for now and suppose all gases are ideal.

For additional information, visit qr.njatcdb.org Item #2493

The ideal gas law formula is defined as follows:

$$PV = nRT$$

where

P = Pressure (in pascals)

V = Volume (in cubic meters)

n = Number of moles of the gas (Avogadro's constant defines 1 mole as $6.02214179 \times 10^{23}$ molecules of the substance)

R = Gas constant

T = Temperature (in Kelvin)

Assume a process vessel contains a gas. The gas constant R and the number of moles n are fixed in this example, so the remaining three variables can be observed. If the temperature inside the vessel increases, the pressure must also increase. For this example, the volume of the chamber will remain fixed as shown. **See Figure 5-34.**

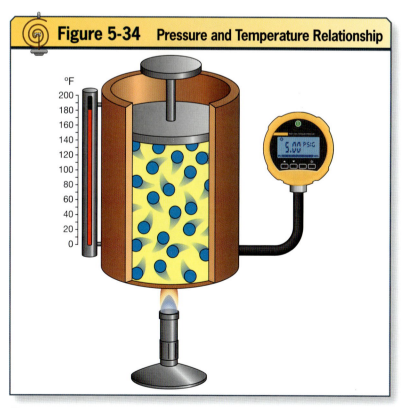

Figure 5-34. *The relationship between pressure and temperature is directly proportional.*

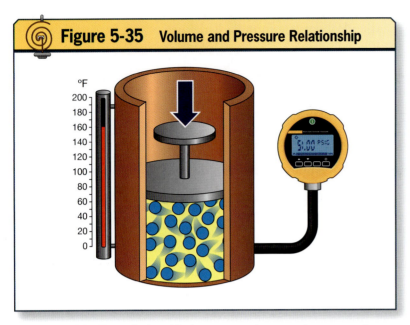

Figure 5-35. *The relationship between volume and pressure is indirectly proportional.*

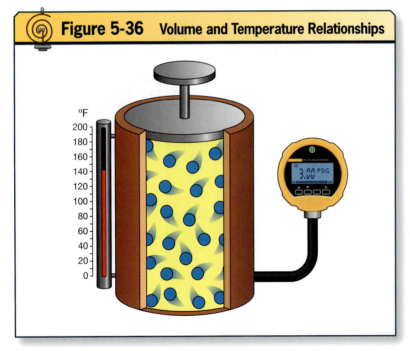

Figure 5-36. *As the container volume is increased, the pressure and temperature will decrease.*

Similarly, if the volume is decreased in the vessel, the pressure must also increase. By decreasing the volume that the gas can occupy, but not releasing molecules of gas to compensate, the gas is compressed. By compressing a gas, the molecules are pressed together, which results in friction. The friction generates heat, which is indicated by the temperature gauge. Additionally, this expelled heat continues to cause the compressed gas to expand more, inside an already decreasing chamber. **See Figure 5-35.**

Finally, assume that a container's volume is increased. As the plunger rises, the gas will assume the shape of the container. As with the other examples, the same amount of gas is still present within the container. As the volume increases, the gas molecules become farther apart. This expansion results in a decreased temperature. The pressure inside the container will also become lower in response to the higher volumetric area inside the container. **See Figure 5-36.**

Through these basic examples, it becomes evident that gasses behave in very unique ways. Because of these unique properties, it is often necessary to measure more than one variable at a time, in order to get an accurate reading. For instance, when gas pressure is measured it is not uncommon to also take a temperature reading of the gas that is measured.

The technician must understand these interrelated variables in the context of the ideal gas law.

SUMMARY

Pressure measurement and corresponding relationships to flow, level, and temperature allow a range of applications in the process control industry. Understanding how changes in pressure occur and the methods used to accurately measure these changes is vital to an instrumentation technician.

REVIEW QUESTIONS

1. **Atmospheric pressure is defined as which measurement?**
   a. 0 pounds per square inch absolute
   b. 0 pounds per square inch differential
   c. 14.7 pounds per square inch absolute
   d. 14.7 pounds per square inch gauge

2. **Which type of pressure sensing element uses a one-piece metallic unit with deep folds to convert pressure changes to linear movement?**
   a. Bellows
   b. Bourdon tube
   c. Diaphragm
   d. Piston

3. **A pressure that is observed to change within a sealed, fixed volume vessel is likely indicative of a change in what to the vessel?**
   a. Shape
   b. Size
   c. Temperature
   d. Volume

4. **What is the name of a pressure-sensing device that begins to measure increasing pressure from 0 pounds per square inch absolute?**
   a. Absolute pressure device
   b. Differential pressure device
   c. Gauge pressure device
   d. Vacuum pressure device

5. **A pressure switch tagged PSL 1234 trips at what point?**
   a. A decrease in pressure below 1,234 pounds per square inch
   b. A decrease in pressure below the set point
   c. An increase in pressure above 1,234 pounds per square inch
   d. An increase in pressure above the set point

6. **What is the weight of 1 cubic foot of water at standard temperature pressure conditions?**
   a. 62.43 grams
   b. 62.43 kilograms
   c. 62.43 ounces
   d. 62.43 pounds

7. **What type of pressure transducer generates a voltage as the pressure increases?**
   a. Capacitive
   b. Linear variable displacement
   c. Piezoelectric
   d. Piezoresistive

8. **If a pressure transmitter with an output of 4–20 milliamperes DC and calibration of 0–25 pounds per square inch gauge is transmitting a signal of 6 milliamperes DC with 0 pounds per square inch gauge present on the sensor, this could be a sign of what type of issue?**
   a. Blocked sensing port
   b. Loss of loop power
   c. Span error
   d. Zero error

9. **What is the unit of pressure measure in the SI system of measurements?**
   a. The bernoulli
   b. The charles
   c. The newton
   d. The pascal

10. **The relationship between pressure and temperature of a gas can be examined using which of the following?**
    a. Ideal gas law
    b. Pressure–temperature analyzer
    c. Theory of relativity
    d. Tire method

Principles of Level

A variety of devices are used to measure, control and record levels. The choice of instrument technology used to measure the level is application specific and changes depending on variables such as accuracy, characteristics of the product being measured, environmental concerns, cost, and reliability. Calibration measurement errors can be introduced through improper or uncompensated mounting procedures. If the installation of the device is not performed correctly, the process may operate suboptimally or present risks to personnel and the environment. The details of the process characteristics must be understood to accurately measure and record levels.

Objectives

» Describe level measurement properties and measurement principles.
» Identify methods to measure hydrostatic head.
» Explain how specific gravity affects hydrostatic head measurements.
» Classify level instruments in categories of operation.
» Describe how differential pressure can be used to derive a level measurement.
» Explain Archimedes's law.
» Identify installations that require temperature compensation.

Chapter 6

Table of Contents

Level Fundamentals 112
   Point Level and Continuous Level Measurement 112
   Physics of Level Measurement 112
      Volume 113
      Mass and Density 113
      Specific Gravity 114
      Hydrostatic Pressure 115
Methods of Level Measurement 117
   Manual and Visual Level Instruments 117
   Mechanical Instruments 120
      Ball Float 120
      Chain Float 121
      Magnetic Bond Mechanism 122
      Displacer 122
   Discrete and Level Switches 124
      Conductivity Probe Method 124
      Vibrating Fork Level Switch 124
      Rotating Paddle Level Switches 125
   Hydrostatic Pressure Instruments 125
      Open-Tank Measurement 126
      Closed-Tank, Dry Reference Leg Measurement 126
      Closed-Tank, Filled (Wet) Reference Leg Measurement 127
      Instrument Elevation and Suppression Errors 128
      Bubblers 129
   Contact Level Instruments 130
      Capacitance Level Sensor Measurement 130
      Guided Wave Radar 131
   Noncontact Level Instruments 132
      Ultrasonic Level Measurement 132
      Radar Level Measurement 133
      Nuclear Measurement 133
      Ion Chamber 134
      Scintillation Tube 134
Density Compensation 134
   Specific Volume 134
   Reference Leg Temperature Considerations 135
   Pressurizer Level Instruments 136
   Steam Generator Level Instruments 137
Environmental Concerns 138
Summary 138
Review Questions 139

For additional information, visit
qr.njatcdb.org
Item #2690

LEVEL FUNDAMENTALS

A typical processing plant contains various tanks, vessels, and reservoirs, all of which store product in various stages. It is essential to record accurate measurement of the vessels to maintain an automatic working process. Most tanks contain a liquid of some form, but some may contain solids. Level measurement is an integral part of process control and may be used in a variety of applications.

Level measurement in a process is necessary for the process to operate as designed safely and efficiently. The level of a product inside a process vessel is used to determine the overall volume, a measurement of occupied space. Commonly used units of volume are gallons, liters, and cubic feet. The plant management must measure the current level of product in a tank to determine production scheduling and material ordering data. Inaccurate level readings can lead to poor production efficiency and, in some cases, potentially catastrophic overfilling of tanks that in turn leads to environmental and personnel dangers.

Level can be expressed in terms of percentage filled (67% full), which is commonly called fillage, and in terms of percentage empty (33% empty), called ullage. The level measurement may also be combined with other dimensions of the tank, such as diameter or length, to determine volume. For oddly shaped tanks, tables called strapping tables are used to determine the volume based on a level measurement.

Liquid level measuring devices are classified according to whether they are used for direct or inferred measurements. An example of a direct device is the motor oil dipstick in a car, which measures the height of the oil in the oil pan. An example of an inferred device is a pressure gauge at the bottom of a tank that measures the hydrostatic head pressure from the height of the liquid.

Point Level and Continuous Level Measurement

Level-sensing devices can be grouped into two categories: point level measurement and continuous level measurement. Point level (also called level switch or discrete point) measurement monitors a specific level height and sends a discrete signal when this point is reached. Discrete signals are used for alarm conditions, motor starting or stopping, emergency shutdown signals, and several other purposes.

A continuous level sensor monitors the level height of a liquid over a wider range, usually 0-100%, rather than at the single point monitored using point level measurement. The continuous level instrument (e.g., differential pressure transmitter) provides the system controllers with a continuous stream of information about the level in the tank. In contrast, point level controllers work blind until the discrete point is reached. This is a reactive approach to process conditions that does not provide the opportunity to make corrections until a discrete point is reached. Continuous level monitoring instead offers a proactive process control that adjusts the process as needed before the process variables deviate from the set point. The design parameters for each control system determine the type of control loop that is required.

Physics of Level Measurement

The technician must understand the underlying physics of level measurement to properly install and maintain

Figure 6-1 Volume Calculation

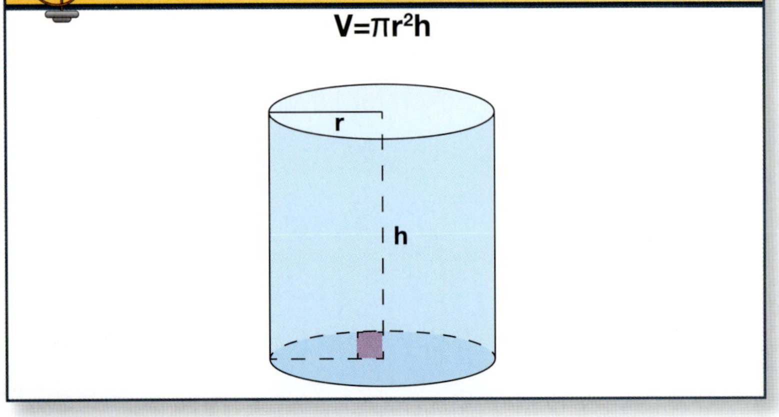

$$V = \pi r^2 h$$

Figure 6-1. Finding the volume of a cylinder is often necessary to determine the amount of process fluid in a tank.

instrument systems. Concepts such as volume, mass, density, specific gravity, and hydrostatic head pressure explain how various level instruments function. Current trends point toward a level measurement device used in industrial applications that is relatively inexpensive and easy to install, calibrate, and maintain.

Volume

Volume is the measurement of occupied space and is given in cubed units, such as cubic feet or cubic centimeters. **See Figure 6-1.**

The volume of a cylinder is provided by the following formula:

$$V = \pi R^2 H$$

where

V = Volume

π = 3.141 (a mathematical constant)

R = Radius (half the diameter)

H = Height

A technician can use this formula to find the volume of a process tank with, for example, a radius of 5 feet and a height of 27 feet:

$$V = \pi \times 5 \text{ ft}^2 \times 27 \text{ ft}$$

$$V = 2{,}120.58 \text{ ft}^3$$

In this case, the tank has a total capacity or volume of 2,120.58 cubic feet. The technician can then determine the volume in gallons if the tank is 13% full. Multiplying the volume by the percentage indicates the tank contains approximately 275.7 cubic feet of product:

$$2{,}120.58 \text{ ft}^3 \times 0.13 = 275.7 \text{ ft}^3$$

The technician can apply the unit conversion factor (see the table in the appendix) to convert this measurement into gallons:

1 cubic foot = 7.4805 gallons

Volume from the previous problem = 275.7 cubic feet

$$275.7 \text{ ft}^3 \times \frac{7.4805 \text{ gallons}}{1 \text{ ft}^3} = 2062.4 \text{ gallons}$$

This tank, filled to 13% of its capacity, contains 2,062.4 gallons.

Mass and Density

Mass is defined as the amount of matter in an object and is commonly provided in kilograms or pounds. Two objects with the same mass but different volumes are said to have a different *density*, or mass per unit volume, given in units such as pounds per cubic foot. The following formula explains this relationship:

$$D = \frac{M}{V}$$

where

D = Density

M = Mass

V = Volume

Two objects can have the same mass but have different densities. A smaller object can be denser yet weigh the same amount as a larger object. **See Figure 6-2.**

Figure 6-2	Density Comparisons Based on Volume	
	Object One	Object Two
Mass	70 pounds	70 pounds
Volume	100 in³	40 in³
Density	0.7 pounds/in³	1.75 pounds/in³

Figure 6-2. *An object with a higher amount of mass per unit volume has a higher density.*

For additional information, visit qr.njatcdb.org Item #2378

Figure 6-3   Specific Gravity Values for Commonly Used Substances

Liquid	Specific Gravity	Gas	Specific Gravity
Water	1.00	Air	1.00
Mercury	13.6	Hydrogen	0.06
Alcohol	0.79	Nitrogen	0.96
Gasoline	0.67	Oxygen	1.10

Figure 6-3. Specific gravity is the measure of density of a fluid in comparison to water in the case of liquids and air in the case of gases.

Specific Gravity

The densities of fluids vary depending on their mass and volume. Measurements must take into account the difference in densities to ensure the level measurement is accurate. *Specific gravity* is the ratio of the mass of any material to the mass of the same volume of water at 4°C. This dimensionless unit has a baseline of 1 and is referenced to pure water at 4°C. A fluid with a density higher than that of water would have a specific gravity higher than 1, such as mercury, which has a specific gravity of 13.6. Fluids with a specific gravity less than 1, such as gasoline, which has a specific gravity of 0.67, are less dense than water. Knowing the specific gravity of a fluid is necessary to equate a pressure measurement to a known level of the fluid in a vessel.

Different specific gravity values are most commonly used for different substances. Almost all substances have been assigned specific gravity values. Fluids that are in a gaseous state are referenced to air, which like water has a specific gravity of 1. **See Figure 6-3.**

Fluid levels can be calculated using a pressure sensor because all fluids create pressure that is directly related to their density and depth. As the depth of a pressure increases, the pressure of the fluid also increases. The head of a fluid can be used to calculate the resulting pressure. **See Figure 6-4.**

Water (chemical symbol: H_2O) weighs 62.43 pounds per cubic foot. Suppose that same cubic foot (volume) contains mercury instead. Mercury (chemical symbol: Hg) has a specific gravity that is 13.6 times heavier than water, and the resulting weight of the mercury for this volume is approximately 850 pounds:

$$62.43 \frac{\text{lbs } H_2O}{\text{ft}^3} \times 13.6 = 849.048 \frac{\text{lbs Hg}}{\text{ft}^3}$$

A manometer is one of the simplest yet most effective instruments for

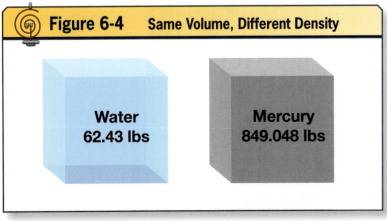

Figure 6-4. Mercury has a higher density than water per unit volume.

measuring pressure. It consists of a U-shaped tube filled with a fluid of a known specific gravity. Consider two manometers: one filled with water and the other filled with mercury. Both manometers have an applied pressure of 1 pound per square inch. The water manometer indicates a pressure difference of 27.7 inches, and the mercury manometer indicates a difference of 2.04 inches. Because mercury is 13.6 times heavier than water (e.g., it has a specific gravity of 13.6), it follows that the mercury will show less displacement than water when subjected to an equal pressure. Using mercury, the range of pressure measurements can be expanded up to 13.6 times an equivalent measurement using water. **See Figure 6-5.**

Although measuring pressure with a manometer is useful, a manometer can practically measure only up to 100 inches of a water column, which is the equivalent measurement of 3.6 pounds per square inch. A larger manometer would be too expensive and unwieldy to use. For pressure measurements greater than 3.6 pounds per square inch, other materials are needed. By using liquids that have a specific gravity greater than that of water, the range of the manometer can be extended. Using mercury, the range of measurement can be extended up to 13.6 times what could be measured with water. The maximum pressure measurement becomes 13.6 multiplied by 3.6 pounds per square inch, or approximately 49 pounds per square inch.

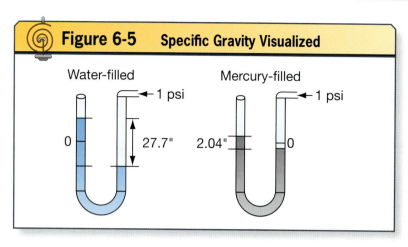

Figure 6-5. U-tube manometers show the density relationship between water and mercury.

Hydrostatic Pressure

Pressure is the result of force acting over a given area, and it is often defined in terms of hydrostatic head, or head for short. A head is a column of fluid that exerts some pressure on a measuring point. The head pressure is a direct representation of the density of the fluid and the height. The volume of the container occupied by the fluid is not a factor in the pressure exerted. To understand head, observe a 30-foot column of water sitting on a pressure sensor. **See Figure 6-6.**

The weight of a column with an area of 1 square foot can be derived by first

TechTip!

When using differential pressure devices to read static head, always compensate for the specific gravity of the process fluid. Water has a specific gravity of 1.00. If a process has a rise of 30 inches and a specific gravity of 0.90, the fluid pressure at the transmitter will be 30.0 inches (rise) multiplied by 0.90 (specific gravity) to equal 27 inches of water.

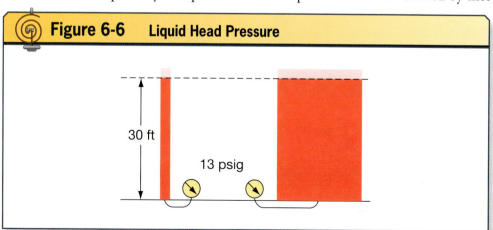

Figure 6-6. The width of a vessel does not affect head measurement. Thirty feet of the same liquid produces an equal head pressure.

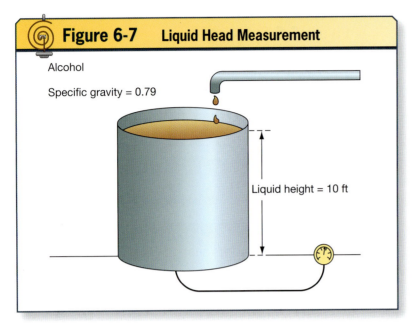

Figure 6-7. The tank contains alcohol to a height of 10 feet.

finding the volume of the column. This is the area of the base times the height of the liquid:

$$1 \text{ ft}^2 \times 30 \text{ ft} = 30 \text{ ft}^3$$

The approximate weight of water is 62.43 pounds per cubic foot. So the total weight of the example column is calculated as follows:

$$30 \text{ ft} \times 62.43 \frac{\text{lbs}}{\text{ft}^3} = 1{,}872.9 \frac{\text{lbs}}{\text{ft}^2}$$

Pressure is commonly measured in pounds per square inch gauge or inches of water. This means the square-foot measurement needs to be reduced into square inches for this example. There are 144 square inches per square foot, and dividing the weight of the water column by this number results in 13 pounds per square inch gauge.

$$1{,}872.9 \frac{\text{lbs}}{\text{ft}^2} \times \frac{1 \text{ ft}^2}{144 \text{ in}^2}$$
$$= 13.0 \frac{\text{lbs}}{\text{in}^2} = 13 \text{ PSIG}$$

The hydrostatic head pressure measured by the gauge is equal to 13 pounds per square inch gauge.

The same type of calculation can be made with substances other than water by incorporating their specific gravity. For example, a tank is filled with alcohol to a level of 10 feet. It has a pressure gauge to read the pressure of the liquid. The specific gravity of alcohol is 0.79. **See Figure 6-7.**

The following formulas show the mathematical approach. First, find the pressure as though the fluid as if it were water. To do this, multiply water's weight (62.43 pounds per cubic foot) by the 10 feet of liquid height to get a result in pounds per square foot:

$$10 \text{ ft} \times 62.43 \frac{\text{lbs}}{\text{ft}^3} = 624.3 \frac{\text{lbs}}{\text{ft}^2}$$

The most common measurement for pressure is in inches, so convert from pounds per square foot to pounds per square inch gauge (there are 144 square inches per square foot).

$$624.3 \frac{\text{lbs}}{\text{ft}^2} \times \frac{1 \text{ ft}^2}{144 \text{ in}^2}$$
$$= 4.34 \frac{\text{lbs}}{\text{in}^2} = 4.34 \text{ PSIG}$$

The result is 4.34 pounds per square inch gauge. Now multiply the equivalent water pressure by the specific gravity of alcohol (0.79):

$$4.34 \text{ PSIG} \times 0.79 = 3.43 \text{ PSIG}$$

Thus, the reading for the equivalent amount of alcohol is 3.43 pounds per square inch gauge.

ThinkSafe!

Always verify the specific gravity of any liquid used for a level indication based on head pressure. Failure to account for the changes in the physical liquid height at a given head pressure because of changes in specific gravity can cause serious process hazards, such as vessel overfillage or underfillage.

This method can be used on any liquid if the specific gravity is known. Specific gravity allows the hydrostatic pressure of any substance to be converted into a level measurement. The primary pressure units are pounds per square inch or inches of water, but the technician should be familiar with other units, such as bars, atmospheres, torrs, and pascals.

Because all liquids can be calculated to express their weight in height and density, the following mathematical relationship is used to express level:

$$P = D \times S \times H$$

where

P = Pressure

D = Density

S = Specific gravity

H = Height

The result is that P can be expressed in inches of a water column or in pounds per square inch as needed. For example, calculate equivalent pressure in both units for a 15-foot-high tank filled with gasoline. **See Figure 6-8.**

To obtain the pressure of the liquid, use the preceding formula. The density of water is 62.43 pounds per cubic foot and gasoline has a specific gravity of 0.67:

$$P = D \times S \times H$$

$$62.43 \frac{\text{lbs}}{\text{ft}^3} \times 0.67 \times 15 \text{ ft}$$

$$= 627.4 \frac{\text{lbs}}{\text{ft}^2}$$

Because most transmitters read in inches, the result must be converted to pounds per square inch gauge:

$$627.4 \frac{\text{lbs}}{\text{ft}^2} \times \frac{1 \text{ ft}^2}{144 \text{ in}^2}$$

$$= 4.36 \frac{\text{lbs}}{\text{in}^2} = 4.36 \text{ PSIG}$$

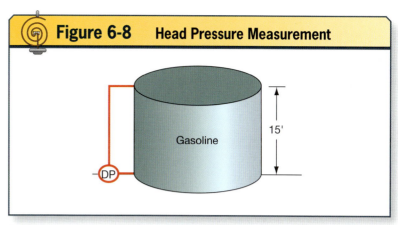

Figure 6-8. *The storage tank contains gasoline at a height of 15 feet.*

Hydrostatic pressure calculations combine the concepts of mass, density, specific gravity, and pressure to infer the level in a tank. Level measurements made using head account for the single largest technology used in process control. The technician must fully understand these concepts before installing or maintaining level instruments.

METHODS OF LEVEL MEASUREMENT

A wide variety of measurement methods exist to determine the level in a process vessel. No single method will produce the optimal result in all applications. When selecting a measurement method, process characteristics such as temperature, accuracy requirements, density, and fluid conductivity must be considered. The instrument's initial cost must be considered, as well as the reliability and cost of maintenance over the life of the instrument. To effectively calibrate and troubleshoot the system, the technician must understand the operating principle of the various level measurements.

Manual and Visual Level Instruments

The gauge glass method is a simple means by which a liquid level is measured in a vessel. In the gauge glass method, a transparent tube is attached to the bottom and top of the tank (the

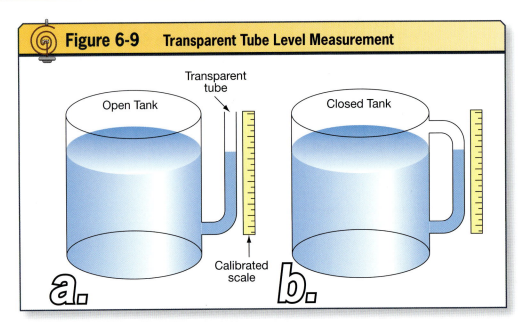

Figure 6-9. (a) The first gauge glass is used for vessels with liquid at ambient temperature and pressure conditions. (b) The second gauge glass is used for vessels with liquid at an elevated pressure or in a partial vacuum.

top connection is not needed in a tank open to atmosphere). The height of the liquid in the tube is equal to the height of the water in the tank. **See Figure 6-9.**

The gauge glasses effectively form a U-shaped tube manometer in which the liquid seeks its own level because of the pressure of the liquid in the vessel. Isolation valves, vents, and drain points should always be installed on gauge glass devices to isolate the process so that maintenance activities may be conducted on or about the gauge glass. **See Figure 6-10.**

Gauge glasses made from transparent tubular glass or plastic are used for

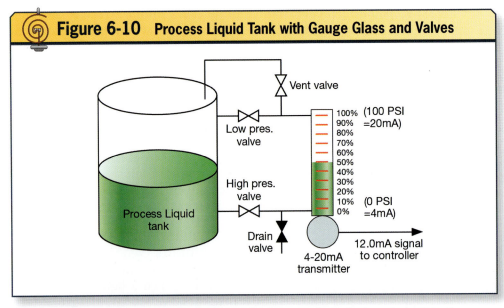

Figure 6-10. Valves are installed when using gauge glass devices to isolate the gauge for maintenance.

processes with pressures up to 450 pounds per square inch and temperatures up to 400°F. Gauges made of special armored glass are used to measure the level of a vessel at temperatures up to 4,000°C and pressures up to 5,800 pounds per square inch. The type of gauge glass used in this instance has a body made of metal with a heavy glass or quartz section for visual observation of the liquid level. The glass section is usually flat to provide strength and safety. **See Figure 6-11.**

In another type of gauge glass, reflex gauge glass, one side of the glass section is prism shaped. **See Figure 6-12.**

A reflex gauge is used primarily for liquids that are clear and therefore difficult to view through regular glass gauges. The glass is molded so that one side has a series of 90-degree angles that run lengthwise. Light rays strike the outer surface of the glass at a 90-degree angle. The light rays travel through the glass, striking the inner side of the glass at a 45-degree angle. The presence or absence of liquid in the chamber determines whether the light rays are refracted into the chamber or reflected to the outer surface of the glass.

When the liquid is at an intermediate level in the gauge glass, the light rays encounter an air–glass interface in one portion of the chamber and a water–glass interface in the other portion of the chamber. Where an air–glass interface exists, the light rays are reflected to the outer surface of the glass because the critical angle for light to pass from air to glass is 42 degrees. This causes the gauge glass to appear silvery white. In the portion of the chamber with the

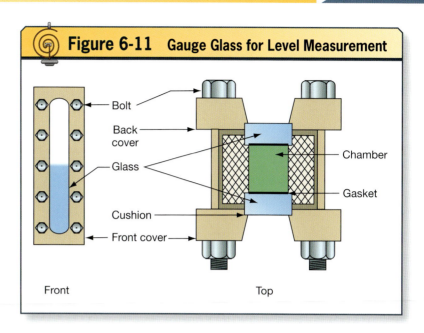

Figure 6-11. A gauge glass is used for level measurement.

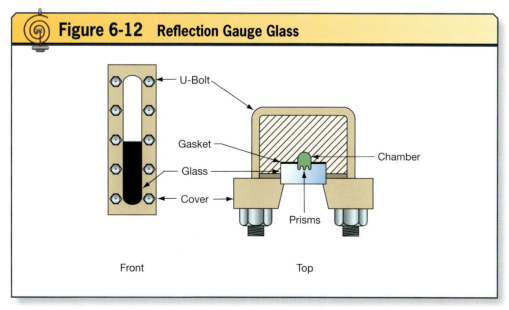

Figure 6-12. The reflex gauge glass has a prism and is useful to measure the level of clear liquids.

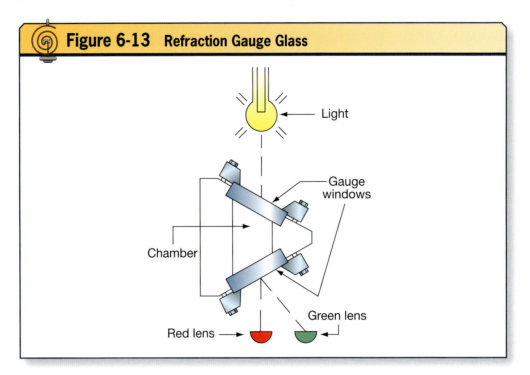

Figure 6-13. *A refraction gauge indicates level based on the light traveling through a fluid.*

water–glass interface, the light is refracted into the chamber by the prisms. Reflection of the light to the outer surface of the gauge glass does not occur because the critical angle for light to pass from glass to water is 62 degrees. This results in the glass appearing black, because it is possible to see through the water to the walls of the chamber, which are painted black.

A third type of gauge glass is the refraction type. This type is especially useful in areas of reduced lighting; lights are usually attached to the gauge glass. Operation is based on the principle that the bending of light, or refraction, is different as light passes through various media. Light is refracted to a greater extent in water than in steam. For the portion of the chamber that contains steam, the light rays travel relatively straight, and the red lens is illuminated. For the portion of the chamber that contains water, the light rays are bent, causing the green lens to be illuminated. Thus, portion of the gauge containing water appears green, and the portion of the gauge from that level upward appears red. **See Figure 6-13.**

Mechanical Instruments

Mechanical measurement techniques use moving parts to measure the level of a process fluid. A series of linkages, chains, cams, and pulleys are used in these instruments. While these instruments may have a transmitter that generates a signal back to the control room, many mechanical instruments serve as local indicators. When mechanical instruments are used as local indicators, they typically do not require electrical or pneumatic connections.

Ball Float

The basic float switch is a simple float that changes level with the level of the fluid. A recording or measuring device measures the change in the level of the float compared with a reference point and transmits the output. The ball float method directly reads the liquid level. The most practical design for the float is a hollow metal ball or sphere. However, there are no restrictions on the

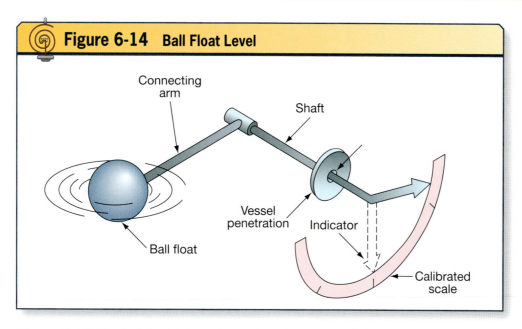

Figure 6-14. The ball float is less dense than the process fluid, which causes it to float on the surface of the liquid.

size, shape, or material used. The design consists of a ball float attached to a rod, which is connected to a rotating shaft that indicates level on a calibrated scale. **See Figure 6-14.**

The operation of the ball float is simple. The ball floats on top of the liquid in the tank. If the liquid level changes, the float follows and changes the position of the pointer attached to the rotating shaft. The travel path of the ball float is limited by its design to be within ±30 degrees from the horizontal plane, resulting in optimal response and performance. The actual level range is determined by the length of the connecting arm. A stuffing box is incorporated to form a watertight seal around the shaft and prevent leakage from the vessel. Ball floats should be calibrated by filling the vessel and observing the operation of the float if possible.

Chain Float

The chain float gauge has a float ranging up to 12 inches in diameter, and it is used when small level measurement range imposed by ball floats must be exceeded. The range of the level measured is limited only by the size of the vessel. The operation of the chain float is similar to that of the ball float except in the method of positioning the pointer and in its connection to the position indication. The float is connected to a rotating element by a chain with a weight attached to the other end to provide a means of keeping the chain taut during changes in level. **See Figure 6-15.**

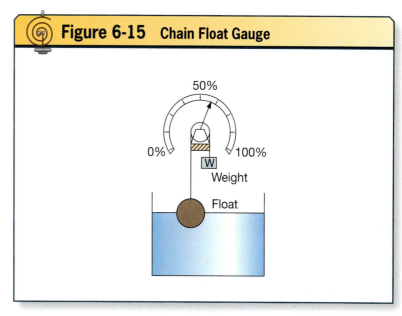

Figure 6-15. The chain float gauge indicates the height of a liquid.

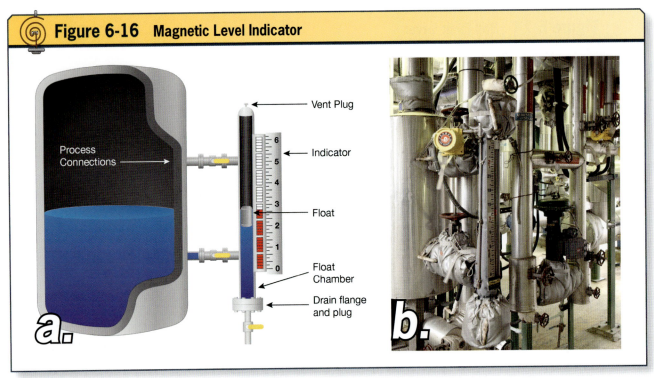

Figure 6-16. (a) The indicator shows the level of the fluid in the chamber based on the position of the float. (b) This setup shows the magnetic float level in the foreground and the condensate receiver tank level.

Magnetic Bond Mechanism

The magnetic bond method is used to isolate the process fluid and the measurement instrumentation. **See Figure 6-16.**

The magnetic bond mechanism consists of a magnetic float that rises and falls with changes in level. The float travels inside of a nonmagnetic tube, commonly made of stainless steel, which houses an inner magnet. This tube is substantially stronger than the material used in sight glasses. A magnetic follower is attached outside the tube. When the internal float rises and falls, the outer follower indicates on a visual scale the level inside the sealed tube. The magnetic bond method offers a safe method for measurement of process fluids that are a release hazard.

Displacer

The displacer method of measuring a level uses the concept of buoyancy, but unlike the float methods listed previously, the displacer never floats. Instead, the displacer is denser than the process fluid and is suspended by a mechanical connection, often a rod or cable. Buoyant force is always directed in a way to force an object out of the liquid in which it is submerged.

Archimedes's law states that when an object is placed into a fluid, the object is subject to a buoyant force equal to the weight of the fluid displaced by the object. An example is the displacement and buoyant force on a 150-pound object fully submerged in a process fluid, with a volume of 1 cubic foot, in an alcohol-filled tank (0.79 specific gravity). **See Figure 6-17.**

ThinkSafe!

The use of the magnetic bond method is advisable when the accidental release of process fluid could pose a threat to personnel or the environment. The magnetic bond method offers more protection than a glass tube device.

The first step is to find the weight of the displaced fluid. Use the weight of an equivalent amount of water (62.43 pounds per cubic foot) to find the buoyant force:

$$62.43 \frac{\text{lbs}}{\text{ft}^3} \times 0.79 = 49.32 \frac{\text{lbs}}{\text{ft}^3}$$

The weight of the displaced fluid is 49.32 pounds per cubic foot and is equal to the buoyant force applied to the displacer according to Archimedes's law. The displacer is 1 cubic foot in volume, so the resultant buoyant force is equal to 49.32 pounds per cubic foot. Because the displacer weighs 150 pounds, the apparent weight of the object is found by subtracting the buoyant force from the real weight when submerged:

$$150 \text{ lbs} - 49.32 \frac{\text{lbs}}{\text{ft}^3} = 100.68 \text{ lbs}$$

This value is only apparent weight; the object still weighs 150 pounds, but the buoyant force of the liquid causes the object to seem lighter. The buoyant force principle can be used to measure a fluid's level. If the fluid level is below or at the bottom of the displacer tube, the displacer will register its full weight on the sensor: 150 pounds. As the fluid level rises, the displacer's apparent weight decreases in exact proportions. Once a displacer is fully submerged, it has reached the maximum level it can measure and subsequent increases in level are not registered by the displacer. The measurement range of a displacer is limited by the length of the displacer.

To determine the apparent weight of a displacer that is partially submerged, the volume of fluid that is displaced must be found. Consider a displacer with a weight of 20 pounds, a diameter of 4 inches, and a height of 30 inches which is 50% submerged in water. **See Figure 6-18.**

Here, 1 cubic inch of water is equivalent to 0.0361 pounds and the apparent weight of the object is found by

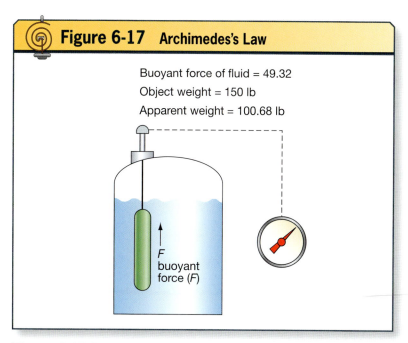

Figure 6-17. The displacer has an apparent weight of 100.68 pounds.

For additional information, visit qr.njatcdb.org Item #1737

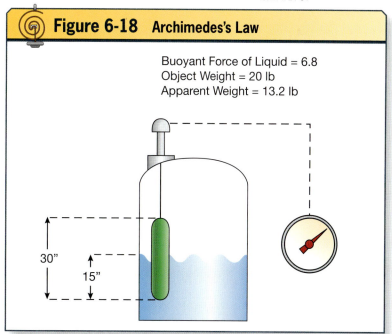

Figure 6-18. The displacer has an apparent weight of 13.2 pounds.

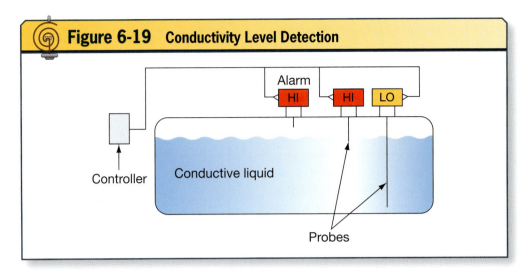

Figure 6-19. *The level switches measure the level at three different points in the tank.*

subtracting the buoyant force from the real weight when submerged:

$$V = \pi R^2 H$$

$$V = \pi \times 2 \text{ in}^2 \times 15 \text{ in} = 188.4 \text{ in}^3$$

$$188.4 \text{ in}^3 \times 0.0361 \frac{\text{lbs}}{\text{in}^3} = 6.8 \frac{\text{lbs}}{\text{in}^3}$$

$$20 \text{ lbs} - 6.8 \frac{\text{lbs}}{\text{in}^3} = 13.2 \text{ lbs}$$

The level can be detected only along the length of the displacer. When the level is below or above the displacer, the output of the transmitter is the maximum or the minimum, depending on calibration. A displacer can also detect the density of a liquid. Liquids with heavier densities cause a more apparent weight change.

Discrete and Level Switches

Point level detection is accomplished by using discrete switches and level switches. These measurement instruments generate a signal that indicates whether the process fluid is above or below a single level point. Some processes can be controlled using only discrete or level switches. In other installations, both point level and continuous level instruments are used to control the process.

Conductivity Probe Method

In a conductivity probe level detection system, a probe in direct contact with the process fluid is used to determine the level. A common design uses two probes separated by an air gap. Air in the vessel acts as an insulator when the process fluid is not in contact with the probe. When the process fluid level reaches the probes, the fluid acts as a conductor, completing the circuit. In response, a sensing unit in the instrument causes relay contacts to open or close. The relay actuates an alarm, a pump, a control valve, or all three. A typical system has three probes: a low-level probe, a high-level probe, and a high-level alarm probe. **See Figure 6-19.**

Vibrating Fork Level Switch

A vibrating fork liquid level switch operates on a principle similar to that of a tuning fork. The sensor is installed in a tank or pipe wall in direct contact with the process. These types of sensors can be used to detect high or low levels of solids or liquids. A piezoelectric crystal causes the fork to oscillate at a set frequency. When the fork comes in contact with a fluid that has a density higher than that of air, the frequency changes. The electronics integral to the device sense this change. By detecting the change in oscillation frequency, the sensor uses a discrete or analog signal to inform the control system of the level change. The devices often have delays to prevent false alarms, such as the sloshing of process fluid when filling a tank. **See Figure 6-20.**

Vibrating level switches require little maintenance, because they have no moving parts and work for a range of fluids. They are also useful in sanitary applications that require frequent cleaning. Unlike most process instruments, vibrating fork level sensors do not require frequent calibration after initial installation.

Rotating Paddle Level Switches
Dry solids such as grains or bulk powders require different types of technology to measure level. One of these types is the rotating paddle level switch. The unit consists of a paddle that is affixed to a shaft, which is turned by a motor with a low number of revolutions per minute. The motor is mounted to a flange through the wall of the vessel, with the paddle mounted inside the vessel. When the level of material is below the rotating paddle, it turns freely. As the level increases and material contacts the paddle, the motor stalls because of the increased resistance. The instrument then sends a signal, often by changing the state of a set of contacts, to indicate the level of material. **See Figure 6-21.**

Hydrostatic Pressure Instruments
One of the most common ways to measure a liquid level is to use a pressure gauge to measure the pressure of a liquid and, knowing the specific gravity of that liquid, to obtain the proper level. This method uses the differential pressure measurement device to detect the desired level.

The differential pressure detector method of liquid level measurement uses a differential pressure detector connected to the bottom of the tank being monitored. The higher pressure caused by the fluid in the tank is compared with a lower reference pressure (usually atmospheric). This comparison takes place in the differential pressure

Figure 6-20. The vibrating fork level detects level based on a change in vibration frequency due to contact with process fluids.

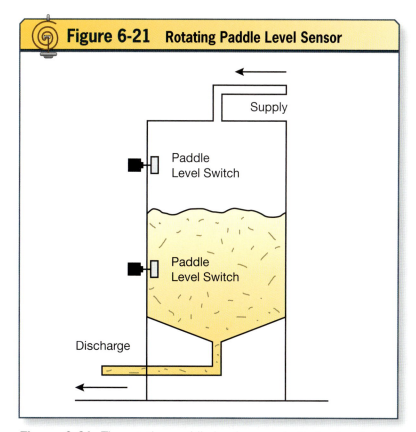

Figure 6-21. The rotating paddle level sensor indicates level when the paddle contacts the process media.

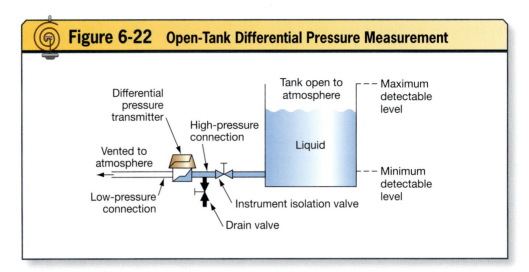

Figure 6-22. The level of an open tank can be measured using a differential pressure transmitter.

detector, which can be attached to an open tank. **See Figure 6-22.**

Open-Tank Measurement

When a tank is open to the atmosphere, it is necessary to use only the high-pressure connection on the differential pressure transmitter. The low-pressure side is vented to the atmosphere. Because the atmospheric pressure is the same on both the low side of the transmitter and the process fluid that creates pressure on the high side of the transmitter, these forces cancel each other out and the net differential pressure because of atmosphere is zero. Care must be taken to ensure the low side pressure connection does not get plugged, because this would result in an erroneous reading.

Closed-Tank, Dry Reference Leg Measurement

Not all tanks and vessels are open to the atmosphere. Many are totally enclosed to prevent vapors or steam from escaping or to allow pressurizing of the contents of the tank.

When measuring the level in a tank that is pressurized or can become pressurized, both the high-pressure and the low-pressure sides of the differential

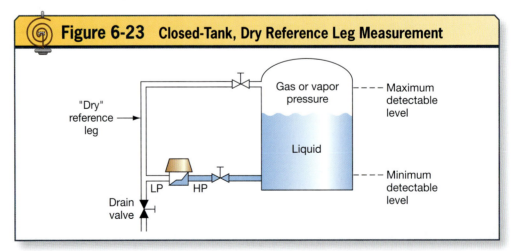

Figure 6-23. The low pressure connection on the differential transmitter is connected to the top of this tank to offset any pressure buildup in the tank.

pressure transmitter must be connected. **See Figure 6-23.**

The most common application is when the high-pressure connection is connected to the tank at or below the lower-range connection point on the tank. The low-pressure side is connected to a reference leg that is attached at or above the upper-range value to be measured.

The reference leg is pressurized by the gas or vapor pressure, but no liquid is permitted to remain in the reference leg. The reference leg must be maintained dry so that there is no liquid head pressure on the low-pressure side of the transmitter. The high-pressure side is exposed to the hydrostatic head of the liquid plus the gas or vapor pressure exerted on the liquid's surface. The gas or vapor pressure is equally applied to the low-pressure and high-pressure sides. These forces oppose and cancel each other out. The output of the differential pressure transmitter is therefore directly proportional to the hydrostatic head pressure, which is the level in the tank.

Closed-Tank, Filled (Wet) Reference Leg Measurement

A closed tank that contains fluids that are likely to condense requires a different compensation system, called a wet leg. A wet leg has a reference leg similar to a dry leg, the only difference being that the wet leg is filled with a liquid. If the compensation leg were allowed to collect various and undetermined amounts of condensed fluid, the transmitter reading would not accurately represent the level in the tank. There would be no way to determine the level in the tank. Therefore, the level of the wet leg must be known to derive meaning from the transmitter indication. The liquid in the reference leg applies a hydrostatic head to one side of the transmitter, and the value of this level is constant as long as the reference leg is maintained full. If this pressure remains constant, any change in differential pressure is caused by a change on the opposite side of the transmitter. **See Figure 6-24.**

The differential pressure transmitter is exposed to equal pressure on the high-pressure and low-pressure sides when the liquid level is at its maximum; therefore, the differential pressure is zero. As the tank level goes down, the pressure applied to one side of the transmitter also goes down, and the differential pressure increases. As a result, the differential pressure and the transmitter output are inversely proportional to the tank level. Either the high-side or the low-side transmitter connection can be connected to the wet leg. For example, assume a tank has a lower range value (LRV) of

For additional information, visit qr.njatcdb.org Item #1738

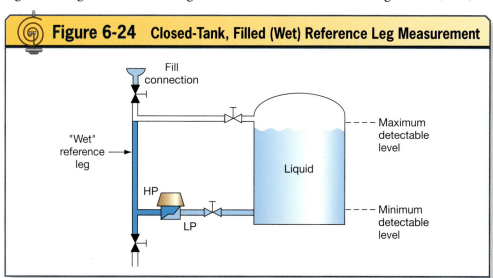

Figure 6-24. A wet reference leg is used in applications that are likely to fill with condensation.

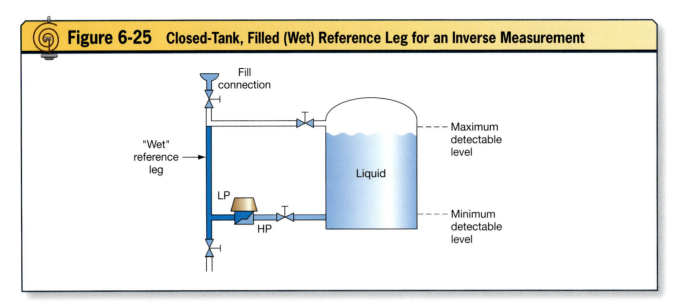

Figure 6-25. *The transmitter high and low connections are reversed in this installation.*

0 "H_2O and an upper range value (URV) of 100 "H_2O. In the case of the high side connected to the wet leg, a full tank would indicate 0 inches of water and an empty tank would indicate 100 inches of water. Conversely, if the low side were connected to the wet leg, a full tank would indicate 0 inches of water and an empty tank would indicate –100 inches of water. **See Figure 6-25.**

Wet legs are useful for eliminating measurement errors in condensable fluids, but the technician must take into account the density of the fill fluid. Temperature affects the density of matter. If the process fluid inside the vessel is held at an elevated temperature and the wet leg is allowed to cool, the density in the wet leg may increase. This can lead to measurement errors, because more fluids with a higher density produce more hydrostatic head. Multivariable transmitters can reduce these errors by factoring density in relation to temperature. Care must also be taken prevent the wet leg from freezing. This is often accomplished by heat tracing the wet leg in cold climates.

Instrument Elevation and Suppression Errors

The mounting location of differential pressure sensors can play a part in the calibrated measurement of the level. If the differential pressure sensor is located below the tank bottom, the fluid in the line connecting the transmitter will exert additional pressure on the sensor. This is called a suppressed zero installation. Whenever the pressure instrument is located below the tank and additional, unwanted pressure is applied, the instrument zero is suppressed and the output is elevated. Transmitters installed below tank connections typically have lower range values that are nonzero, such as a transmitter with a

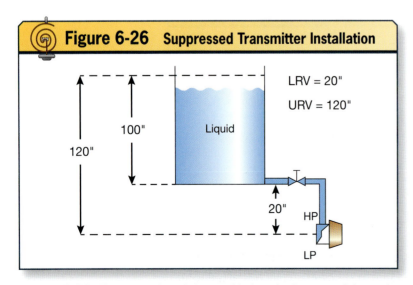

Figure 6-26. *The transmitter is located below the bottom of the tank.*

range of 20–120 inches. **See Figure 6-26.**

Field conditions may require the differential pressure transmitter to be installed above the lower level of the tank. This installation results in a negative pressure exerted on the sensor called an elevated zero and suppressed output. This application requires a special type of connection to the tank, called a remote seal. A remote seal consists of an isolation diaphragm with a capillary tube attached that is factory sealed with a fill fluid. This sealed capillary effectively extends the sensing capability through hydraulic pressure to the point of connection on the tank. The remote seal prevents air or foreign material from being introduced to the transmitter connections. Transmitters installed above the connections to the tank typically have lower range values that are less than zero, such as a transmitter with a range of –100 to –40 inches. **See Figure 6-27.**

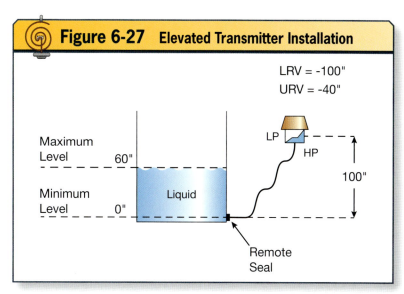

Figure 6-27. *The transmitter is located above the bottom of the tank.*

Bubblers

A bubbler level measurement system uses the concept of hydrostatic head to measure the level in a tank using a gas source and simple physics. **See Figure 6-28.**

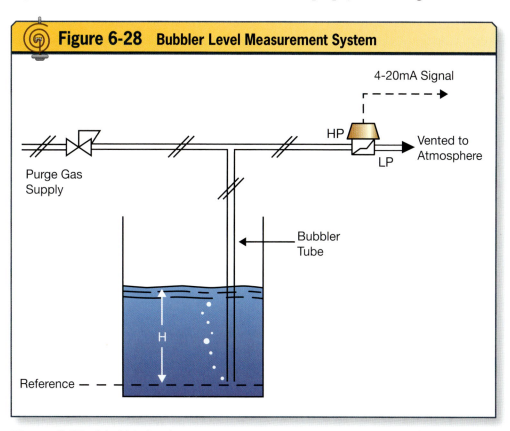

Figure 6-28. *A bubble level measurement system uses hydrostatic head in the form of backpressure to derive a level measurement.*

A gas is forced to flow through a tube inserted in the tank, called a dip tube or bubbler tube. The tube opening is located near the bottom of the tank. The pressure that builds up in the tube is measured using a differential pressure transmitter or gauge. If the tank is closed or pressurized, the transmitter must measure the pressure inside the tank by connecting the low side of the transmitter to the tank. This additional connection cancels out unwanted pressure and results in a measurement of hydrostatic head.

For the bubbler system to work, the dip tube must have a source of gas under more pressure than that of the hydrostatic head. This flow is commonly verified by a flow indicator called a rotameter. The bubbler is a simple measurement system that is self-cleaning. The one disadvantage of a bubbler is the requirement for a source of gas flow and the cost associated with maintaining the gas source, although this may be overcome by using a hand pump in remote locations.

Contact Level Instruments

Contact level instruments that use capacitance or guided radar to measure level utilize a probe in direct contact with the process fluid. These applications have no moving components, in contrast to the mechanical measurement instrument methods mentioned previously. This reduces failure rates due to wear or breakage of mechanical parts. Contact level measurement instruments can be mounted on the top of a process vessel, allowing the technician to remove and replace the instrument without draining the vessel. To ensure safety, the technician should use proper fall protection when working on the top of the process vessel.

Capacitance Level Sensor Measurement

Another method for measuring level is to use a capacitance level sensor. A capacitor is defined as two conductors, commonly called plates, that are separated by a nonconductive medium, called a *dielectric*. The capacitor stores an electrical charge that is equal to the

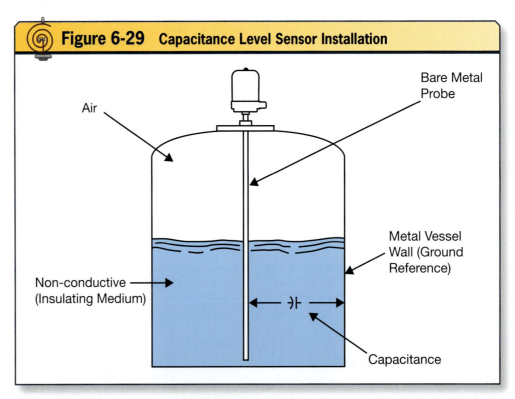

Figure 6-29. *Capacitance level measurement uses a probe and the tank wall to form a capacitor whose capacitance value varies with level height.*

applied voltage from the power supply or battery. **See Figure 6-29.**

Capacitance is the property that may be expressed as the time integral of flow rate (heat, electric current, etc.) to or from storage divided by the associated potential change. Thus, it is the storing of an electrical charge and is measured in farads. Because one farad is a large number, the unit picofarad (10^{-12} farads) or microfarad (10^{-6} farads) is more commonly used to measure capacitance. The ability of the capacitor to store charge is defined by the following formula in the units of picofarads:

$$\text{Capacitance} = 0.225 \times K \frac{\text{Area}}{\text{Distance}}$$

Here, K is the dielectric constant of the material. Because the area of the plates and the distance between the plates of the capacitor are fixed, the only variable is this dielectric constant. Two common process fluids are air, with a dielectric constant of 1, and water at 68°F, which has a dielectric constant of 80.4. Materials that are good insulators have low dielectric constants, while materials that are poor insulators have high dielectric constants. Standard dielectric constants are published by the National Institute of Standards and Technology (NIST).

A capacitance sensor probe is inserted from the top down into the tank and senses the level change by measuring capacitance. The result is an accurate level measurement. In one form of a capacitance sensor, a metal rod (or plate) is inserted into a tank. This rod serves as one of the capacitance plates, and the tank wall serves as another. Another type of sensor has two rods or plates separated by air. This type is necessary in plastic tanks and other nonmetallic vessels.

When the tank is empty, the dielectric is air. When the level in the tank rises, the air is displaced by the fluid. The fluid can have a better or worse dielectric rating. A change in the dielectric value of the material directly changes the capacitance value of the sensor. This change in capacitance is read by the transmitter electronics and affects the output signal.

Calibration procedures for capacitance sensors require measurement devices to be calibrated in place and only with the process fluid to be measured. Failure to do so is likely to result in erroneous measurements. The concepts of zero and span are the same as for other transmitters.

Guided Wave Radar

Guided wave radar is a technology that uses microwaves traveling through a probe to determine the distance and therefore the level in a tank. **See Figure 6-30.**

The probe and transmitter are mounted to a connection on the top of the tank. The basic operating principle employs a microwave signal generator and a receiver. The generator sends a

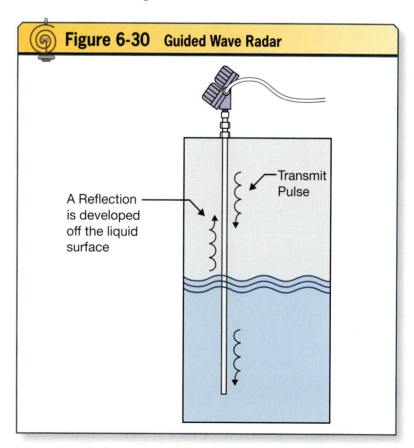

Figure 6-30. *A guided wave radar transmitter sends a signal along a probe and measures the amount of time required for the signal to return after reaching the surface of the liquid.*

For additional information, visit qr.njatcdb.org Item #2380

microwave pulse, which travels at the speed of light, down the probe. When the signal encounters a material with a dielectric constant greater than air, a portion of the signal is bounced back to the receiver. The time the signal takes to travel down the probe and back up the probe is calculated. Since the speed of the signal is constant, the distance is measured between the receiver and the top of the material in the tank.

The parameters of the tank are calibrated into the device during installation, which allows an accurate measurement of level. The guided wave radar level measurement systems have no moving parts and are commonly used to replace older displacement technology. During the instrument setup, the tank parameters need to be configured. The probe selected must be compatible with the process material and should be kept away from internal agitators or heating coils that could damage the probe. The measurement distance is limited by the probe length.

Noncontact Level Instruments

Noncontact level measurements using radar and ultrasonic applications to derive level measurement are unique applications. Each application is based on a process material's tendency to absorb or reflect radiation based on the dielectric constant. These applications have no moving components and in most cases require no physical contact with the process measured. The main advantage of a noncontact level measurement is the absence of moving parts and the ability to detect a level without making physical contact with the process fluid.

Ultrasonic Level Measurement

The ultrasonic level transmitter uses an ultrasonic pulse, typically in the range of 20–70 kilohertz. For contrast, the frequency range most humans can detect is below 20 kilohertz. The velocity of an ultrasonic pulse varies with the substance through which it travels and the temperature of that substance. This means that if the speed of sound is to be used in measuring a level (distance or position), the substance through which it travels must be known and its temperature variations must be measured and compensated.

The transmitter can be mounted in a top-down configuration or a bottom-up configuration. When mounted on the top of the tank, the sensor detects the distance to the surface of the process fluid. Accurate knowledge of the shape of the tank's cross-section is required to determine the volume of liquid. This information comes from strapping tables for the individual tank. When mounted on the bottom of the tank, the signal travels upward through the process fluid and is reflected by the interface between two liquids. The time it takes for the echo to return is an indication of the location of the interface. **See Figure 6-31.**

Ultrasonic waves are sound waves, and sound waves are affected by the temperature and density of the medium in which they travel. In addition, sound waves disperse at a rate consistent with the square of the distance. A disadvantage of

Figure 6-31 Time-of-Flight Radar Measurement

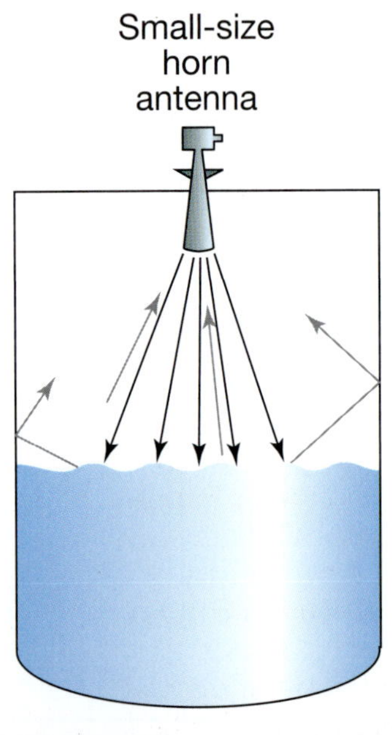

Figure 6-31. The time of flight radar measures the amount of time required for the signal to travel to the surface of the liquid and return to the transmitter.

ultrasonic transmitters is that foamy surfaces and steam tend to attenuate the signal and reduce accuracy. In addition, internal components of the tank such as agitators or additional sensors can interfere with the reading.

Radar Level Measurement

Radar measurement consists of a transmitter, an antenna, a receiver, and an operator interface. The transmitter is mounted on top of the vessel. Its solid-state oscillator sends out an electromagnetic wave (using a selected carrier frequency and waveform) aimed downward at the surface of the process fluid in the tank. The frequency used is typically between 6 and 28 gigahertz. This signal has advantages over ultrasonic level measurement because of the nature of sound wave propagation. While ultrasonic signals are affected by the chemical composition of the atmosphere in the tank, the electromagnetic wave generated by a radar level transmitter is unaffected. The electromagnetic wave travels at a consistent speed. A portion of the transmitted wave is reflected to the antenna, where it is collected and routed to the receiver. A microprocessor calculates the time of flight and calculates the resulting level from the time-of-flight measurement.

Time of flight is the period between the transmission of the radar pulse and the reception of the return echo. Radar beams can penetrate plastic and fiberglass; therefore, noncontact radar gauges can be isolated from the process vapors by a seal. The accuracy of radar measurement devices is affected by materials that tend to form layers of foam on their surface. The foamy surface causes scattering or absorption of the signal. Tanks with agitators can also cause interference that affects the accuracy of the transmitter. Depending on manufacturer and type, radar level transmitters may require too much power to be loop powered and need an additional power source. Radar transmitters are better suited than ultrasonic transmitters for measuring levels in high vacuum environments.

Nuclear Measurement

A form of level measurement that is used to measure hazardous or toxic chemicals is a nuclear level device. These measurement devices are most commonly ion chambers and scintillation tubes. These two devices are used for continuous level measurement and point level measurement. Mounting methods for ion and scintillation devices are similar. It is important to understand the basic operation of each type. Radiation measurement is a detailed field, and additional training is needed before the technician interacts with these devices.

Gamma rays are electromagnetic energy that can pass through materials. These rays are not capable of causing the materials through which they pass to become radioactive. However, exposure to gamma rays in sufficient quantity is dangerous to the technician's health. Just as lockout/tagout procedures are used for electrical hazards, nuclear sources can and should be locked out to protect workers during maintenance. The two most common isotopes used for a nuclear source are cesium-137 and cobalt-60. These materials are pelletized and contained within an enclosure with a shutter that directs the gamma rays in a specific direction. The gamma rays are absorbed, or attenuated, by solids or liquids at a different rate than the attenuation rate through the empty tank. The detector, composed of either a single ion chamber or a scintillation tube assembly, is mounted on the opposite side of the tank to measure the gamma rays.

Gamma rays, in effect, see through solid tank walls, nuclear radiation gauges are perhaps the ultimate in noncontact sensing devices. While nuclear devices sound dangerous, and can be without observing proper procedures, they offer a great benefit in that they can measure level, irrelevant of temperature or density, directly through a tank wall with no penetrations that could possibly leak. However, their licensing, testing, and inspection requirements make them more costly, especially from

For additional information, visit qr.njatcdb.org
Item #2381

TechTip!

Never install more than one ultrasonic level sensor in the same tank. The signals may affect each other.

an administrative perspective. Nuclear devices are expensive and require a gamma radiation source, they are often employed only as a last choice.

Ion Chamber

Ion chambers are tubes that contain plates, much like a capacitor, and are filled with a high-pressure gas. They are mounted next to the vessel wall being measured. The nuclear source is mounted next to the opposite vessel wall. A voltage is applied to the plates of the tube, and as radiation produced by the source passes through the tube, a small current passes between the plates. Material in the vessel blocks some portion of the radiation from reaching the ion tube. This current increases when more radiation passes through the vessel and into the tube, indicating a lower level in the vessel. Current decreases when less radiation passes through the vessel and into the tube, indicating a higher level in the vessel. The current is measured, and a microprocessor sets a current output range between 0% and 100% based on the current readings at the calibration ranges.

Scintillation Tube

A scintillation tube consists of a crystal that emits light pulses when subjected to radiation. The light pulses are measured by a photomultiplier, and the information is sent to a microprocessor for signal conditioning. More radiation passing into the scintillation tube produces a higher pulse rate. Less radiation passing into the scintillation tube produces a lower pulse rate. Less material in the vessel results in a greater photomultiplier current. More material in the vessel results in less photomultiplier current. The scintillation tubes are more sensitive, which is seen as an advantage because of the reduced amount of radiation needed from the gamma source.

Both ion chambers and scintillation tubes are calibrated in place after all process equipment is in place. If the instrument was calibrated before tank insulation was installed, the calibration would not be accurate. The unique nature of nuclear level instruments requires the manufacturer's directions and procedures to be followed to ensure safe and reliable operation.

DENSITY COMPENSATION

The density of a material is defined by its mass and volume. More specifically, the density of an object is the amount of mass per the amount of volume the object occupies. Level measurement technologies such as hydrostatic head pressure transmitters require the density to be known to determine an accurate level. As the temperature of a fluid changes, the volume also changes. This fundamental principle must be understood as the volume, and subsequently the density, of a fluid changes in relation to temperature changes.

The difference between the reference legs and the temperature of the process often requires compensation for accurate level measurement. To compensate for density differences, the technician must understand why the compensation is needed and know how to accomplish this task.

Specific Volume

Before considering an example that shows the effects of density, "specific volume" must be defined. Whereas density is given by the following formula:

$$D = \frac{M}{V}$$

where

D = Density

M = Mass

V = Volume

the specific volume of a substance is defined as the volume per unit mass, as shown in the following formula:

$$\text{Specific volume} = \frac{V}{M}$$

> **ThinkSafe!**
> Depending on the size of the radioactive source, specific safety requirements must be followed to protect workers in the area. Always check the Nuclear Regulatory Commission guidelines and specific state regulations to determine source safety and regulatory requirements.

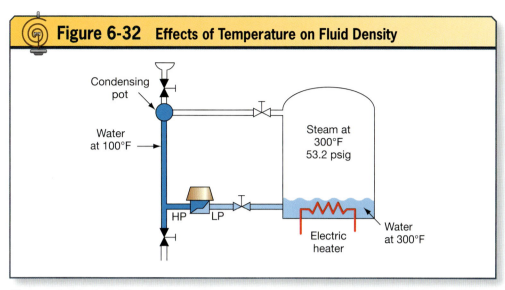

Figure 6-32. *A fluid's density varies as the temperature changes.*

Specific volume is also the reciprocal of density, as shown in the following formula:

$$\text{Specific volume} = \frac{1}{D}$$

Specific volume is the standard unit used in working with vapors and steam that have low values of density. For the applications that involve water and steam, specific volume can be found using saturated steam tables, which list the specific volumes for water and saturated steam at different pressures and temperatures. The density of steam (or vapor) above the liquid level has an effect on the weight of the steam or vapor bubble and the hydrostatic head pressure. As the density of the steam or vapor increases, the weight increases and causes an increase in hydrostatic head even though the actual level of the tank has not changed. The larger the steam bubble, the greater the change in hydrostatic head pressure.

A vessel can hold water at saturated boiling conditions. A condensing pot at the top of the reference leg is incorporated to condense the steam and maintain the reference leg at a full level. The effect of the steam vapor pressure is canceled at the differential pressure transmitter because this pressure is applied equally to the low-pressure and the high-pressure sides of the transmitter. The differential pressure to the transmitter reflects only the difference in hydrostatic head pressure between the low-pressure and the high-pressure connections. **See Figure 6-32.**

Reference Leg Temperature Considerations

Measuring the level of a pressurized tank that has differing temperatures throughout the system, such as a boiler unit, requires several consequences to be considered. As the temperature of the fluid in the tank is increased, the density of the fluid decreases. As the fluid's density decreases, the fluid expands, occupying more volume. Even though the density is lower, the mass of the fluid in the tank is the same. The problem encountered is that as the fluid in the tank is heated and cooled, the density of the fluid changes but the reference leg density remains relatively constant. This causes the reference leg hydrostatic pressure to remain constant. The density of the fluid in the reference leg depends on the ambient temperature of the room in which the tank is located; therefore, it is relatively constant and independent of tank temperature.

Figure 6-33. The water level in a steam drum must be maintained accurately.

If the fluid in the tank changes temperature and therefore density, some means of density compensation must be incorporated to produce an accurate indication of tank level. This is the problem encountered when measuring a pressurized water level or steam generator water level in pressurized water reactors and when measuring a reactor vessel water level in boiling water reactors. Failure to use density compensation in these examples would produce and not account for an excessive liquid level in the tank. The result could be unsafe process conditions.

Pressurizer Level Instruments

A pressurizer is used to maintain the process pressure at a controlled level within a nuclear reactor coolant system. The pressurizer operates by controlling the density of the water and steam contained inside. The surge line on the bottom of the unit is connected to the reactor coolant line and creates an interface to the rest of the system. If the pressure in the system increases above the set point, the spray line discharges cold water inside the vessel, condensing the steam and lowering the volume, and density, of the steam. If the pressure decreases below the set point, the heaters located in the bottom of the vessel are energized, increasing the temperature and therefore the pressure. **See Figure 6-35.**

An example is the level detection instrumentation for a boiler steam drum. The water in the reference leg is at a lower temperature than the water in the steam drum. It is, therefore, denser and must be compensated for to ensure the indicated steam drum level is accurate. **See Figure 6-33.**

The different densities of water can be compared based on temperature. **See Figure 6-34.**

Figure 6-34 Water Density Versus Temperature

Water Temperature (°F)	Water (lbs/ft³)	Water (lbs/gal)
32	62.42	8.34
80	62.22	8.32
160	61	8.15
212	59.83	7.99

Figure 6-34. The density of water varies in relationship to temperature.

Pressurizer temperature and pressure are held fairly constant during normal operation, commonly 575°F and 2,100 pounds per square inch gauge. The differential pressure detector 1 is calibrated for use when the pressurizer is at operational temperature. However, the pressurizer is not always hot. It may be cooled down for maintenance conditions, in which case differential pressure detector 2, calibrated for ambient temperatures, replaces the normal differential pressure detector. Differential pressure detector 3 measures both the level and the density of the water in the reference leg. The level of the water in the pressurizer is determined by the difference in pressure between the low side and the high side of differential pressure detector 1. If the pressure difference is zero, the pressurizer vessel is at the same level as the reference leg.

Steam Generator Level Instruments

A steam generator is a piece of equipment that turns water into steam. Maintaining an accurate water level is necessary. A level that is too high could result in liquid water entering the steam lines; a level that is too low could expose the heat exchanger tubes and cause them to sustain damage. **See Figure 6-36.**

The differential pressure detector measures actual differential pressure. A separate pressure detector measures the pressure of the saturated steam. Because saturation pressure is proportional to saturation temperature, a pressure signal can be used to correct the differential pressure for density. An electronic circuit uses the pressure signal to compensate for the difference in density between the reference leg water and the steam generator fluid. As the saturation temperature and pressure increases, the density of the steam generator water decreases. The differential pressure detector should indicate a higher level, even though the actual differential pressure has not changed. The increase in pressure is used to increase the output of the differential pressure level detector in proportion to

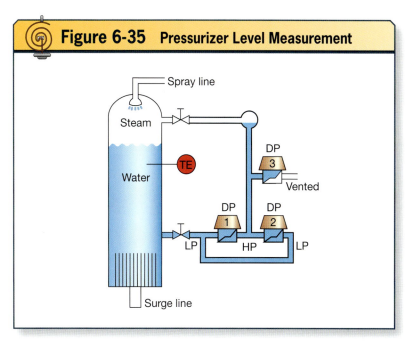

Figure 6-35. A pressurizer maintains a consistent pressure inside a steam loop.

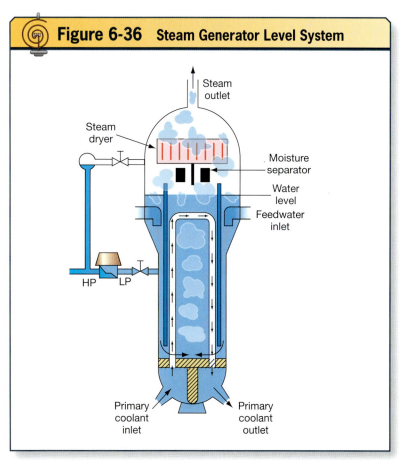

Figure 6-36. The level of water inside this steam generator is measured using a differential pressure transmitter with wet leg.

Figure 6-37. A remote diaphragm seal transmitter is useful to prevent slurries from clogging the impulse tubing.

saturation pressure to reflect the change in the actual level.

ENVIRONMENTAL CONCERNS

Ambient temperature variations affect the accuracy and reliability of level detection instrumentation. Variations in ambient temperature can directly affect the resistance of components in the instrumentation circuitry and therefore affect the calibration of electronic equipment. The effects of temperature variations are reduced by the design of the circuitry and by maintaining the level detection instrumentation in the proper environment.

Humidity also affects most electrical and electronic equipment. High humidity causes moisture to collect on the equipment, causing short circuits, grounds, and corrosion, which can damage components. The effects of humidity are controlled by maintaining the equipment in the proper environment.

In some cases, a process fluid can cause damage to the pressure transmitter. An example is a *slurry*, a fluid containing insoluble matter (e.g., mud), which can easily block or harden at the transmitter, rendering it inoperative. Remote diaphragm seals and sealed capillary tubes can be used to avoid this problem.

A remote seal may be flush mounted or may come equipped with an extended diaphragm. Extended diaphragms may be necessary to obtain a valid reference reading of the impulse pressure. Extended diaphragms allow the diaphragm to extend into the flange nozzle or the inner tank wall. **See Figure 6-37.**

Fluid pressure against the remote diaphragm must be transferred to the transmitter. This is accomplished by flexible tubing filled with an incompressible fluid. The sealed capillary connects the remote diaphragm at one end to the pressure transmitter at the other end. Sealed capillaries may be obtained in a variety of lengths, allowing the transmitter to be mounted at a safe and convenient location.

SUMMARY

Level measurement uses physical properties involving pressure, Archimedes's law, conductivity, resistance, and many other physical relationships to determine accurate levels. These devices require the installer to adequately understand the principles involved and be able to determine whether the installation and operation of the devices are affected. Physical concerns, installation requirements, and proper physical parameters confront each measurement application. When initially specifying a device, extreme care must be taken that all of these physical parameters have been included and that all safety and environmental concerns have been addressed.

REVIEW QUESTIONS

1. A level measurement device monitoring the level of glycol in a process tank reports a level of 37%. After some time, the level measurement device reports a level of 42%. What type of instrument is the level measurement device?
   a. Buoyancy
   b. Continuous
   c. Hydrostatic head
   d. Point

2. A cylindrical process vessel has a diameter of 13.5 feet and a height of 50 feet. What is the volume of the vessel?
   a. 675 ft³
   b. 2,121 ft³
   c. 7,157 ft³
   d. 28,628 ft³

3. How many gallons would be contained in a process vessel with an internal volume of 1,293 cubic feet?
   a. 173 gal
   b. 1,293 gal
   c. 7,102 gal
   d. 9,672 gal

4. What is the density of an object with a mass of 361 pounds that occupies a volume of 100 cubic inches?
   a. 3.61 lb/in³
   b. 36.1 lb/in³
   c. 3,610 lb/in³
   d. 361,000 lb/in³

5. What is the specific gravity of a liquid that weighs 156.49 pounds per cubic foot?
   a. 0.399
   b. 1.00
   c. 2.51
   d. 13.6

6. A tank contains 304.7 inches of water. What is the pressure at the bottom of the tank?
   a. 2.16 psi
   b. 8.44 psi
   c. 11.0 psi
   d. 25.4 psi

7. A column of mineral oil with a specific gravity of 0.905 generates a hydrostatic head of 7 pounds per square inch. How high is the column of mineral oil?
   a. 175.5"
   b. 194"
   c. 201"
   d. 214"

8. What is the buoyant force exerted on a displacer with a volume of 200 cubic inches fully submerged in a tank filled with water (a specific gravity of 1.00)?
   a. 5.37 lbs
   b. 6.98 lbs
   c. 7.22 lbs
   d. 8.45 lbs

9. The capacitance value of a capacitor can change in relation to a change in which of the following?
   a. The area of the plates
   b. The dielectric
   c. The distance between plates
   d. All of the above

10. Find the specific volume of a gas with a density of 0.00356 pounds per cubic foot.
    a. 2.809 lb/ft³
    b. 3.56 lb/ft³
    c. 280.9 lb/ft³
    d. 356 lb/ft³

Principles of Flow

One of the basic process variable measurements performed by industrial instrumentation is the measurement of flow. However, the principles of flow and fluid flow measurement may be among the most complex topics within the field of industrial instrumentation. To understand these principles, a detailed study of the associated variables needs to be performed.

Most industrial processes rely on the accurate measurement of process flow to operate safely and efficiently. The inability to make accurate flow measurements can result in unsafe conditions for people, the environment, and process equipment.

Objectives

» Describe flow measurement properties and measurement principles.
» Describe how differential pressure can be used to derive a flow rate.
» Understand flow relationships and express them mathematically.
» Describe the types of orifice plates and their features.
» Solve for the value of appropriate flow variables given area and mass flow rates.
» Explain the various methods used for open channel flow measurement.

Chapter 7

Table of Contents

Flow Fundamentals	142
Fluids	142
Component Variables	142
Process Energy	143
Volumetric Flow Equation	144
Law of Conservation of Energy	145
Law of Continuity	145
Bernoulli's Principle	146
Square Root Signal Extraction	147
Flow Characteristics	148
Fluid Viscosity	148
Density	148
Flow Characteristics Across a Restriction	148
Reynolds Number	149
Elements of Flow Meters and Relevant Process Variables	150
Methods of Flow Measurement	151
Flow Switches	152
Displacement Meters	152
Differential Pressure and Flow	153
Orifice Plates	153
Pressure Taps	155
Pipe Elbows	156
Flow Tubes	157
Pitot Tube	158
Flow Nozzles	159
Target Meters	159
Rotameters	160
Mass Flow Meters	161
Coriolis Meter	161
Hot-Wire Anemometer	161
Velocity Meters	161
Electromagnetic Flow Meter	162
Ultrasonic Flow Equipment	162
Open-Channel Meters	163
Flumes	163
Weirs	164
Environmental Considerations	165
Summary	166
Review Questions	167

For additional information, visit qr.njatcdb.org
Item #2690

FLOW FUNDAMENTALS

The accurate measurement and control of fluid flow inside process piping is essential for a process to function. **See Figure 7-1.** To gain an understanding of flow and its importance in industrial process systems, the terms associated with the fundamentals of flow need to be defined.

Fluids

Fluids are substances, such as liquids, gases, or vapors, that have no fixed shape of their own and yield easily to external pressure. In other words, fluids take on the shape of the vessel in which they are contained. For example, liquid water, on its own, has no discernible shape and is capable of flowing readily along any given pathway. However, when placed in a glass, the water takes on the shape of that glass. This holds true for any substance in a liquid state.

Both gases and vapors exhibit this same ability. *Gases* are fluids, such as air, that have neither independent shape nor volume but tend to expand indefinitely. An example would be the air people breathe daily, which takes on the shape of the lungs once inhaled and expands back out and into the atmosphere when exhaled.

Component Variables

For a fluid to flow, it must be moving. To accurately calculate the rate of flow, the technician must be able to determine how fast the fluid is moving through whatever contains it—water in a pipeline, for example. One such component in determining the rate of flow is velocity. *Velocity (v)* is the rate at which a substance moves from its original position in a given amount of time. More specifically, velocity is the distance, in a given direction, that the substance moves over time. A common unit used in this measurement is inches per second.

Another component used in determining the rate of flow is area. *Area* is the measure of a surface expressed in units squared within a set of boundary lines. For example, a square is 4 inches by 4 inches. **See Figure 7-2.** The area, A, of the square can be determined by multiplying the length, L, by the height, H, of the square:

$$A = L \times H$$
$$A = 4 \text{ in.} \times 4 \text{ in.}$$
$$A = 16 \text{ in}^2$$

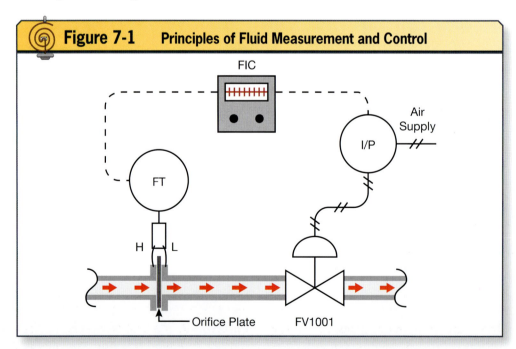

Figure 7-1. A typical control loop is used to measure and control flow in a process line.

However, within the field of industrial process control, process lines tend to be circular. In the formula necessary for calculating the cross-sectional area of a circle, the area is equal to pi multiplied by the square of the radius, r, of the circle:

$$A = \pi r^2$$

Most commonly, the unit square inches is used to express this cross-sectional area. **See Figure 7-3.**

Mass and volume are important values used in process measurement, allowing for the quantitative measurement of process flow. Flow rates are frequently expressed in volumetric terms, such as cubic feet per hour, or mass terms, such as pounds per hour. Mass flow rates require the density, the ratio between mass and volume, to be known.

An additional component used in determining the rate of flow is differential pressure, represented as *h*. Differential pressure is the difference in static pressure between two points of a system, such as the difference between two pressure taps on a differential pressure transmitter. This is highly relevant for day-to-day liquid-level measurement. However, differential pressure, as it relates to fluid flow, can be complex compared to a liquid-level measurement application.

When using a differential pressure transmitter in a liquid-level application, the high- and low-pressure sensing elements of the transmitter are used to measure the hydrostatic head pressure, based on the height of a fluid, and compare this measured value to atmosphere or some other reference. In this way, an inferred measurement of the fluid level in a vessel can be obtained. In contrast, fluid flow measurement obtained using differential pressure relies on mathematical relationships that vary greatly from head pressure calculations. This is largely because a moving fluid contains energy.

Process Energy

There are two main types of energy within a process system: potential energy and kinetic energy. *Potential energy*

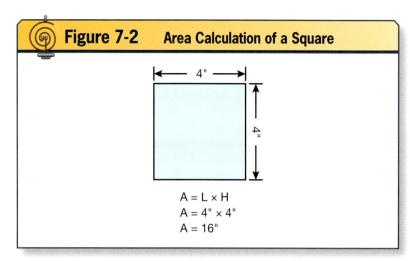

Figure 7-2. *The area of a square is calculated by multiplying the length by the height of the square.*

is the energy that is stored in an object because of its positional relationship to another object. An object at rest at some position above another object is said to have potential energy. With respect to fluid flow within a process, potential energy can be viewed as the static energy of the fluid within the process lines. Potential energy within a system can be expressed by the following equation:

$$PE = \rho g H$$

where *PE* is the potential energy, ρ is

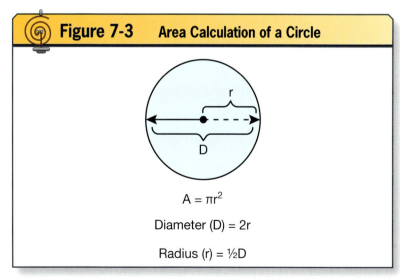

Figure 7-3. *The area of a circle is calculated by multiplying pi and the radius of the circle squared. The radius of the circle is equal to one-half the diameter of the circle.*

the fluid density, g is the force of gravity, and H is the relative height of the object.

Kinetic energy is the energy possessed by an object because of its motion, or the work necessary to accelerate an object with a given mass from a resting position to some velocity. Kinetic energy is present as a direct result of the fluid flowing within the process lines. This is largely because the fluid can be seen as an object with a given mass accelerating to a stated velocity within the process lines due to the pressure applied. Kinetic energy within a process can be expressed by the following equation:

$$KE = \frac{1}{2}\rho v^2$$

where KE is the kinetic energy, ρ is the fluid density, and v is the velocity.

Total energy in a flow consists of the sum of the potential energy, the kinetic energy, and the pressure, P, as expressed by the following equation:

$$\text{constant flow} = \rho g H + \frac{1}{2}\rho v^2 + P$$

Applying this equation, the energy within a flow is constant at any point along the streamline. This can be seen by looking at two different points along a process line. **See Figure 7-4.** Using this information, some general observations can be made about the fundamentals of flow.

VOLUMETRIC FLOW EQUATION

The standard volumetric flow equation is the product of the cross-sectional area of the process line times the velocity of the fluid flowing through it. It is expressed as follows:

$$Q = vA$$

where Q is the volumetric flow rate, v is the velocity of flow in inches per second, and A is the cross-sectional area in square inches of the process line. Applying this equation, the rate of flow, in cubic inches per second, is directly related to the velocity of the flow and the diameter of the process line. *Volumetric*

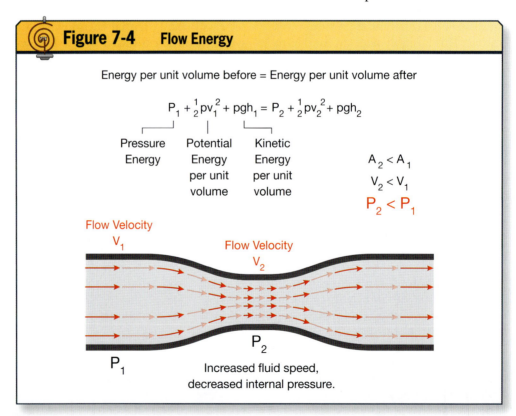

Figure 7-4. The sum of the pressure energy, kinetic energy, and potential energy is equal at any point within a fluid flow.

flow rate (Q) is often defined as a given quantity moving past a given point in a specified period. Velocity is the rate at which a substance moves from its original position in a given amount of time. It equates only to how fast a substance is moving. By knowing, or being able to calculate, volumetric flow rate, the technician can ensure the safe and efficient operation of a process within its design parameters.

When calculating volumetric flow rate, through the use of the equation $Q = vA$, the unit is often expressed in cubic inches per second. This value can be converted to other common units of volumetric flow rate, such as gallons per second, gallons per minute, and gallons per hour, using common unit conversion formulas, tables, or both. Mass flow rate represents the true mass of the fluid instead of its volume, which varies by pressure and temperature. However, volumetric flow rate is technically the more convenient measurement and is often the preferred measured value. Changes in temperature result in changes in volume, requiring correction in some cases.

Measurement devices often do not reflect flow rates in the same units that operations personnel use to control the process. It is helpful to have an understanding of conversion tables and how to use them to obtain the desired measured variable. Many conversions can be obtained within the programming of the flow device electronics. There are also numerous software programs available that accomplish the required conversions. Still, it is good to have a basic knowledge of how these conversions can be manually completed. For example, assume a sensor measures differential pressure using the units inches of water.

The following process shows how a conversion to gallons per hour proves its equivalent measuring units. Consider a process with a metered fluid flow rate of 2,748 cubic inches per second. The technician desires to solve for an equivalent gallons per hour. Unit conversion tables show there are 1,728 cubic inches and 7.4805 gallons in 1 cubic foot:

$$2{,}748 \text{ in}^3/\text{s}$$
$$2{,}748 \text{ in}^3/\text{s} \times 60 \text{ s/min} = 164{,}880 \text{ in}^3/\text{min}$$
$$164{,}880 \text{ in}^3/\text{min} \times 60 \text{ min/hr} = 9{,}892{,}800 \text{ in}^3/\text{hr}$$
$$\frac{9{,}892{,}800 \text{ in}^3/\text{hr}}{1{,}728 \text{ in}^3/\text{ft}^3} = 5{,}725 \text{ ft}^3/\text{hr}$$
$$5{,}725 \text{ ft}^3/\text{hr} \times 7.4805 \text{ gal/ft}^3 = 42{,}825.86 \text{ gal/hr}$$

It is sometimes necessary to convert from a mass flow rate (in pounds per hour) to its equivalent gallons-per-hour variable. The following example uses a process reading of 3,240 pounds per hour and converts it to gallons per hour. Assume 1 cubic foot of water weighs 62.43 pounds:

$$\frac{62.43 \text{ lb/ft}^3}{7.4805 \text{ gal/ft}^3} = 8.35 \text{ lb/gal}$$
$$\frac{3{,}240 \text{ lb/hr}}{8.35 \text{ lb/gal}} = 388 \text{ gal/hr}$$

LAW OF CONSERVATION OF ENERGY

The law of the conservation of energy, a fundamental law of physics, bears mentioning because it proves to be useful when considering flow measurement. *Energy* is the capacity for doing work. Simply stated, the law of the conservation of energy is that energy can be neither created nor destroyed; it can only be altered in form. An example of this principle is when electrical energy flows through a resistor and transforms to heat energy. Regardless, the total energy of an isolated system remains constant. By extending this principle into the study of fluid flow, it can be observed that the total energy at any point in a process line does not change because energy can be neither created nor destroyed.

Law of Continuity

The law of continuity is a fundamental law of physics that has a bearing on the

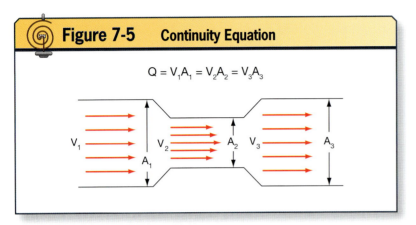

Figure 7-5. Flow rate (Q) stays consistent in a piping system in accordance with the continuity equation.

Bernoulli's Principle

The conservation of energy principle is at the foundation of many scientific principles, including *Bernoulli's principle*. In the 18th century, Daniel Bernoulli found that the sum of pressure energy and velocity energy in a line is a constant throughout the system if potential energy and friction are ignored. Bernoulli's principle is simply a variation of the conservation of energy as it relates to fluid flow and fluid flow measurement. Bernoulli's principle shows that as the velocity in a given line is increased, the pressure must decrease; conversely, when the velocity in a given line is decreased, the pressure must increase. So, what happens to a fluid flowing in a process line when the diameter decreases and then increases? The flow rate, Q, does not change at any point in the process line. When the cross-sectional area, A, of the process line decreases, the velocity, v, must increase; conversely, when the cross-sectional area of the process line increases, the velocity must decrease. **See Figure 7-6.**

The velocity at v_2 is greater than that at v_1, and because energy remains the same, the pressure at P_2 is less than the pressure at P_1. By the same reasoning, the velocity at v_3 is less than that at v_2, and the pressure at P_3 is greater than the pressure at P_2. The result is that the energy balances at all points along the line.

This illustrates an important relationship that can be observed between velocity, v, and differential pressure, h. The change in pressure is not directly proportional to the change in velocity but is instead proportional to the square of the change in velocity. From this observation, the velocity is proportional to the square root of the differential pressure. It may be easier to visualize the differential pressure change through a restriction by placing a manometer across the restriction in the line and observing the effects of the process pressure on it. **See Figure 7-7.**

This behavior, labeled the *Bernoulli effect*, is the lowering of fluid pressure in regions where the flow velocity is increased. This lowering of pressure through

For additional information, visit qr.njatcdb.org Item #2637

principle of flow within a process system. Fluid moving through a process line must follow the law of continuity, which states that the product of the average velocity and pipe cross-sectional area must remain constant at any point within the streamline. **See Figure 7-5.**

Flow calculations commonly assume a flowing fluid is incompressible, meaning the density does not change substantially; therefore, the continuity equation simplifies to the following equation:

$$Q = v_1 A_1 = v_2 A_2 = v_3 A_3$$

where Q is the flow rate, v is the velocity, and A the cross-sectional area.

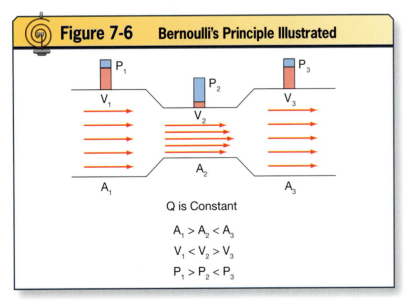

Figure 7-6. The flow rate does not change, but the velocity increases as the cross-sectional area decreases, and vice versa.

a restriction in a flow path may seem counterintuitive. However, as fluid flows through the restriction, kinetic energy must increase at the expense of pressure energy, therefore the pressure lowers.

By taking this relationship into consideration, differential pressure can be used in place of velocity in the equation for flow rate:

$$Q \propto \sqrt{h} \times A$$

where flow rate, Q, is proportional to the square root of differential pressure, h, multiplied by the cross-sectional area, A.

A flow measurement can be obtained by measuring the differential pressure across a known restriction in a process line, most commonly an orifice plate, and calculating the cross sectional area of the pipeline's internal diameter.

Square Root Signal Extraction

The relationship between flow rate and differential pressure is nonlinear. This is because the relationship between flow rate, Q, and differential pressure, h, is proportional based on the following equation:

$$Q \propto \sqrt{h} \times A$$

Through algebraic manipulation of this proportionality, it becomes apparent that differential pressure is proportional to the square of the flow rate as represented by the proportionality:

$$h \propto Q^2$$

This relationship is more easily understood by graphing differential pressure as a function of flow rate. **See Figure 7-8.**

A flow transmitter connected across a differential pressure flow element does not directly measure flow rate, but it does measure the square of the flow rate. As a result, the instrument transmits the proper output values corresponding to inputs of 0% and 100% (when properly calibrated to these inputs), but does not respond linearly to all other intermediate inputs. As a result of this nonlinear response, any

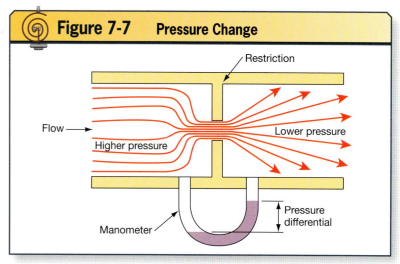

Figure 7-7. A change in differential pressure can be visualized through the placement of a manometer across a line restriction.

controller, indicating device, or recording device receiving the output from the transmitter registers these nonlinear values. This is because the output signal from the transmitter is not in direct correlation with the flow rate in the process line, which is often undesirable in a process control environment.

An accurately calibrated device should

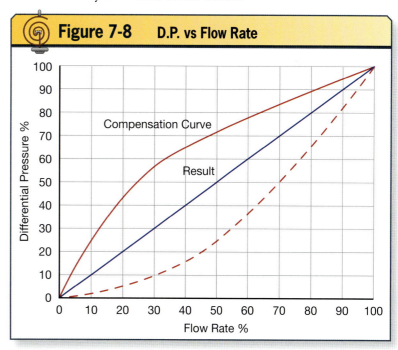

Figure 7-8. The red dotted line plots the differential pressure in relation to the flow rate. To achieve a linear output signal, as depicted by the blue line, a compensation curve is necessary.

have a percent output signal that is directly proportional to a given input signal. In the case of a differential pressure transmitter connected across a flow element, a mathematical function must be employed to extract or condition the required output signal: the square root. This is often performed at one of two locations: inside the transmitter or inside the controller. Regardless of where the math is performed, it must occur at one point in the loop such that the flow rate can be accurately registered across its full range of operation in the process.

FLOW CHARACTERISTICS

Additional factors which affect the flow of a fluid include the fluid's viscosity and density, as well as the friction of the fluid in contact with the inner walls of the process line.

Fluid Viscosity

Flow in a pipe is not always ideal. *Viscosity* is the extent of friction between two adjacent layers of a fluid; the greater the viscosity of a fluid, the more energy required to cause it to slide past itself. In other words, viscosity is the inherent resistance of a substance to flow. The more viscous a fluid is, the thicker it seems to be; the less viscous a fluid is, the thinner it seems to be. Clean, fresh water is an example of a low-viscosity fluid, whereas molasses at room temperature is a high-viscosity fluid. The unit used to express dynamic viscosity, commonly the Greek letter mu (μ), is known as *poise*, where 1 poise is equal to 0.1 Pascal-second. Both of these units are generally too large for common use, so the *centipoise* is used most often when discussing viscosity. Water has a dynamic viscosity close to 1.000 centipoise, whereas glycerin has a dynamic viscosity of around 950.000 centipoise, which makes it more viscous than water.

Density

Density is a contributing factor to fluid flow. Density is the mass per unit volume of a fluid, with the density of fresh water at 4°C being 62.43 pounds per cubic foot. The density of a fluid flowing within a process line affects how the fluid moves through the line. Combined with velocity and viscosity, it is a contributing factor to the fluid's momentum and drag force, i.e. friction, as it flows through the process. *Friction* is energy lost to heat dissipation when a fluid flows through a pipe. Friction results when a moving fluid comes into contact with the inner walls of a pipe. Viscosity and fluid motion result in heating of the process lines, which reduces the total energy of the moving fluid. This is undesirable, because this "lost" energy results in process inefficiency, requiring higher pump and pump motor capacity, which in turn leads to higher process costs.

Flow Characteristics Across a Restriction

One of the means of reducing the undesirable aspects of fluid flow is to control flow rate within a process system. By controlling the rate of flow, the characteristics of the fluid flow can be made more viscous, or laminar. *Laminar flow* is characterized as the straight-line flow of a fluid. **See Figure 7-9.** This typically occurs at a slow rate of flow along the inner walls of the process line, where the flow of fluid approaches a near-zero rate (because of friction), compared to the rate of flow at the center of the streamline, where it can be twice the average velocity in the line. Due to the lack of a uniform flow, laminar flow is not considered desirable for most flow measurement applications.

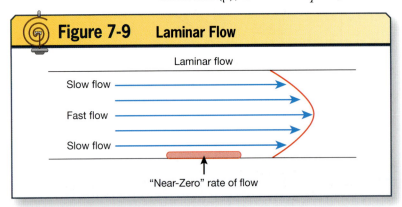

Figure 7-9. *Fluid flows faster in the center of the line and more slowly along the wall.*

The contrast to laminar flow is turbulent flow. *Turbulence*, violent or unsteady movement of air or water, occurs because the inertial forces of the fluid flow are greater than the viscosity forces, holding the fluid together. As the liquid flow rate increases, the formation of an *eddy* or eddies, swirling motion of the liquid as it flows past an obstacle, becomes greater. Eddies also occur whenever flow direction changes, such as at a line elbow or joint and adjacent to pipe walls. Though the term turbulent commonly invokes a negative connotation, turbulent flow is necessary for many flow meters to operate properly. For example, a venturi tube requires a turbulent flow pattern to accurately determine flow rates. **See Figure 7-10.**

When the fluid flow is of a gaseous nature, a similar phenomenon occurs, known as a *vortex*. A common example of a vortex in action would be a tornado, or a mass of whirling air. A variety of methods to control flow profiles exist, such as increasing straight runs of piping or utilizing *flow straighteners*, which are devices that have special blades to reduce turbulence. In many cases simply slowing the rate of flow of the fluid can have a positive effect on smoothing a flow profile.

Reynolds Number

A British scientist, Sir Osborne Reynolds, observed a relationship among the four basic factors that affect flow: the velocity, viscosity, and density of fluid combined with the diameter of the line through which the fluid flows. This relationship is expressed as *Reynolds number (RN)*, which is a unitless number used to gauge the amount of fluid flow resistance in a line. The equation for this relationship is expressed as follows:

$$RN = \frac{\rho v d}{\mu}$$

where ρ is the fluid density, v is the average velocity of the fluid, d is the geometric length (usually the diameter of a round process line), and μ is the dynamic viscosity of the fluid.

RN is a ratio of a fluid's inertial forces (i.e., flow rate and density) and its drag force (i.e., line diameter) to the viscosity of the fluid. In other words, it is a ratio of the fluid's momentum to its viscosity, which aids in predicting how a fluid will flow. It is common for turbulence to build quickly after a certain RN has been exceeded. The RN is unitless, but how it is calculated depends on the units of measure used for the other values in the equation.

With regard to laminar flow, which occurs when viscosity is high, the force of friction tends to dictate the manner in which a fluid flows. However, turbulent flow is the result of there being almost no friction within the fluid, and the fluid tends to flow smoothly and freely. RN can be a predictor of whether flow is more laminar or more turbulent. Generally, if RN is less than 2,000, flow is considered to be laminar; if RN is between 2,000 and 4,000, flow is said to be transitional; and when RN is greater than 4,000, flow is considered to be turbulent.

Consideration must be given to flow with regard to its ability to create wear and tear on components used in flow measurement applications. Over time, abrasive, erosive, and corrosive effects are realized on the various elements within a process flow loop, especially the primary elements. As a result, care and attention must be given because the primary sensing elements are in direct contact with the fluid stream. Loss of accuracy, dependability, and repeatability can occur due to these effects.

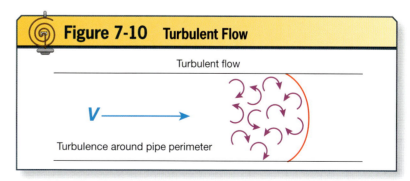

Figure 7-10. Turbulent flow occurs when viscous forces are overcome, resulting in mixing across the fluid flow profile.

The basic equation for flow, $Q = vA$, can be modified to more accurately reflect real flow rates rather than ideal flow rates. Instead of the basic equation, the following equation can be used:

$$Q = C_d A \sqrt{\frac{2h}{\rho\left(1-\left(\frac{d}{D}\right)^4\right)}}$$

where Q is the volumetric flow rate, C_d is the discharge coefficient, A is the cross-sectional area of the restriction in square inches, h is the differential pressure, d is the diameter of the restriction, and D is the inside diameter of the pipe.

The discharge coefficient is dependent on the RN and is determined by laboratory tests or field verification. The *beta ratio* is the ratio of the diameter of a pipeline constriction (i.e., the dimension of the bore opening of an orifice plate) to the diameter of an unconstricted pipe. The dimension of the bore opening is normally given on the handle of the orifice plate as a decimal or a fraction. To check for accuracy, simply divide the measured diameter of the hole in the orifice plate by the measured inner diameter of the pipe.

For example, observe the following scenario. A water pipe with an inside diameter of 1.049 inches has an orifice plate with a diameter of 0.5 inches inserted into a set of flanges. A differential pressure transmitter measures 127.7 inches of water pressure across the restriction. Assume a discharge coefficient of 0.59 for the orifice plate. The density of water is 0.0361 pounds per cubic inch at 4°C:

$$Q = 0.59 \times 0.1963 \text{ in}^2 \times \sqrt{\frac{2 \times 127.7 \text{ "H}_2\text{O}}{0.0361 \text{ lb/in}^3 \left(1-\left(\frac{0.5 \text{ in}}{1.049 \text{ in}}\right)^4\right)}}$$

$$Q = 10.00 \text{ GPM}$$

The beta ratio, density, and discharge coefficient are fixed values in the equation. To simplify flow calculations, the following equation is used:

$$Q = k\sqrt{h}$$

In this simplified equation, k represents a constant involving RN, the discharge coefficient, and the density of the fluid. Using the previous example, the k value can be found as follows:

$$10.00 \text{ GPM} = k\sqrt{127.7 \text{ "H}_2\text{O}}$$

$$k = \frac{10.00 \text{ GPM}}{\sqrt{127.7 \text{ "H}_2\text{O}}}$$

$$k = 0.8849$$

Now that the technician knows the k value, the gallons per minute value can be calculated for any differential pressure, h. For example, if a differential pressure of 11.5 inches of water were measured, the gallons per minute would be found as follows:

$$Q = 0.8849\sqrt{11.5 \text{ "H}_2\text{O}}$$

$$Q = 3.00 \text{ GPM}$$

When a fluid moves through an orifice plate, the fluid forms a concentrated flow, with the lower pressure area immediately downstream of the plate. The term used to describe the point downstream of a line restriction, such as an orifice plate, where the fluid velocity is greatest and pressure is lowest due to the inertia of the moving fluid, is called the *vena contracta*. **See Figure 7-11.** This area of low pressure is an important aspect related to the installation of certain types of flow instruments.

Elements of Flow Meters and Relevant Process Variables

Process measurement and control demand accuracy, repeatability, and dependability while interacting with the various process variables that must be considered in each instance. Monitoring and control of a flow loop often is contingent on the line size and velocity. The characteristics of the flow within the process may directly influence accuracy, repeatability, and dependability of flow measurement. In addition, conditions such as operating temperatures, ambient pressures, exothermic process reactions, endothermic process reactions, and line surges may cause the measurement devices to no longer perform as desired.

Although these factors are design considerations for engineers, field technicians are confronted with such conditions, and others, that may affect the functioning of equipment within the process environment. Many process problems are directly related to the measuring equipment selected, installed, or implemented in a flow control loop. A basic understanding of the advantages and disadvantages of the different types of flow measurement devices can help in diagnosing problems and implementing solutions. Those who understand device requirements and limitations can appreciate the selection of flow measurement instruments based on the importance of accuracy, dependability, cost, and other factors. After these factors have been determined, several physical factors come into play as well, including line size, range of flow rate (e.g., minimum and maximum), fluid characteristics (e.g., liquid, slurry, suspended solids, operating temperature, and operating pressure), corrosive effects, and steady or surging flow.

The process industry deals with the rate of flow for process reactions, storage, and profit. It is estimated that flow measurement is one of the more common measurements performed in many industrial process systems, but some technicians claim that many of the flow measurement devices installed do not perform satisfactorily. Faulty specifications by designers and engineers account for some of these problems, but improper installation accounts for some of this subpar performance. Instrument technicians who understand the flow instrument's overall function and limitations can contribute more accurate flow data, resulting in increased profitability of industrial process systems.

METHODS OF FLOW MEASUREMENT

In general, there are many methods of flow measurement. These include flow switches (discrete measurement), displacement meters, differential pressure flow meters, a range of primary

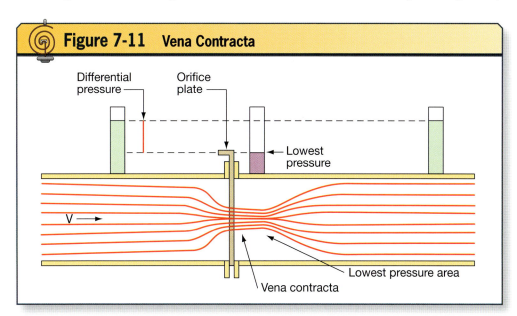

Figure 7-11. Concentrated flow forms a point called the vena contracta while fluid moves through an orifice plate.

Figure 7-12. A paddle flow switch is one of the many available types of flow switches.

elements, target meters, rotameters, mass flow meters, velocity meters, and open-channel meters.

Flow Switches
Flow switches are used to provide discrete measurement of flow within a process system. They are designed and configured to operate at a specific flow rate and offer a simplified method for determining the presence of fluid flow. Their discrete, or "on" or "off," signaling is commonly used to signal abnormal or undesired conditions. **See Figure 7-12.**

Displacement Meters
Operation of displacement meters such as gear or nutating disc type involves separating liquids into accurately measured increments and then moving them along. Each segment is connected to a register that counts each segment as a volume amount. These units are popular with automatic batch processes and accounting applications. The meters are particularly good for applications that measure viscous liquids using a simple mechanical meter system. **See Figure 7-13.**

A nutating disc meter is commonly used for water service. The movable element is a circular disc that is attached to a central ball. A shaft is fastened to the ball and held in an inclined position by a cam or roller. The disc is mounted in a chamber that has spherical side walls and conical top and bottom surfaces. The fluid enters an opening in the spherical wall on one side of the partition and leaves through the other side. As the fluid flows through the chamber, the disc wobbles, or executes a nutating motion. Because the volume of fluid required to make the disc complete one revolution is known, the total flow through a nutating disc can be calculated by multiplying the number of disc rotations by the known volume of fluid.

 **TechTip!**

Gear meters should not be used when the process contains media that could become lodged in the gears and prevent gear rotation.

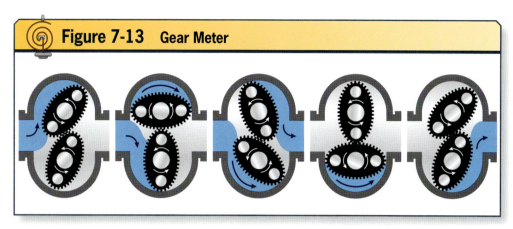

Figure 7-13. *A gear meter utilizes the force of the fluid flow to rotate a set of gears. Each rotation of these gears corresponds to a volume of fluid, and as these rotations are counted in relation with time, volumetric flow rate is determined.*

Differential Pressure and Flow

Differential pressure flow meters may be the most common units in use for flow measurement. Like most other flow meters, differential pressure flow meters have primary and secondary elements. The primary element is responsible for a change in the kinetic energy, which causes the pressure drop across the element. An orifice plate is common in this application. This unit is precisely sized for correct line size, flow rate, and liquid properties and allows an accurate measurement over a reasonable range. The secondary element analyzes the primary element's information and provides a signal or readout that is converted to the actual flow rate. A multitude of primary elements are used to provide a measurable difference in pressure across a line restriction. These flow elements include orifice plates, pressure taps, pipe elbows, flow tubes, Pitot tubes, and flow nozzles.

Orifice Plates

The most common liquid flow meter in use today is the orifice plate. The orifice plate is the simplest of the flow path restrictions used in flow detection, and it is the most economical.

Orifice plates are flat plates that are typically $1/16$ to $1/4$ inch thick. **See Figure 7-14.** They are normally mounted between a pair of flanges and are installed in a straight run of smooth pipe. The outside diameter of the plate is cut so that it fits concentrically within the bolt circle of a flange, with suitable

For additional information, visit qr.njatcdb.org Item #2550

TechTip!

Always verify the proper installation position on a meter before installation. Some meters require installation positioning above the process line, and others require installation positioning below the process line.

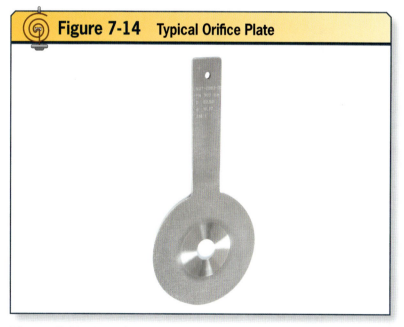

Figure 7-14. *An orifice plate offers a simple and relatively inexpensive method of creating a pressure drop for flow measurement.*

Courtesy of Endress+Hauser

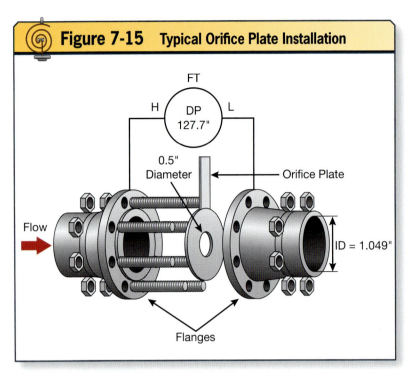

Figure 7-15. An orifice plate is mounted between two pipe flanges, or other specially designed apparatus.

orifice is equidistant (concentric) to the inside diameter of the pipe. Flow through a sharp-edged orifice plate is characterized by a change in velocity. As the fluid passes through the orifice, the fluid converges, and the velocity of the fluid increases to a maximum value. At this point, the pressure is at a minimum value. As the fluid diverges to fill the entire pipe area, the velocity decreases to the original value. The pressure increases to about 60% to 80% of the original input value. The pressure loss is irrecoverable; therefore, the output pressure is always less than the input pressure. The pressure levels on both sides of the orifice plate are measured, resulting in differential pressure that is proportional to the square of the flow rate.

Segmental and eccentric orifice plates are functionally identical to the concentric orifice plate. A segmental orifice plate is designed to measure the flow of liquids containing solids. The circular section of the segmental orifice plate is concentric with the pipe. The segmental section eliminates damming of foreign materials on the upstream side of the orifice plate when mounted in a horizontal pipe. Depending on the type of fluid being measured, the segmental section is placed on the top or bottom of the horizontal pipe.

Eccentric orifice plates have a center that does not coincide with the centerline of the pipe. Instead, they shift the edge of the orifice to the inside of the pipe wall. This design prevents upstream damming and is used in the same way as the segmental orifice plate. The eccentric orifice plate is used to promote the passage of entrained water or gas, rather than allowing buildup in front of the orifice plate.

For additional information, visit qr.njatcdb.org Item #2551

gaskets upstream and downstream between the plates and the flanges. **See Figure 7-15.** The orifice constricts the flow of liquid to produce differential pressure across the plate. The opening in the orifice plate must be carefully calculated for size, according to the flow to be measured.

Three kinds of orifice plates are typically used: concentric, eccentric, and segmental. **See Figure 7-16.** The concentric orifice plate is the most common and is recommended for use in clean liquid and gas applications. The

ThinkSafe!
Never remove a high- or low-pressure sensing line while the flow is in service. The flow must be isolated from the device before servicing the sensing lines.

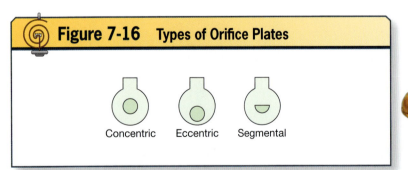

Figure 7-16. Three kinds of orifice plates are concentric, eccentric, and segmental.

Orifice plates have two disadvantages. First, they cause a high, permanent pressure drop (outlet pressure is 60% to 80% of inlet pressure). Still, the amount of pressure drop varies depending on the type of orifice plate used; that is, the drop through segmental orifice plates is only about half that of conventional orifice plates. Second, they are subject to erosion, which eventually causes inaccuracies in the measured differential pressure.

Orifice plates for steam or gas may have a small hole above or below the measuring opening. The drain hole is designed to prevent the buildup of condensate when measuring condensable gases or vapors, and it must be installed on the bottom of the orifice plate. **See Figure 7-17.** The vent hole is used to relieve the accumulation of gases when measuring liquids with entrained gases, and it must be installed at the top of the orifice plate. **See Figure 7-18.**

In almost all cases, when installing an orifice plate, the bevel or recess must face downstream. Types of orifice plates in which the bevel is installed on the upstream side of the plate are called conical and quadrant orifice plates. They are relatively new and are used to yield a more constant and predictable discharge coefficient at low flow velocity.

Pressure Taps

The piping connections by which the pressure drop across a restriction is measured are commonly called pressure taps. When a stream of fluid passes through an orifice, a jet is formed similar to the one that forms beyond the nozzle of a hose. After passing the sharp upstream edge of the orifice, this jet continues to contract to a point approximately one-half of the pipe diameter downstream from the orifice. At its most constricted point, it reaches its greatest velocity and lowest pressure. This narrowest point is the vena contracta. Beyond the vena contracta, the jet expands until it finally equals the pipe's ID, but full pressure is never recovered because of friction losses.

Pressure taps may be placed at several locations, each requiring a different calculation for the size of the orifice. Orifices for different tap locations are not interchangeable.

Three of the taps commonly used in the United States are the flange tap, the vena contracta or radius tap, and the pipe tap. Whether any of these taps are located in the vertical or horizontal plane depends upon the physical state of the fluid: gas, vapor, or liquid.

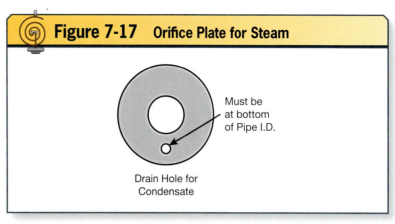

Figure 7-17. An orifice plate for steam has a drain hole for condensate.

TechTip!
Always verify the direction of an orifice plate upon installation in the process line. Backward installation of the plate is possible with many styles of orifice plates.

ThinkSafe!
Always verify that the gasket is the proper size, material, and pressure rating for the orifice plate and flange assembly.

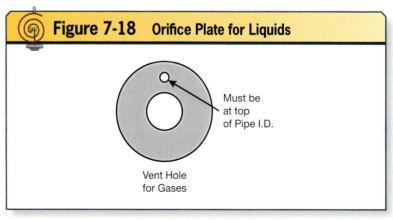

Figure 7-18. An orifice plate for liquids has a vent hole for the gases.

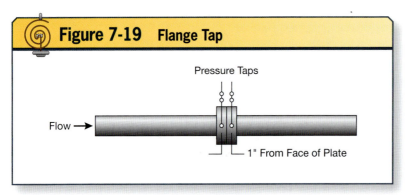

Figure 7-19. Pressure ports installed through a conventional flange measure the pressure one inch in each direction of the orifice plate. These ports are attached to the HI and LO of a differential pressure transmitter.

connections, the female tap connection is used most frequently.

The installer must drill the hole first, making sure that it is parallel to the flange face so that the socket can be installed in the proper plane. When welding an orifice flange to the end of a pipe, root penetrations, tack welds, and so on must be removed from the inside of the pipe. Burrs also must be removed from the inside of the pipe where the hole was drilled. Scale deposits that resulted from welding the socket to the pipe must also be removed. Ultimately, the inside surface must be smooth.

A special tool can be used for installing pressure tap sockets on a pipe. This tool can be used to position the socket correctly (parallel to the face of the flange and at the proper plane) by inserting it in the hole with the socket attached. The socket can be tacked onto the pipe true and square and then welded in place.

It is of extreme importance no serious disturbance occurs in the flow pattern near the orifice. Therefore, the pipe sections immediately upstream and downstream are considered parts of the primary measuring device. They are known as the metering runs or sometimes as the measuring section. To avoid swirls or crosscurrents, a certain amount of straight pipe is necessary on either side of the pressure taps. On the downstream side, five pipe diameters are usually sufficient. Pipe used for metering runs should be smooth inside. Commercially available pipe is usually satisfactory, but reconditioned or cheaply constructed pipe should be scrutinized carefully.

Pressure taps can be built directly into a pair of flanges by the manufacturer. Holes are drilled on specified centers from the face of the flange, or in such a manner that the center of the hole is one inch from the face of the orifice plate when the flange is made with gaskets. The flange is also drilled by the manufacturer at the location or locations of the holes large enough and of the proper depth to be tapped for national pipe thread (NPT) pipe connection. These holes and tapped threads are parallel to the face of the flange. **See Figure 7-19.**

Pipe taps and radius taps are made by drilling holes at the proper location in a pipe and welding sockets over the hole. **See Figure 7-20.** Of the two types (male and female) of pipe tap

Pipe Elbows

The pipe elbow can be used as a flow measurement device. **See Figure 7-21.** By placing a pressure tap at both the inner and the outer radii of an elbow and measuring the pressure differential caused by the differences in flow velocity between the two flow paths, any 90° pipe elbow can serve as a liquid flow meter. All that is required is the placement of two small holes in the elbow located at either 45° or 22 1/2° points

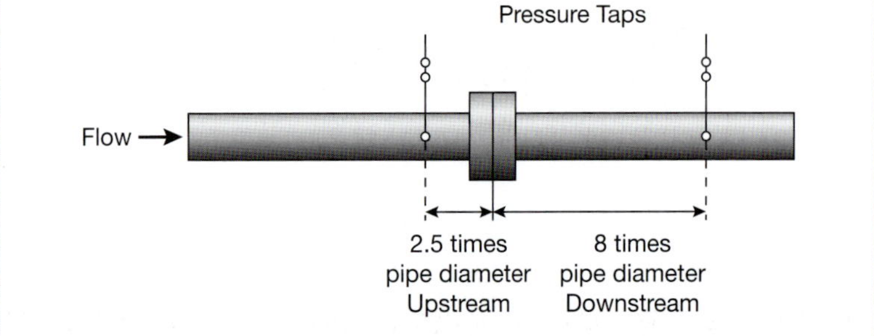

Figure 7-20. Accurate placement of pipe taps is essential to achieving a proper pressure reading, depending on the application.

on the elbow. Differential pressure is created by the centrifugal force occurring when a fluid changes direction in passing through the elbow. The differentiated pressure is proportional to the flow rate, and the pressure is transmitted by a differential pressure cell transmitter. The advantage of this measuring device is that it can measure flow in either direction. The disadvantage is its poor accuracy.

Flow Tubes

Where pressure recovery is of great importance, the Venturi tube is often used. The Venturi tube has the advantage of being able to handle large flow volumes at low pressure drops. The Venturi tube provides the excellent accuracy in measuring turbulent fluids, as well as streams with solids in suspension.

The Venturi tube has a converging conical inlet, a cylindrical throat, and a diverging recovery cone. **See Figure 7-22.** It has no projections into the fluid, no sharp corners, and no sudden changes in contour. The inlet section decreases the area of the fluid stream, causing the velocity to increase and the pressure to decrease. The low pressure is measured in the center of the cylindrical throat, where the pressure is at its lowest value and neither the pressure nor the velocity changes. The recovery cone enables the recovery of pressure such that total pressure loss is only 10% to 25%. The high pressure is measured upstream of the entrance cone.

Instead of single pressure taps, the Venturi tube may use piezometer rings in gas measurement applications. These are hollow rings around the entrance (high pressure) and throat (low pressure) sections of the venture tube. A series of pressure taps are installed within the unit to average the measured pressure. The Venturi tube should be installed between companion flanges, in a straight section of pipe, as far downstream as possible from sources of disturbance. The major disadvantages of this type of flow detection are the difficulty of installation and inspection once it is installed.

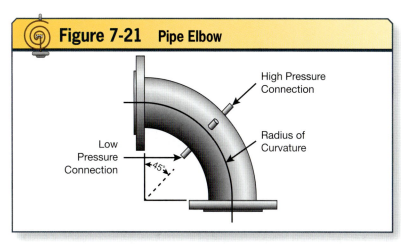

Figure 7-21. Pressure taps installed in an elbow can offer a method of flow measurement when piping is restricted and accuracy is not critical.

The Dall flow tube has a higher ratio of pressure developed to pressure lost than seen in the Venturi flow tube. It is more compact and is commonly used in large flow applications. The tube consists of a short, straight inlet section followed by an abrupt decrease in the inside diameter of the tube. This section, called the inlet shoulder, is followed by the converging inlet cone and a diverging exit cone. The two cones are separated by a slot or gap. The low pressure is measured at the slotted throat, which is the area between the

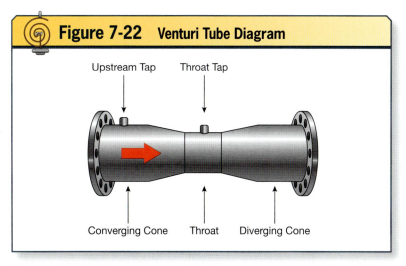

Figure 7-22. The upstream or high pressure tap occurs just before the taper begins, while the low pressure tap occurs at the point of constriction within the Venturi tube. The red arrow indicates flow direction.

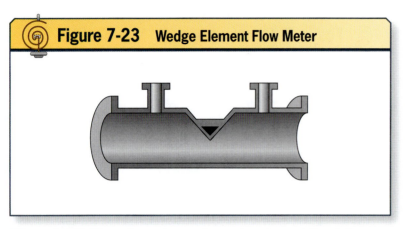

Figure 7-23. The unique shape of a wedge element flow tube allows it to be used bi-directionally.

two cones. The high pressure is measured at the upstream edge of the inlet shoulder.

The Dall flow tube is available in medium to very large sizes. The cost of the large sizes is normally less than that of a similarly sized Venturi flow tube. This type of flow tube has a pressure loss of about 5%.

Wedge element flow meters are designed mainly for use in slurry applications, and they exhibit characteristics that are similar to those of a segmental orifice plate. The main difference is that the obstacle presented to fluid flow happens more gradually. **See Figure 7-23.** The principal advantage of this type of flow meter is its use in applications where a very low RN occurs (500–1,000). In contrast, when orifice plates, Venturi tubes, and flow nozzles are used, an RN greater than 10,000 may be necessary to achieve the required square root relationship between differential pressure and flow rate.

Pitot Tube
The Pitot tube is a primary flow element used to produce differential pressure for flow detection and is used in a range of applications. In its simplest form, it consists of a tube with an opening at the end that detects the flow velocity at one point in the flow stream. **See Figure 7-24.** This provides the high-pressure input to a differential pressure detector. A pressure tap provides the low-pressure input.

The dual-walled, or jacketed, Pitot tube is constructed with one tube mounted within the other. **See Figure 7-25.** The inner tube senses the impact pressure, while the annular space between the tubes transmits the static pressure. The static pressure is the operating pressure of the environment upstream of where the tube is installed. The impact pressure is the total of the static and kinetic pressures of the environment and is detected at the Pitot opening as the flow stream affects the opening.

The operation of the Pitot tube differs from that of other primary devices in that it does not set up differential pressure by restricting the flowing stream. The Pitot tube measures fluid velocity instead of fluid flow rate but still provides a differential pressure measurement.

The Pitot tube is sometimes used where no line restrictions can be tolerated or in large lines where the cost of restrictive devices becomes prohibitive. Use of most Pitot tubes is limited to a single point measurement. The units are susceptible to plugging by foreign material and cannot be used to measure the flow of dirty liquids or liquids with solids in suspension. Some advantages of Pitot tubes are low cost, absence of moving parts, easy installation, and minimum pressure drop.

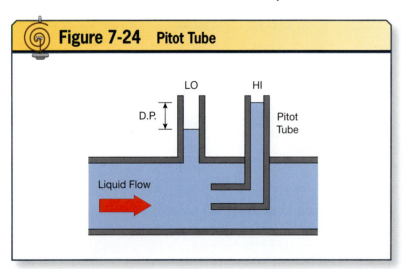

Figure 7-24. A Pitot tube measures differential pressure.

ThinkSafe!
Media containment considerations should always be reviewed before installing any flow device that uses a glass or plastic flow tube. Never install glass or plastic tubes in hazardous services.

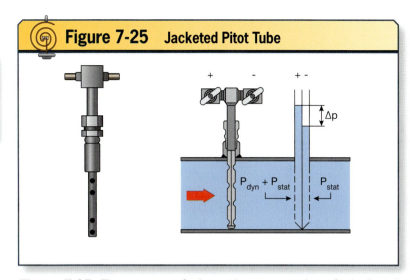

Figure 7-25. The upstream facing or impact opening of a jacketed pitot senses both the dynamic (Pdyn) and static (Pstat) pressure, while the downstream side senses only the static pressure.

Flow Nozzles

The flow nozzle is commonly used for the measurement of steam flow and other high-velocity fluid flow measurements in which erosion may occur. Steam flow detection is normally accomplished with the use of a steam flow nozzle. Because steam is considered a vapor, changes in pressure and temperature can greatly affect its density. Flow nozzles can be used for both clean and dirty liquids; however, their use is not recommended for highly viscous liquids or those containing large amounts of sticky solids.

The flow nozzle offers almost all of the advantages of a Venturi tube but does not require as much space. It is capable of measuring approximately 60% higher flow rates than an orifice plate with the same diameter because of the streamlined contour of the throat, which is a distinct advantage for the measurement of high-velocity fluids. The flow nozzle also requires less straight-run piping than an orifice plate. However, the pressure drop is about the same for both. **See Figure 7-26.**

Flow nozzles can be mounted between flanges and have a lower permanent pressure loss than an orifice plate. The pressure taps for a flow nozzle are commonly located one pipe diameter upstream and one-half pipe diameter downstream. A flow nozzle should be installed in a smooth, straight section of pipe, as far downstream from sources of disturbance as possible.

For additional information, visit qr.njatcdb.org Item #2552

Target Meters

A target meter has a physical paddle or target placed directly in the fluid flow

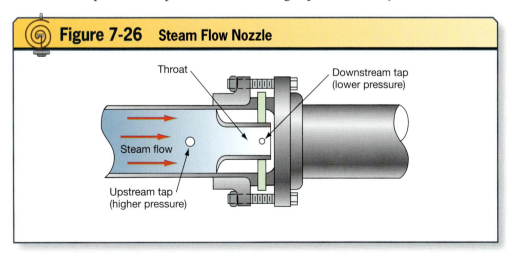

Figure 7-26. The tapered inlet cone allows for a smooth flow of steam or vapor laden gases, which could easily condense at an obstruction.

> **ThinkSafe!**
> Many steam flow applications operate at extremely high pressure. Use proper protection, and never disconnect any portion of a steam meter unless the steam flow is isolated on both sides of the meter.

stream. **See Figure 7-27.** As the target is deflected within the fluid flow, the force bar senses the pressure exerted on the target. The body of the instrument houses the transducer, which converts the degree of mechanical deflection into an output signal that is then sent to a controller. Advantages of a target meter include the ability to be used in any type of fluid flow application (e.g. liquid, gas, or steam), high reliability because it has few moving parts that can fail, use in a variety of line sizes of 1/2 inch and larger, and ability to make bidirectional flow measurements.

Rotameters

The rotameter is a variable area flow meter whose indicating element is a float, and is typically used to measure low flow rates. The rotameter consists of the float and a conical glass tube which is constructed such that the internal diameter increases with the tube height. **See Figure 7-28.** When there is no fluid passing through the rotameter, the float rests at the bottom of the tube. As fluid enters the tube, the higher density float remains on the bottom. The space between the float and the tube allows fluid flow past the float. As flow increases in the tube, the pressure drop increases. When the pressure drop is sufficient, the float rises to indicate the amount of flow. The higher the flow rate, the greater the pressure drop. The higher the pressure drop, the farther up the tube the float rises. The float should stay at a constant position at a constant flow rate. With a smooth float, fluctuations appear even when

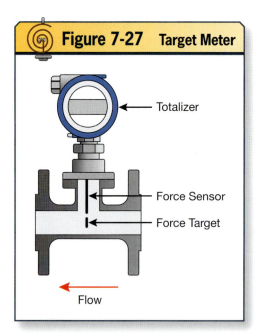

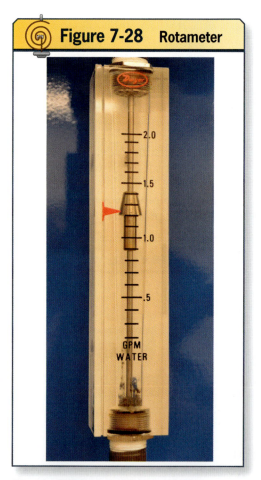

Figure 7-27. A target flow meter measures an impact force that is proportional to the flow rate.

Figure 7-28. A rotameter displays volumetric flow rate.

flow is constant. A float with slanted slots cut in the head maintains a constant position with respect to flow rate. Because the rotameter relies on gravity, it must be installed vertically with the inlet at the bottom of the housing.

Mass Flow Meters

Mass flow meters determine the actual mass flow rate by compensating for density and temperature. Mass flow rates are useful for the measurement of gases and steam due the highly variable effects of temperature changes. Remember that mass is the relationship between density and volume. Volume measurement alone neglects the effects of temperature dependent density.

Coriolis Meter

The *Coriolis meter* is a recent technological development to measure mass flow rates with a high degree of accuracy. A U-shaped tube is enclosed in a housing and serves as the flow path. Two sensors, one located on the inlet side of the tube and one located on the outlet side of the tube detect movement of the tube. An oscillation is induced in the tube via a driver coil. When the fluid is not flowing, the tube oscillates evenly, resulting in an equal and in-phase signal received at the two sensors. **See Figure 7-29.** When fluid begins to flow, the Coriolis Effect creates a "wiggling" motion and results in a phase shift between the two sensors signals.

In addition to flow rate, the Coriolis meter can measure mass directly by measuring the oscillation frequency. A fluid with a high density, such as molasses (93.65 lbs/ft³), will oscillate at a slower frequency than a lower density fluid such as water (62.43 lbs/ft³). Many Coriolis meters also incorporate temperature measurement that is used to calculate total mass flow.

Hot-Wire Anemometer

The hot-wire *anemometer*, an instrument for measuring or indicating the velocity of air flow that is principally used in gas flow measurement, consists of an electrically heated, fine platinum wire that is immersed into the flow. As the fluid velocity increases, the rate of heat flow from the heated wire to the flow stream increases. A cooling effect on the wire electrode causes its electrical resistance to change. In a constant-current anemometer, the fluid velocity is determined from a measurement of the resulting change in wire resistance. In a constant-resistance anemometer, fluid velocity is determined from the current needed to maintain a constant wire temperature and therefore the resistance constant.

Velocity Meters

Velocity meters operate with linearity with respect to the flow volume. There is no square-root relationship, and their rangeability is greater. Velocity meters have minimum sensitivity to viscosity changes when applied to a flow with a RN greater than 10,000. Most velocity meters come with flanges, making them suitable piping arrangements to allow installation directly into pipelines.

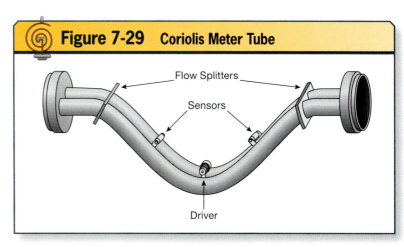

Figure 7-29. *The oscillation of a Coriolis tube is necessary for the mass flow measurement.*

For additional information, visit qr.njatcdb.org Item #2553

ThinkSafe!
Use caution and protective gloves when dealing with a hot-wire anemometer. While power is on, some devices reach hot temperatures that can cause burns if contacted.

ThinkSafe!
Always wear eye, face, and hand protection when working with pressurized flow devices that must be installed or removed from their process service.

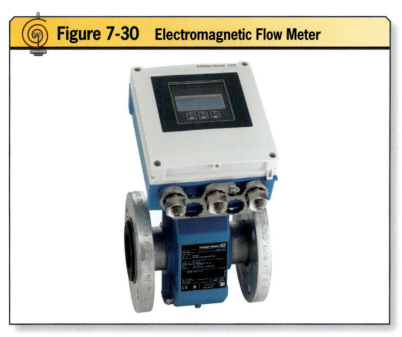

Figure 7-30. An electromagnetic flowmeter can be used to measure the flow rate of conductive fluids.

Courtesy of Endress+Hauser

Electromagnetic Flow Meter

The electromagnetic flow meter is similar in principle to the generator. The rotor of the generator is replaced by a pipe placed between the poles of a magnet so that the flow of the fluid in the pipe is exposed to the magnetic field. As the fluid flows through this magnetic field, an electromotive force is induced on the process fluid. This electromotive force can be measured with the aid of electrodes attached to the pipe and connected to a galvanometer or an equivalent device. For a given magnetic field, the induced voltage is proportional to the average velocity of the fluid. However, the water-based fluid should have some degree of electrical conductivity for the electromagnetic flow meter to operate properly. Additionally, a solid ground connection between the electronics and the up and down stream piping is essential to ensure continuity with the process fluid. If the piping is plastic lined steel, fiberglass, plastic or some other non-conductive material, ground rings must be installed to complete the circuit through the process fluid. **See Figure 7-30.**

Ultrasonic Flow Equipment

Devices such as ultrasonic flow equipment use the Doppler frequency shift of ultrasonic signals reflected from discontinuities in the fluid stream to obtain flow measurements. These discontinuities can be suspended solids, bubbles, or interfaces generated by turbulent eddies in the flow stream. The

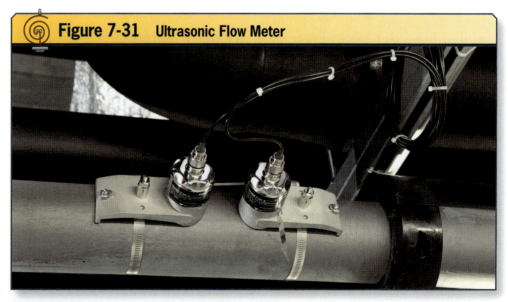

Figure 7-31. A clamp on ultrasonic flow meter can measure flow rates without penetrating the pipe wall.

Courtesy of Endress+Hauser

> **TechTip!**
> Ultrasonics do not work well in applications that have very low turbulence, solids, or entrained gases.

sensor is mounted on the outside of the pipe, and an ultrasonic beam from a piezoelectric crystal is transmitted through the pipe wall into the fluid at an angle to the flow stream. Signals reflected off flow disturbances are detected by a second piezoelectric crystal located in the same sensor. Transmitted and reflected signals are compared in an electrical circuit, and the corresponding frequency shift is proportional to the flow velocity.

Another form of ultrasonic measurement is the transit-time meter. This type uses two piezoelectric elements. These elements transmit pulses and measure the upstream and downstream transmission times through the fluid. The energy moves faster when liquid is flowing than it does when liquid is stopped. The processor on the meter calculates the speed of the energy pulsed into the liquid by measuring the time required for the pulses to travel between the two sensors. When the liquid being measured is flowing, energy moves faster with the flow than it does against the flow. The processor converts this time differential to a fluid flow rate. **See Figure 7-31.**

Open-Channel Meters

The term *open channel* refers to any fluid routing that allows the fluid to flow with a free surface, such as in tunnels, non-pressurized sewers, and rivers. Depth-related measurements are the most common type of measurement. This technique assumes that the instantaneous flow rate can be obtained by measuring the depth of the channel. The depth at a specific flow is determined by the width, depth, and slope of the fluid routing device. Open-channel meters use an obstruction to obtain their measurement. These obstructions are called primary measuring devices and can be divided into two categories: flumes and weirs. Flumes and weirs are two of the oldest and most common open-channel measurement types.

Flumes

A *flume* is a specially shaped, fixed hydraulic structure that under free-flow conditions forces a fluid to accelerate in such a manner that the flow rate through the flume can be measured. This is achieved by measuring the depth at a single specified location in the flume. Acceleration is accomplished through a convergence of the sidewalls, a change in floor elevation, or a combination of the two. All flumes must be installed level with the proper conditions at the converging (entrance) and diverging (exit) sections. In addition, flow must be smooth and non-turbulent when entering the flume to allow for accurate measurement.

The most common flume is called the Parshall flume, and it is widely used in permanent installations such as sewage and wastewater treatment facilities. **See Figure 7-32.** This type of flume is characterized by throat width, which ranges in value from 1 inch to 50 feet. This dimension, as well as the dimensions for slope, converging section, and

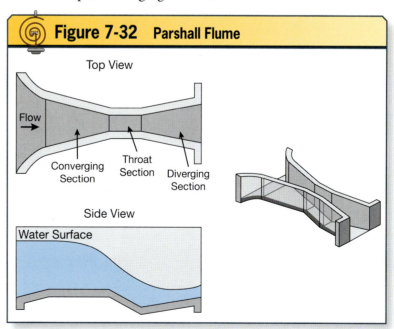

Figure 7-32. *Parshall flumes are used to measure a range of flow rates.*

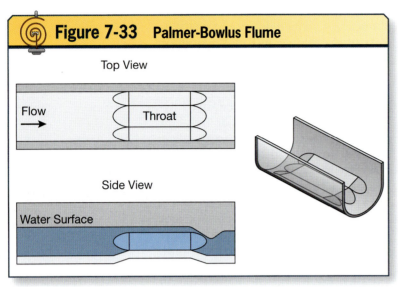

Figure 7-33. A Palmer-Bowlus flume with a flat bottom and a trapezoidal throat is the standard design configuration when used in circular pipes.

diverging section, must be followed to allow the use of standard discharge tables.

In contrast to the Parshall flume, the Palmer-Bowlus flume is designed for installation in an existing channel. Primarily used in the sanitary field, this flume has also proved useful in runoff studies and for measuring flows in manholes. **See Figure 7-33.** The flow rate through this flume type is measured by determining the depth at a point one-half of a pipe diameter upstream of the throat. The Palmer-Bowlus flume ranges in sizes from 4 to 42 inches and, as a result, works well in portable and temporary installations.

Additional flumes exist in the marketplace with a variety of uses across differing industries. Two such types are the H flume and the Cutthroat flume. The H flume was originally used almost exclusively by agricultural researchers when monitoring field runoff. More recently, this type of flume has seen use in the sanitary field to measure sewage and has proved handy for portable installations. **See Figure 7-34.** Similar to the Parshall flume in design, the Cutthroat flume differs in that it has a flat bottom that maintains its elevation throughout the length of the flume. **See Figure 7-35.** The advantage of the Cutthroat flume over the Parshall flume is in applications in which head loss is limited, because the former can provide increased accuracy where excessive grit exists.

Weirs

A *weir* is a damlike obstruction placed within an open channel such that the liquid flows over it, often through a specially shaped opening. The most common types of weirs, named according to the shape of the opening, are the triangular (V-notch) weir, the rectangular weir, and the trapezoidal (Cipolletti) weir. The flow rate over a weir is determined from the measured depth of the liquid in the pool upstream of the weir. Weirs are inexpensive and simple in construction; they are commonly made of metal, fiberglass, and even wood. They are not suitable for measuring flows containing solids.

The head on a weir is measured by first measuring the depth of the liquid passing over the weir upstream of the notch, where drawdown at the surface is at a minimum. This can be achieved by driving a stake into the pool at a distance at least five or six times the depth of flow over the weir. The top of the stake should be level with the bottom

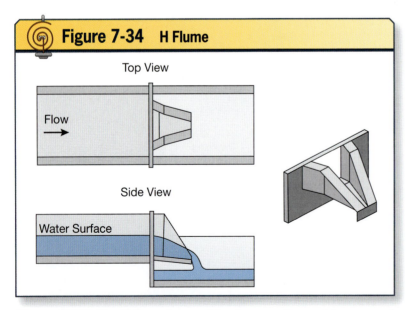

Figure 7-34. The H flume is capable of measuring a wide range of flow rates.

of the weir notch. The distance from the surface of the liquid to the top of the weir is the head, and this distance is easily measured with a scale.

ENVIRONMENTAL CONSIDERATIONS

A technician should never operate equipment outside its rated environment. The density of the fluid whose flow is to be measured can have a large effect on flow-sensing instrumentation. The effect of density is most important when the flow-sensing instrumentation is measuring gas flows, such as steam. Because the density of a gas is directly affected by temperature and pressure, changes in these parameters directly affect the measured flow. Changes in fluid temperature or pressure must be compensated for to achieve an accurate measurement of flow.

Ambient temperature variations affect the accuracy and reliability of flow-sensing instrumentation. Variations in ambient temperature can directly affect the resistance of components in the instrumentation circuitry and therefore affect the calibration of electrical or electronic equipment. The effects of temperature variations are reduced by the design of the circuitry and by maintaining the flow-sensing instrumentation in the proper environment.

Humidity also affects most electrical and electronic equipment. High humidity causes moisture to collect on the equipment, causing short circuits,

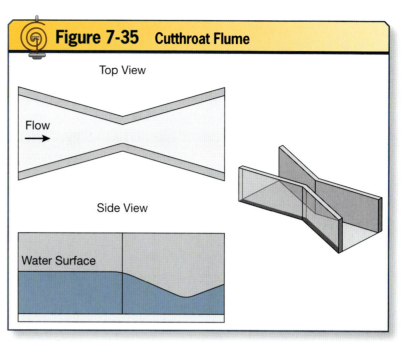

Figure 7-35. Various flume types are used in various applications for the purpose of flow measurement in open channels.

grounds, and corrosion, which can damage components. The deleterious effects of humidity are controlled by maintaining the equipment in the proper environment. The density of the fluid, ambient temperature, and humidity can affect the accuracy and reliability of flow-sensing instrumentation. All instrumentation should be operated within its rated limits for temperature and atmospheric conditions. When used within rated limits, most instrumentation is not greatly affected by these changes and poses no environmental concerns.

SUMMARY

Flow meters rely on many physical and mathematical properties to accurately measure flow. The measured properties can be altered by introducing errors into the measurement loop through installation or calibration errors. Flow-measuring devices must be installed and calibrated correctly to perform their intended functions.

REVIEW QUESTIONS

1. The pressure drop across a restriction is equal to the velocity.
   a. True
   b. False

2. What type of flow is characterized as a straight-line flow of a fluid?
   a. Dense
   b. Inviscid
   c. Laminar
   d. Turbulent

3. In which type of flow meter does all of the fluid pass through the meter in nearly isolated quantities?
   a. Differential pressure
   b. Displacement
   c. Mass
   d. Velocity

4. RN is a unitless number that is used to determine the resistance to fluid flow in a process line.
   a. True
   b. False

5. Which of the following instruments operates on the principle of oscillating tube frequency driven by an electromagnetic drive coil?
   a. Anemometer
   b. Coriolis meter
   c. Rotameter
   d. Target meter

6. Because steam is considered a gas, pressure and temperature changes affect its density.
   a. True
   b. False

7. Which of the following statements defines vena contracta?
   a. The point downstream of a line restriction where fluid velocity is highest.
   b. The point downstream of a line restriction where the fluid streamline recovers its flow profile.
   c. The point upstream of a line restriction where pressure is at its lowest.
   d. The point where the streamline makes contact with a line restriction.

8. What happens when velocity is increased in a flow line?
   a. Pressure always drops to zero
   b. Pressure decreases
   c. Pressure increases
   d. Pressure remains the same

9. An orifice plate is commonly installed so that the beveled edge faces upstream.
   a. True
   b. False

10. Which of the following is a type of flow meter classification?
    a. Density
    b. Mass
    c. Viscosity
    d. Volume

Principles of Temperature

Temperature measurement is the way in which the kinetic energy of the molecules in a material is measured and expressed. Temperatures must be derived indirectly. By measuring a temperature's effects on a given material, the actual temperature is determined. Many devices are commonly used in industry, and it is important for the technician to know about their function, operation, and potential errors. After the physical parameters of temperature measurement are understood, the art of calibration and troubleshooting can begin. There are several ways to introduce errors into a temperature measurement loop, so it follows that a technician must be familiar with the fundamentals to ensure temperature measurement devices are accurate.

Objectives

» Describe the purpose of temperature measurement.
» List the four temperature scales used in temperature measurement.
» Explain the reason for a temperature measurement standard.
» Describe the characteristics and operation of a thermocouple.
» Describe the circuit requirements and compensations needed for a thermocouple circuit.
» Explain the operation of a resistance temperature detector (RTD) measurement circuit.
» Describe how a thermistor measures temperature indirectly.
» Solve for the accurate output of a temperature transmitter.

Chapter 8

Table of Contents

Purpose of Temperature Measurement..........170
- History of Temperature Measurement......171
- Temperature Measurement Units and Standards..172

Methods of Temperature Measurement........173
- Bimetal Thermometers, Mercury Thermometers, and Alcohol-Filled Thermometers...173
- Temperature Switch..............................175
- Continuous Measurement......................176

Temperature Sensors...............................176
- Thermocouple.....................................176
- Resistance Temperature Detector......180
- Thermistor...184
- Transmitters..186

Thermowell Basics......................................188
Summary..189
Review Questions.......................................189

For additional information, visit qr.njatcdb.org
Item #2690

PURPOSE OF TEMPERATURE MEASUREMENT

Industries require their energy management systems to perform at optimal efficiency. When industries control their temperature processes efficiently, they produce their finished products at reduced expense because of lower costs for fuel and power. With the demand for natural resources becoming greater, industries are placing greater emphasis on energy management systems, resulting in an increased demand for knowledgeable control system technicians. One of the most important methods for saving energy is to monitor and control all processes related to *heat*, defined as thermal energy expressed in units of calories, or British Thermal Units (BTUs). Because heat is the transfer of energy, the quantity of energy that is transferred has a direct effect on profits. As industries place greater emphasis on the importance of temperature measurement, it follows that installation and understanding of the fundamentals of temperature measurement will grow in importance for people who work directly with temperature sensors.

Temperatures measurements are classified as performing in three basic functions: indication, alarm, and control. An *alarm* is a device or function that signals the existence of an abnormal condition by means of audible or visible discrete changes intended to attract attention. The temperatures monitored may normally be displayed in a central location, such as a control room, and they may have audible and visual alarms associated with them when specified, preset limits are exceeded. These temperatures may have control functions associated with them so that equipment is started or stopped to support a given temperature condition or so that a protective action occurs.

The transfer of heat energy within a process is understood by examining thermodynamics, a field of science devoted to the study of heat and temperature. The laws of thermodynamics explain interaction between heat and process material. The temperature of a piece of plastic, wood, metal, or other material depends on the molecular activity of the material. Kinetic energy is a measure of the activity of the atoms that make up the molecules of any material. Temperature is therefore a measure of the kinetic energy of a material. **See Figure 8-1.**

> **ThinkSafe!**
>
> Working with temperature-measuring devices can expose the technician to extremely hot surfaces and process piping in certain situations. Always verify the expected operating temperature of the device, and always wear the appropriate personal protective equipment to prevent burn injuries.

Figure 8-1   Molecular Movement in Relation to Temperature

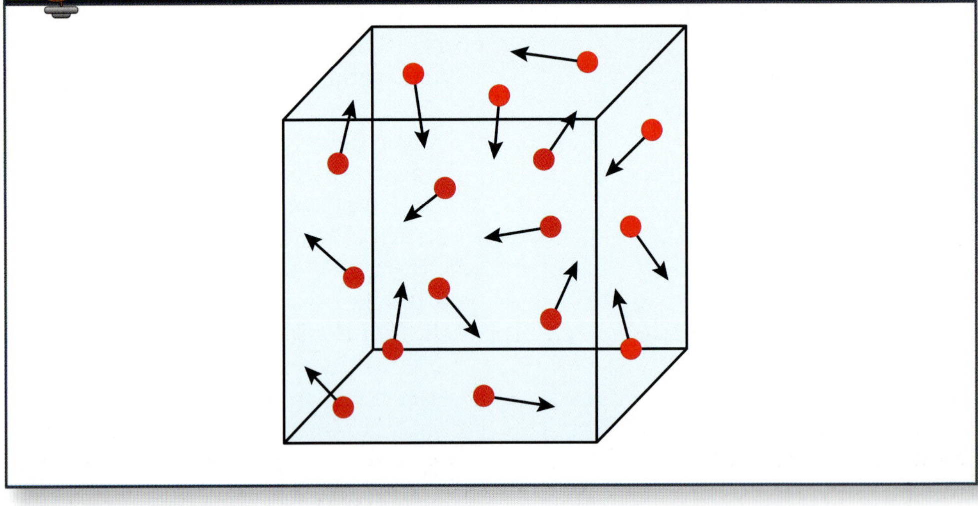

Figure 8-1. Temperature is the measure of the kinetic energy of molecules within a substance.

Take the radiator in a liquid-cooled automobile as an example of heat transfer. The internal combustion that occurs inside the engine block causes heat to transfer from the burning air-and-fuel mixture to the surrounding metal engine block. The process requires the heat to be transferred away from the metal engine block to limit the temperature. Liquid coolant is circulated inside the engine block at a lower temperature, causing the metal engine block to transfer heat to the coolant. The coolant is then pumped to a liquid-to-air *heat exchanger*, a vessel in which heat is transferred from one medium to another; it is often called the radiator. The increased surface area of the fins on the radiator allows the heat to transfer from the higher-temperature coolant to the lower-temperature air flowing through the radiator. This entire system neither creates nor destroys energy; it merely transfers heat to areas of lower energy. The concept of thermal equilibrium explains why the heat transfers from areas of higher energy to areas of lower energy. **See Figure 8-2.**

History of Temperature Measurement

There are different scales for measuring temperature. *Fahrenheit* (F), a temperature scale defined by 32°F at the freezing point and 212°F at the boiling point of water at sea level, is the most common in the United States. *Celsius* (C), a temperature scale defined by 0°C at the freezing point and 100°C at the boiling point of water at sea level, is the next most prevalent temperature scale. Other temperature scales may be used because they record temperature in absolute terms, and it is sometimes helpful to think of the absolute temperature scales as beginning at an absolute zero temperature. Regardless of the technique or scale used, temperature is the measurement of the presence of heat or energy.

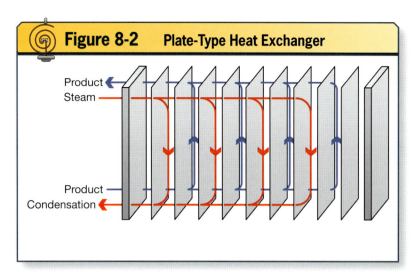

Figure 8-2. This plate-type heat exchanger uses high-temperature steam to transfer heat to a process fluid.

Galileo Galilei is credited with inventing a method to indicate differences in temperature in approximately 1592. He observed a container filled with alcohol and in which a long, narrow glass tube with a reservoir at the upper end was placed. As the temperature was increased, the air trapped in the reservoir expanded, because of the transfer of heat, and was forced out of the tube. As the temperature was decreased, the remaining trapped air contracted and allowed some of the alcohol to begin advancing up the hollow glass tube. This upside-down form of a thermometer was the first one that could be produced in quantity, and it was the first that recorded the same results when subjected to the same temperature change.

Over the years, many forms of temperature measurement scales were used. But it was not until 1724 that Daniel Fahrenheit established a scale with 0°F as the lowest temperature he could record with his mixture of ice, water, and salt (ammonium chloride). He chose 32°F as the temperature of ice and water mix without the salt. Fahrenheit chose 212°F as the temperature of boiling water, a span of 180°F between freezing and boiling. Fahrenheit's temperature scale grew in popularity and was widely adopted.

 TechTip!

Many temperature devices come with multiple scale units (e.g., F and C) on their displays. It is easy to confuse one unit for another. Always verify that the scale unit being read is the one appropriate to the application.

Temperature Measurement Units and Standards

Approximately one-half century after Fahrenheit's work, Anders Celsius proposed a temperature scale that contained a 100°C difference between water's freezing point and its boiling point. This was the beginning of the Celsius scale.

Around 1800, Lord Kelvin proposed a universal thermodynamic scale based on the expansion coefficient of an ideal gas. The *Kelvin* (K) scale uses the unit of absolute temperature in which 0°K represents the complete absence of heat, with 273.15°K corresponding to 0°C and 373.15°K corresponding to the boiling point of water at sea level. For this scale, Kelvin created a model that could theoretically establish the absence of heat as a point of *absolute zero*, the temperature at which thermal energy is at a minimum and all molecular activity ceases, reported as 0°K or –273.15°C. He used the Celsius temperature scale to record the changes that his working model could produce. The resulting Kelvin temperature scale uses the concept of 100°C between water's freezing point and its boiling point.

A fourth type of temperature measurement, the *Rankine* (R) scale, is an absolute temperature scale based on the Fahrenheit scale, with 180°F between the freezing point and the boiling point of water (459.67°R = 0°F). This means that the Rankine scale is another form of measuring absolute temperature. A relationship exists among the four temperature scales. **See Figure 8-3.**

Temperature measurement standards are required to determine the accurate calibration of temperature devices. The industry standard for temperature scales is determined by the International Temperature Scale of 1990 (ITS-90). This standard ensures consistency of temperature measurements across the globe. The ITS-90 standard is referenced on many temperature calibrators and instrument specification sheets.

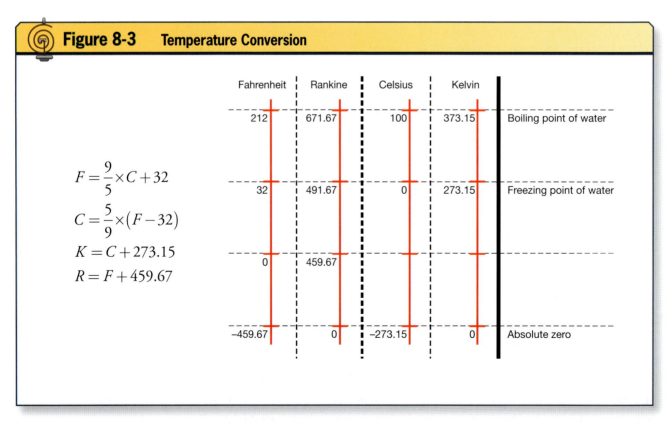

Figure 8-3. Converting among various temperature scales is accomplished using conversion formulas.

METHODS OF TEMPERATURE MEASUREMENT

A technician needs to understand the methods of temperature measurement to install, maintain, and troubleshoot temperature loops. Temperature measurement can provide a visual indication, a discrete signal for alarms, or a continuous signal to monitor temperature over a range of values.

Temperature cannot be measured directly. The technician can measure temperature only by observing changes in other objects or materials. A thermometer displays temperature change by observing the effect temperature has on mercury. As the temperature increases or decreases, the mercury changes with respect to the heat change. The technician observes the change in mercury, not the change of temperature. Temperature change causes all solids, liquids, and gases to expand or contract. In addition to expansion, other physical properties of materials, such as thermoelectric properties or change in resistive properties, are used to determine the amount of heat energy of a process material.

The continuous temperature measurement devices used in process control commonly consist of two distinct components: sensors and transmitters. The sensor, often called a temperature element (TE), measures the temperature and generates a signal to the transmitter. The sensor may be integral to the transmitter or mounted remotely. The temperature transmitter (TT) reads the signal from the sensor and generates an output signal, typically in milliamperes direct current or pounds per square inch, to the controller or indicator. Of the two, the temperature transmitter signal is better suited to travel long distances and is less likely to be compromised by outside factors, such as electromagnetic interference.

Bimetal Thermometers, Mercury Thermometers, and Alcohol-Filled Thermometers

Thermometers are used to provide visual indication of the temperature of a process. A bimetallic thermometer consists of two metals bonded together, each with different coefficients of expansion in relation to temperature. Copper, steel, nickel, and brass are commonly used to form the temperature-sensing element. This results in a piece of metal, called a bimetallic element, that can sense the temperature of a process. Elements are shaped in coils or strips, depending on the design of the thermometer. The element begins to deform as the metals expand at different rates. **See Figure 8-4.**

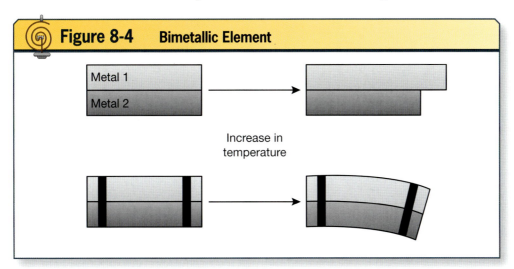

Figure 8-4. A bimetallic element consists of two metals that expand at different rates when heated. When the two metals are bonded together, the assembly curves as it expands.

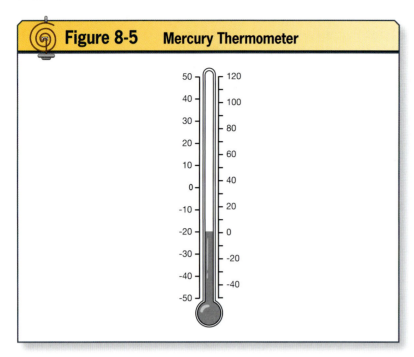

Figure 8-5. Mercury thermometers are accurate but rarely used because of toxicity concerns.

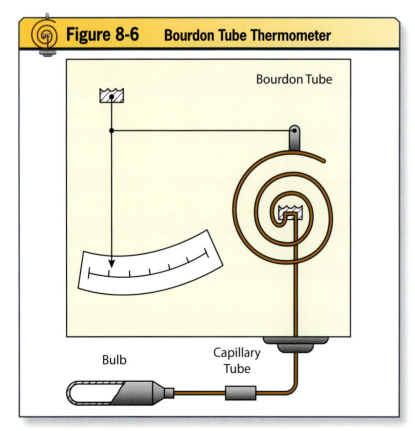

Figure 8-6. A Class 1 filled bulb thermometer uses the expansion of the fluid inside the capillary to deform a bourdon tube, causing the indication needle to display the bulb temperature.

A predetermined amount of deflection is then transferred to an indicating needle, which is calibrated to indicate the temperature of the bimetallic element.

Mercury and alcohol thermometers employ the principle of fluid expansion in relation to temperature because of an increase in kinetic energy to determine the temperature. The immersion depth of a thermometer is an important characteristic. Thermometer designs are classified as full immersion, partial immersion, and complete immersion. This classification determines how the thermometer should be inserted in the process fluid. In the case of the partial immersion thermometer, the immersion depth is marked on the thermometer stem.

The mercury glass stem thermometer is a familiar example. By observing how far the liquid rises in the tube, it is possible to tell the temperature of the measured object. The expansion of mercury is relatively linear, which leads to repeatable, accurate readings. However, the toxicity of mercury has led to fewer applications in process control. **See Figure 8-5.**

Filled bulb thermometers are sealed pressure systems that work based on the principle of material expansion in relation to temperature. They come in several classes, which determine the fill type. Classes 1 and 5 use a fill liquid, Class 2 uses a liquid-and-vapor mixture, and Class 3 uses a gas fill (no Class 4 exists). In all styles, the fill material expands and the pressure inside the sealed system increases. This increased pressure causes a bellows or circular bourdon tube element to deform, generating movement. The movement operates an indication needle, causing the indicator to point to the temperature scale, which represents the temperature of the bulb. **See Figure 8-6.**

All filled bulb systems that use a liquid suffer from liquid head pressure, which requires compensation. The elevation of the bulb in relation to the readout must be calculated and accounted for in the calibration of the instrument. All filled systems are also affected by ambient temperature, which requires compensation by a secondary capillary tube or a bimetallic element in the case.

Temperature Switch

Discrete temperature control is a control method that uses a temperature switch to generate alarms or control a temperature process. Discrete control is also called point control. Temperature switches generally have one or more sets of electrical contacts that change states when a predetermined temperature is reached. When the contacts change state in relation to the temperature, the electrical circuits may energize or de-energize pumps, heaters, alarms, or other control devices.

Temperature switches are often constructed using bimetallic elements or filled bulb-type temperature sensors.

A temperature switch is used in a variety of scenarios. For example, a tank with an electrical heating element may be controlled by a temperature switch. The set point for the process is determined by the process design and the calibration of the temperature switch is performed by the technician. When the temperature exceeds the set point, the heating element is de-energized. Over time, the process loses heat to outside disturbances and therefore needs to be heated to maintain an acceptable temperature. The electrical contacts inside the switch close, allowing electrical current to flow to the heating element. **See Figure 8-7.**

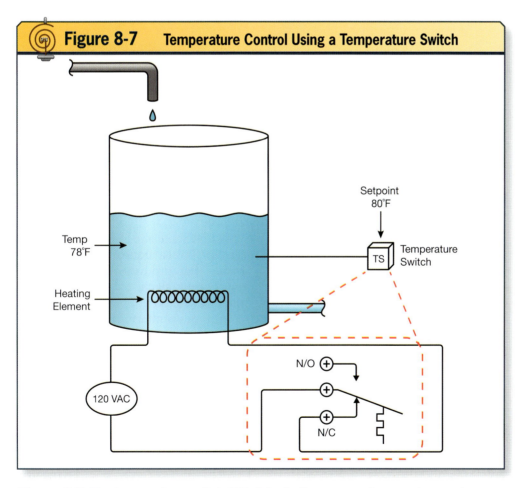

Figure 8-7. The temperature switch (TS) detects the process temperature and controls the heating element. The common, normally open (N/O) and normally closed (N/C) contacts are typically housed inside the temperature switch body.

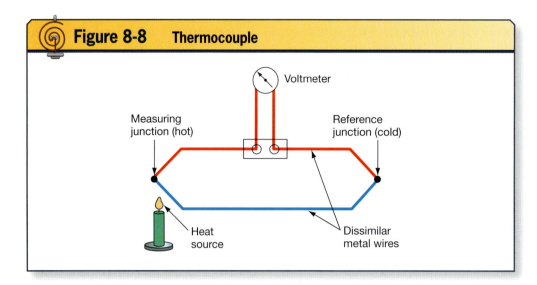

Figure 8-8. A simple thermocouple circuit can be created using a voltmeter.

Continuous Measurement

Maintaining process material at a desired temperature set point is essential to ensure quality and safety. The level of allowable deviation from the set point is determined by the process design. In contrast to a temperature switch, which measures only whether the process temperature is above or below a certain temperature set point, a continuous temperature measurement instrument provides a signal that represents a range of values. The need to "see" a wider range of temperature values results in the implementation of a continuous temperature measurement method.

Temperature Sensors

The three basic types of temperature sensors are thermocouples, resistance temperature detectors (RTDs), and thermistors.

Thermocouple

A *thermocouple* is a device constructed of two dissimilar metals that generates a millivoltage as a function of the temperature difference between measuring and reference junctions. The millivolt signal can be measured, and its magnitude can be used as a measure of the temperature in question.

The fundamental theory that explains the operation of the thermocouple was discovered by Thomas Seebeck in 1821. He found that when two dissimilar metals are joined at both ends and one end is heated, a continuous current flows in the circuit. If this circuit is broken at one end, the resulting open-circuit voltage is a predictable reading that depends on the temperature and the metal types used. **See Figure 8-8.**

All dissimilar metals produce this effect. By knowing the combination of metals used, the resulting temperature can be measured from the millivolt reading.

Standard thermocouple types, such as a type J or K, are manufactured using metals that produce consistent millivolt output signals. There are several standard thermocouple types. The different types of thermocouples are color coded for quick identification. The American National Standards Institute (ANSI) standard color code always uses red to identify the negative lead. **See Figure 8-9.**

TechTip!

Unlike most electrical conductors, the red wire on all thermocouples in U.S. standards is the negative (−) lead. This standard varies by country.

Chapter 8 Principles of Temperature 177

Figure 8-9. Thermocouple Reference Chart

Figure 8-9. Thermocouple color coding and operating ranges are standardized. Notice the differing color codes of ANSI and IEC (International Electrotechnical Commission).

The millivolt output is directly proportional to temperature and is referenced on a thermocouple table. These tables are available for each type of thermocouple. For example, a temperature of 957°F would correspond to an output of 28.172 millivolts. **See Figure 8-10.**

To understand the operation of a thermocouple, the technician can think of the thermocouple as a temperature differential detector. The thermocouple produces a voltage that relates to the difference in temperature between the *measuring junction*, the point in a thermocouple where the two dissimilar metals are joined, and the *reference junction*, the cold junction in a thermocouple circuit that is held at a stable known temperature. If the thermocouple is sitting on a bench, with both junctions at the same temperature, 0 millivolts is read by a voltmeter. **See Figure 8-11A.** Because there is no difference of temperature, no millivoltage exists. If the technician then raises the temperature of the measuring junction 15°F above room temperature, a corresponding millivoltage may be read with a voltmeter. **See Figure 8-11B.**

This measurement method has a fundamental problem. If both the measuring and the reference junction vary in temperature, the reading from the thermocouple merely provides a signal that relates to the difference between the two temperatures. **See Figure 8-11C.** The signal output, 0.394 millivolts, is the same for both examples. This condition can lead to measurements that do not accurately report the temperature of the process.

To overcome this condition, the reference junction is typically held at a stable known temperature. The industry standard used is called the ice point, which is 32°F or 0°C. The use of an ice bath results in a repeatable, stable

For additional information, visit qr.njatcdb.org Item #2123

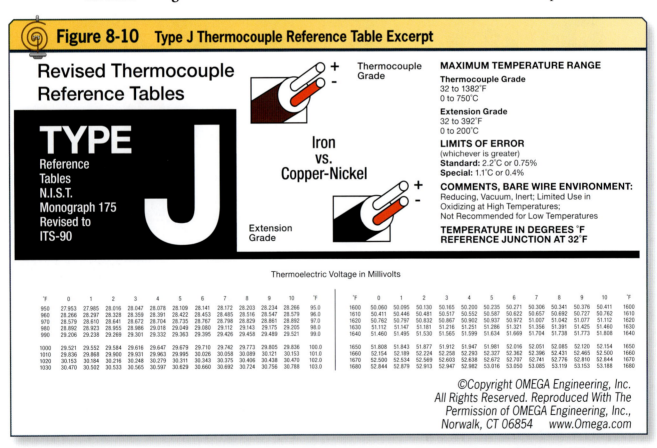

Figure 8-10. Thermocouple charts show the relationship between voltage and temperature. The Type J Thermocouple Reference Table is based on NIST (National Institute of Standards and Technology) and ITS-90 standards.

measurement reference. Instead of using an ice bath, many temperature transmitters use compensation circuitry to adjust the output signal based on the reference junction temperature. **See Figure 8-11D.**

Taking a millivolt reading from thermocouples presents another issue for the technician. The leads of the voltmeter are made of metal. The connections between the voltmeter leads and the thermocouple create additional junctions, which can lead to an inaccurate reading. The law of intermediate metals is useful to understand in this context. The law states that any additional junctions made in a thermoelectric circuit will not affect the millivolt output, contingent upon the additional junctions being held at the same temperature. To ensure the junctions are kept at the same temperature, an isothermal block is often used. The physical construction of an isothermal block ensures the connections are held at the same temperature and that the true temperature signal is not affected. **See Figure 8-12.**

Consider the internal construction of a typical thermocouple. The leads of

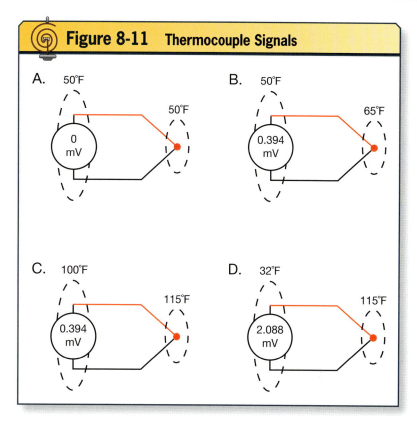

Figure 8-11. Type J thermocouples respond differently to various temperature differences.

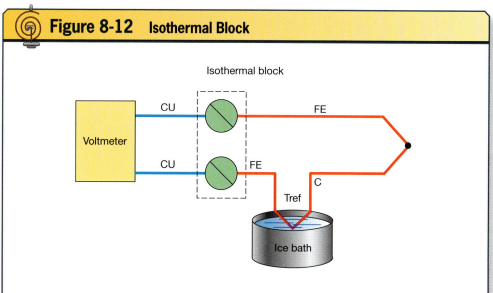

Figure 8-12. An isothermal block in a circuit is used to ensure junction temperatures match. This circuit consists of wires composed of iron (FE), constantan (C), and copper (CU). The ice bath serves as a stable temperature reference (Tref).

TechTip!

If an instrument does not provide ambient temperature compensation, the reference temperature must be measured and converted to a millivoltage based on a thermocouple chart. This additional millivoltage must be added to the measured signal.

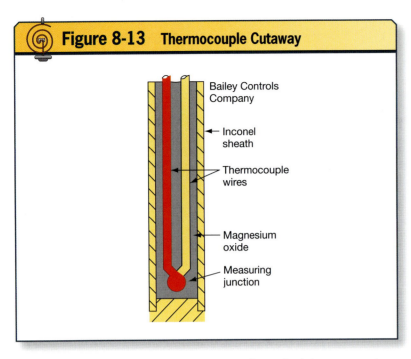

Figure 8-13. The internal construction of a typical thermocouple includes material that enhances heat transfer, such as magnesium oxide.

TechTip!
The most common failure points in temperature accuracy are bad wiring connections. Always verify proper connections when investigating suspected reading errors.

the thermocouple are encased in a rigid metal sheath. The measuring junction is normally formed at the bottom of the thermocouple housing. A material such as magnesium oxide surrounds the thermocouple wires to prevent vibration that could damage the fine wires and enhances heat transfer between the measuring junction and the medium surrounding the thermocouple. **See Figure 8-13.**

The installation of thermocouples and thermocouple wiring requires consideration of several factors. Location, both of where the thermocouple is mounted and of where resultant wiring is run, must be considered. Because of the low levels of thermocouple wiring voltages, stray currents or voltages (i.e., noise) can greatly affect the final reading. Thermocouple lead wiring should always be placed away from high-voltage or high-current conductors. Separate conduit or raceways are always recommended. Shielded twisted-pair wiring is another useful precaution, because the shield protects from noise and the twisting of the pair negates additional induced electromotive forces.

Extending thermocouple wire can cause error in the measurement signal if the wire is improperly installed. The best practice to extend a thermocouple is to use thermocouple-grade extension wire that matches the type of thermocouple being used. While copper wire is sometimes used to extend thermocouples, care must be taken to ensure the reference junction is measuring the appropriate point in the circuit. Several manufacturers make thermocouple-specific connectors and terminal blocks. These connectors are made of materials that are less subject to measurement errors and poor connections. The installation of the transmitter as closely as possible to the thermocouple is recommended. This allows the weak millivolt signal to be converted to a robust 4- to 20-milliampere direct current signal.

Resistance Temperature Detector

The RTD is a temperature measurement device. The RTD incorporates pure metals or certain alloys that increase in resistance as temperature increases and decrease in resistance as temperature decreases. RTDs act somewhat like an electrical transducer, converting changes in temperature to voltage signals by the measurement of resistance. The metals that are best suited for use as RTD sensors are pure, are of uniform quality, are stable within a given range of temperature, and are able to give reproducible resistance–temperature readings. Only a few metals have the properties necessary for use in RTD elements. RTD elements are normally constructed of platinum, copper, or nickel. These metals are best suited for RTD applications because of their relatively linear resistance–temperature characteristics, their high coefficient of resistance, and their ability to withstand repeated temperature cycles. **See Figure 8-14.**

The *coefficient of resistance* is the constant defining the incremental change in resistance of an RTD with respect to change in temperature. It is usually expressed as a percentage per degree of

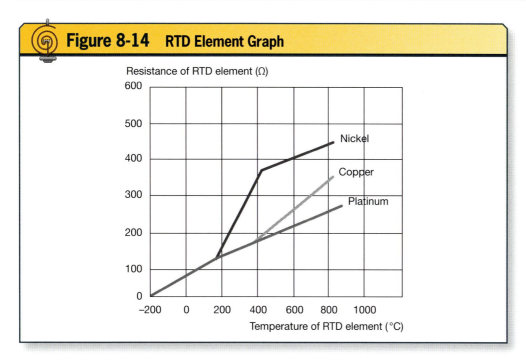

Figure 8-14. The relationship between temperature and resistance of various metals determines their suitability for RTD applications.

temperature change. RTD specification sheets often refer to the coefficient of resistance as the alpha coefficient (e.g., α 0.385).

RTD elements are usually long, spring like wires surrounded by an insulator and enclosed in a sheath of metal. This design has a platinum element surrounded by a porcelain insulator. The RTD material used must be capable of being drawn into fine wire so that the element can be easily constructed. The insulator prevents a short circuit between the wire and the metal sheath. **See Figure 8-15.**

Inconel, a nickel–iron–chromium alloy, is normally used in manufacturing the RTD sheath because of its inherent corrosion resistance. When placed in a liquid or gas medium, the inconel sheath quickly reaches the temperature of the medium. The change in temperature causes the platinum wire to heat or cool, resulting in a proportional change in resistance. This change in resistance is measured by a precision resistance-measuring device that is calibrated to give the proper temperature reading. This device may be a bridge circuit or an electronic transmitter. Both perform the same function.

An RTD has a protective well and a

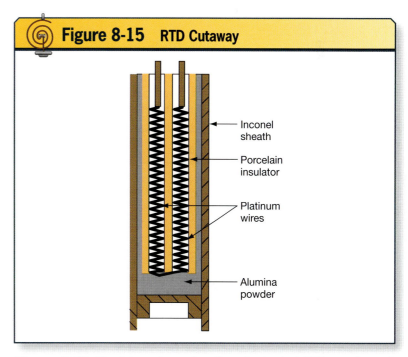

Figure 8-15. The internal construction of a typical RTD consists of a metal RTD element surrounded by insulating material.

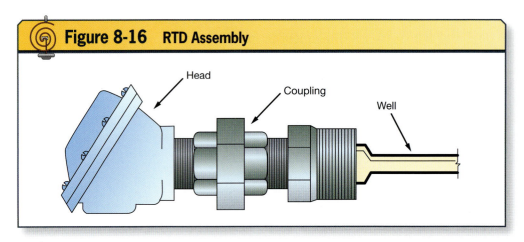

Figure 8-16. An RTD includes a protective well, a coupling, and a head to protect against damage, temperature, and pressure.

terminal head. The well protects the RTD from damage by the gas or liquid being measured. Protecting wells are normally made of stainless steel, carbon steel, inconel, or cast iron, and can be designed for very high temperatures and pressures. **See Figure 8-16.**

The common values of resistance for RTDs range from 10 ohms to more than several thousand ohms; the most common is 100 ohms at a temperature of 0°C. **See Figure 8-17.**

Connecting a resistance measurement device, such as an ohmmeter, to the RTD may require a significant amount of wire to reach the device. This length of wire can introduce substantial error in the measurement circuit. With the leads connected to an RTD, the total resistance of the circuit may be several ohms more than the RTD.

For the following example, assume a lead resistance of 1.5 ohms:

RTD used:
Platinum 100 α 0.385 Ω

Actual resistance of RTD:
154.7 Ω or 143°C

Resistance measurement:
156.2 Ω or 147°C

Error due to lead:
147°C − 143°C = 4°C error

To counteract this lead resistance, industry has adopted the use of 3- and 4-wire RTDs. The 3-wire RTD uses

Figure 8-17 RTD Table

Deg. C	Resistance	Deg. C	Resistance	Deg. C	Resistance	Deg. C	Resistance	Deg. C	Resistance
-45.00	82.29	-15.00	94.12	15.00	105.85	45.00	117.47	75.00	128.99
-40.00	84.27	-10.00	96.09	20.00	107.79	50.00	119.40	80.00	130.90
-35.00	86.25	-5.00	98.04	25.00	109.73	55.00	121.32	85.00	132.80
-30.00	88.22	0.00	100.00	30.00	111.67	60.00	123.24	90.00	134.71
-25.00	90.19	5.00	101.95	35.00	113.61	65.00	125.16	95.00	136.61
-20.00	92.16	10.00	103.90	40.00	115.54	70.00	127.08	100.00	138.51

Figure 8-17. The relationship between temperature (in degrees Celsius) and resistance (in ohms) for a 100-ohm RTD is positively correlated.

three lead wires, two in parallel on one side of the resistive element and one on the opposite side. **See Figure 8-18.**

Typically, the two leads that are on the same side have the same color of insulation and the opposite lead is a different color of insulation.

The following equation illustrates the relationship between lead resistance and overall resistance of the element:

$$\text{Total } \Omega = \Omega_{L_1} + \Omega_{RTD} + \Omega_{L_2} - (\Omega_{L_2} + \Omega_{L_3})$$

$$\text{Total } \Omega = \Omega_{L_1} + \Omega_{RTD} - \Omega_{L_3}$$

$$\text{Total } \Omega = \Omega_{RTD} \text{ (if } \Omega_{L_1} = \Omega_{L_3}\text{)}$$

This equation can be applied to an example. A 3-wire RTD is installed remotely, and each lead has a resistance of 4.2 ohms:

$$\text{Total } \Omega = 4.2\,\Omega_{L_1} + \Omega_{RTD} + 4.2\,\Omega_{L_2} - (4.2\,\Omega_{L_2} + 4.2\,\Omega_{L_3})$$

$$\text{Total } \Omega = 4.2\,\Omega_{L_1} + \Omega_{RTD} - 4.2\,\Omega_{L_3}$$

$$\text{Total } \Omega = \Omega_{RTD}$$

The technician should notice that when the lead resistances are equal, they effectively cancel each other out and a true resistive measurement is read. The leads must be kept as close to the same length as possible to prevent errors in a 3-wire RTD circuit.

The 4-wire RTD has four leads, two on each side of the resistive element. **See Figure 8-19.**

The two leads have the same insulation color to identify which leads share the same side of the element. The measurement principle of the 4-wire RTD is the use of two leads as an excitation current loop, typically in the order of microamperes. The remaining two leads are connected to a voltmeter with very high impedance. The voltmeter measures the voltage drop across the element, which is used in conjunction with Ohm's Law to determine the precise resistance of the element.

A characteristic common to all RTDs is that they are not self-powered. Unlike the thermocouple, which produces a millivolt signal, the RTD requires a current source to measure the resistance of the element. A concern when installing an RTD is that this current source will create a *self-heating* effect, or internal heating of a transducer as a result of power dissipation, because of the current flow through the resistive element. Some transmitters use a pulsed current

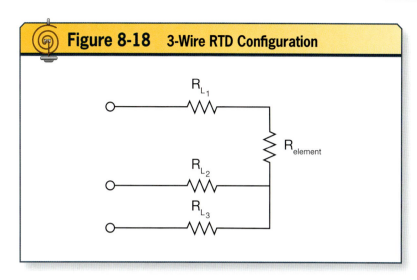

Figure 8-18. The 3-wire RTD uses three lead wires to reduce lead resistance errors. Lead wire 1 (R_{L_1}) is connected to one side of the resistive element ($R_{element}$). Lead wires 2 and 3 (R_{L_2} and R_{L_3}) are connected to the other side of the resistive element.

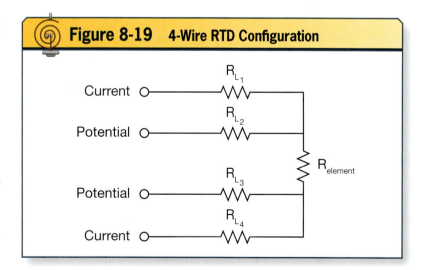

Figure 8-19. A 4-wire RTD uses two leads for the excitation current and two leads for the voltage measurement. Lead wires 1 and 2 (R_{L_1} and R_{L_2}) are connected to one side of the resistive element ($R_{element}$). Lead wires 3 and 4 (R_{L_3} and R_{L_4}) are connected to the other side of the resistive element.

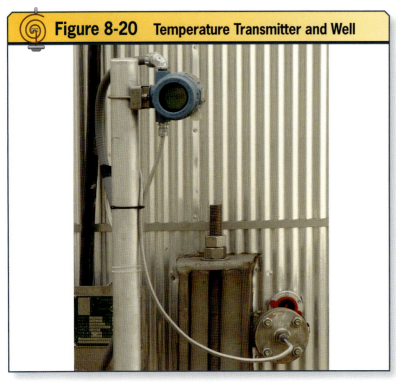

Figure 8-20. A temperature indicating transmitter may include a remote temperature well.

Figure 8-21. Temperature elements are mounted near a process tank.

instead of a constant current to reduce self-heating effects.

The same practical precautions that apply to thermocouples must be observed for RTDs: Use shields where possible, use twisted-pair wiring, avoid stressing the conductors, use a larger wire for long pulls, and avoid high-temperature areas. The RTD is more fragile than the thermocouple, and care must be taken when installing it to minimize damage. During installation or maintenance, a technician may encounter an RTD sensor with either six or eight lead wires. Dual-sensor RTDs, which contain a second, separate sensor, are used to provide redundant temperature readings and prevent the loss of temperature readings in the event of a sensor burnout or short circuit.

Controller microprocessors can make the resistance-to-temperature conversions if required. High-quality digital multimeters (DMMs) may be used to measure RTD resistance to a high degree of accuracy (assuming the DMM is within calibration standards). Calibrators allow easy troubleshooting and calibration of most RTDs. The platinum to copper connection that is made when the RTD is measured can introduce some additional voltage because of the Seebeck effect. By referencing the resistance readings of the RTD at a known temperature, these effects can be eliminated.

Consider a typical installation. The temperature transmitter is often mounted a few feet away from the vessel to limit exposure to excessive heat. **See Figure 8-20.**

Groups of temperature elements may be installed to measure multiple points in a vessel. **See Figure 8-21.**

Typical transmitter wiring diagrams illustrate various RTD and thermocouple sensors. **See Figure 8-22.**

Thermistor

The thermistor, like the RTD, is a temperature-sensitive resistor. If the thermocouple is considered the most versatile temperature sensor and the RTD is considered the most stable, the thermistor could be considered the most

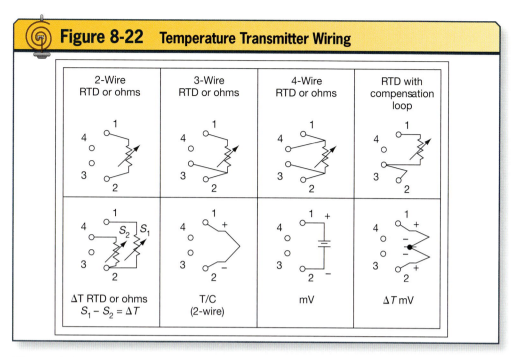

Figure 8-22. A temperature transmitter wiring diagram can be created for both RTD and thermocouple sensors. Some connections use two sensors, normally called Sensor 1 (S_1) and Sensor 2 (S_2). With two sensors, the temperature difference (ΔT) between two points can be measured.

sensitive sensor. An additional advantage of the thermistor is that its large resistance value effectively eliminates concern with wire lead resistance. For example, a 10,000-ohm thermistor that is extended with lead wires that add an additional 2 ohms of resistance would be subject to an error of only 0.02% (2 ohms divided by 10,000 ohms).

The thermistor is made chiefly of semiconductor materials that allow it to be many times more sensitive than thermocouples or RTDs. One disadvantage is that thermistors do not maintain a direct proportional resistance change with temperature. **See Figure 8-23.**

The nonlinear curve that determines resistance can react in two manners: positively or negatively. A *positive temperature coefficient thermistor* is a thermistor that increases in resistance when measured temperature increases. Conversely, if an increase in temperature results in a decrease in resistance, the thermistor is said to have a negative temperature coefficient.

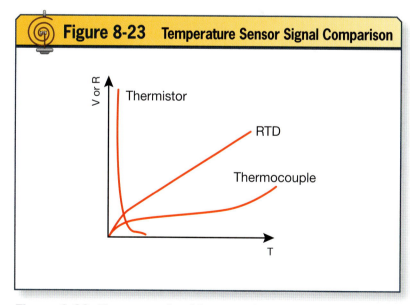

Figure 8-23. The output signal linearity of a thermistor, RTD, and thermocouple can be compared graphically. The X-axis is given in unit temperature (T). The Y-axis is given is units of voltage (V) in the case of the thermocouple and resistance (R) in the case of the RTD and thermistor.

Figure 8-24. Different styles of temperature transmitters are available to suit the process environment.

A thermistor works the same way as an RTD, but the nonlinearity of the response demands a different approach to determine the temperature of the sensor. The nonlinearity of the thermistor requires an equation or table to determine the temperature based on the thermistor resistance value. Most modern controllers can accept and interpret the thermistor signal and translate the resistive value to a temperature value.

Though thermistors are considered the most sensitive temperature sensors, they have several disadvantages. They are susceptible to high-temperature damage because they are made of a semiconductor material. They are usually limited to a few hundred degrees Celsius. Continued exposure to high temperatures well within the range specified can still lead to *drift*, a change of a reading or a set point value over long periods caused by several factors, including changes in ambient temperature, time, and line voltage.

Thermistors are often small, which means their response time is fast; this also means that they are susceptible to self-heating errors.

Transmitters

The signals produced by RTDs, thermocouples, and thermistors are relatively weak and subject to error over long wire distances. Therefore, the temperature transmitter is used to convert the signal from either a millivolt (thermocouple) or an ohm (RTD or thermistor) value to a more robust signal, such as 4 to 20 milliamperes direct current. Electrical magnetic interference (EMI) concerns are reduced when using a 4- to 20-milliampere direct current signal, and wire lengths can be greater between the sensor and the controller.

Temperature transmitters come in multiple configurations and are selected based on the environment and mounting requirements. Direct-mount transmitters are mounted at the process connection and result in the shortest signal leads. Transmitters can also be mounted in cabinets that protect the transmitter from the process hazards, such as excessive heat or corrosive vapors. **See Figure 8-24.**

The technician must understand how the temperature transmitter's input and output signals relate to the process. Proper calibration of the transmitter is essential for proper operation of the process. As an example, assume a temperature transmitter has a thermocouple input and a temperature range from 30°C to 600°C. The transmitter's output is the standard 4 to 20 milliamperes direct current. **See Figure 8-25.**

In this example, what is the output in milliamperes direct current when the sensor temperature is 167°C? The output is found using the following formula:

Output = (% of input range × output span) + output LRV

The following values are then applied:

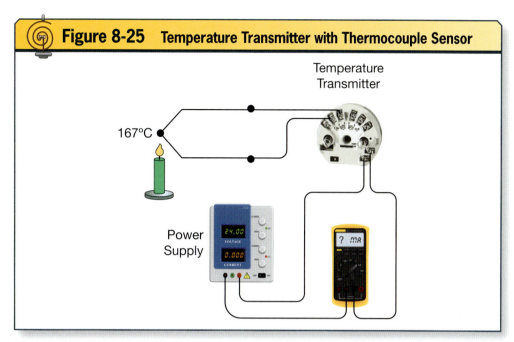

Figure 8-25. *The loop current can be calculated by measuring the sensor temperature.*

$$\text{Measurement span} = \text{URV} - \text{LRV} = 600°C - 30°C = 570°C$$

$$\text{Input value} = 167°C - 30°C = 137°C$$

$$\% \text{ of input range} = \left(\frac{137°C}{570°C}\right) \times 100 = 24\%$$

$$\text{Output} = (24\% \times 16 \text{ mADC}) + 4 \text{ mADC} = 7.84 \text{ mADC}$$

For the temperature, the upper-range value (URV) is 600°C and the lower-range value (LRV) is 30°C in this example. The input measurement span is 570°C. The output span is found by subtracting transmitter's output LRV (4 milliamperes direct current) from its output URV (20 milliamperes direct current). This results in an output span of 16 milliamperes direct current. The final output is 7.84 milliamperes direct current.

The technician could also take a milliampere direct current reading in a temperature loop and convert the signal to the process temperature. As an example, assume the temperature transmitter has an RTD input and a temperature range from −20°F to 300°F. The transmitter's output is the standard 4 to 20 milliamperes direct current. **See Figure 8-26.**

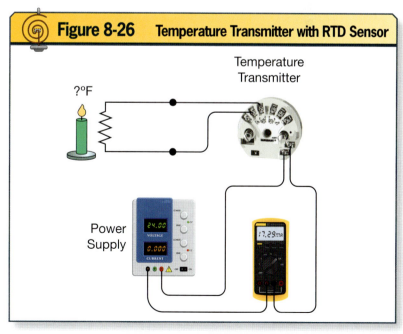

Figure 8-26. *The temperature can be calculated by measuring the loop current.*

For this example, what is the temperature when the loop current is measured at 17.29 milliamperes direct current? This input is found using the following formula:

$$\text{Input} = \left(\left(\frac{\text{Output} - \text{Output LRV}}{\text{Output span}}\right) \times \text{Input measurement span}\right) + \text{LRV}$$

The following values are then applied:

Measurement span $= \text{URV} - \text{LRV} = 300°F - (-20°F) = 320°F$

Output value $= 17.29 \text{ mADC} - 4 \text{ mADC} = 13.29 \text{ mADC}$

Percentage of output range $= \left(\dfrac{13.29 \text{ mADC}}{16 \text{ mADC}}\right) \times 100 = 83\%$

Input $= (83\% \times 320°F) + (-20°F) = 245.6°F$

The input measurement span is defined as the URV minus the LRV, in this case, 320°F. The output LRV is 4 milliamperes direct current, and the output span is 16 milliamperes direct current.

Temperature transmitters enable process signals to travel long distances and maintain their accuracy. Properly calibrated, temperature transmitters are essential components in temperature loops. Care should be taken to always follow the design parameters in regards to environment, mounting, and hazardous conditions.

THERMOWELL BASICS

Temperature sensors are rarely installed directly into the process. They are placed into a protective covering called a *thermowell*, a closed-end tube designed to protect temperature sensors from harsh environments. **See Figure 8-27.**

The thermowell, in turn, is placed into the process. Thermowells can be screwed, welded, or held in place with a flange on the vessel or line. Thermowells are often made of 316 or 304 stainless steel, but other alloys are available. Thermowells can slow down the response time of the thermocouple, because the thermowell acts as an insulator, which can isolate the sensor from the process. If this thermal lag is problematic, heat-conducting fluids or gels can be used to fill the air space inside the thermowell.

ThinkSafe!
Never install a process temperature-measuring device directly into the process media. Always use a protective thermowell. This protects the technician from exposure to process media while performing maintenance activities on the device.

Figure 8-27. A screw-in-type thermowell is used to isolate the temperature sensor from the process material.

SUMMARY

Temperature measurement is a broad field that involves numerous conditions and various measurement devices. It is necessary to understand all principles of operation, physical properties, circuit layouts, and sensor types to achieve the optimal process operation.

REVIEW QUESTIONS

1. Temperature is the measure of the __?__ of molecules in a substance.
   a. ionic concentration
   b. kinetic energy
   c. potential energy
   d. specific gravity

2. This temperature scale is based on the freezing point of water at 0° and the boiling point at 100°.
   a. Celsius
   b. Fahrenheit
   c. Kelvin
   d. Rankine

3. Convert 762°R to Fahrenheit.
   a. 150.18°F
   b. 302.33°F
   c. 1,035.15°F
   d. 1,221.67°F

4. The __?__ temperature element works on the principle of two metals that expand at different rates when heated.
   a. bimetallic
   b. RTD
   c. thermistor
   d. thermocouple

5. Temperature switches use this type of signal to control a process.
   a. Discrete
   b. Milliampere direct current
   c. Millivolt
   d. Variable resistance

6. A __?__ element works on the principle of two dissimilar metals that generate a small voltage as a function of temperature difference.
   a. bimetallic
   b. RTD
   c. thermistor
   d. thermocouple

7. What is the color of the negative lead of an ANSI-compliant type J thermocouple?
   a. Green
   b. Red
   c. White
   d. Yellow

8. The __?__ temperature sensor measures temperature based on a change in resistance and is commonly constructed of a fine platinum wire because of the inherent linearity.
   a. bimetallic
   b. RTD
   c. thermistor
   d. thermocouple

9. The most sensitive temperature sensor is the __?__.
   a. bimetallic element
   b. RTD
   c. thermistor
   d. thermocouple

10. A temperature transmitter ranges from 25°F to 1,190°F. The output of the transmitter is measured at 14.96 milliamperes direct current. What is the process temperature?
    a. 798°F
    b. 823°F
    c. 1,089°F
    d. 1,114°F

11. A temperature transmitter ranges from 0°C to 100°C. The process temperature is 23.8°C. What should the transmitter output read?
    a. 3.8 mADC
    b. 5.9 mADC
    c. 7.8 mADC
    d. 23.8 mADC

Principles of Smart Instrument Communication and Calibration

The term "smart" refers to process instruments that possess enhanced capabilities in comparison to analog devices, which are commonly called "legacy," "conventional," or "dumb" devices. Still, both legacy and smart instruments are used for instrumentation projects. A smart instrument requires a communicator to properly set up, configure, and calibrate the smart device. A smart communicator does not calibrate the device; instead, it provides a working interface to the device so that a calibration procedure may be performed.

Objectives

- Explain the similarities and differences between a conventional and a smart device.
- Describe how a smart communicator is used to interact with a smart device.
- Explain how information is exchanged from a smart device to a smart communicator.
- Explain the calibration process for a HART device.
- Describe smart device switches and state their purpose.

Chapter 9

Table of Contents

Basics of Smart Instruments 192
Types and Methods of Communication 192
    HART 192
    Foundation Fieldbus 193
Smart Instrument Communicators 194
Systems and Applications 198

Using the Communicator to Calibrate a
Smart Instrument 201
    Connect the HART Communicator 201
    Verify the Sensor Input Section 202
    Configure the Conversion Section.......... 203
    Verify the AO Section 203
    Example of a Typical Installation
    Sequence for a Smart Device 204
Summary.. 206
Review Questions 207

For additional
information, visit
qr.njatcdb.org
Item #2690

BASICS OF SMART INSTRUMENTS

Smart instruments have increased capabilities in comparison to legacy devices. The basis of a smart instrument is the inclusion of a digitally operated microprocessor. The microprocessor-based instrument allows digital communication to be established with the device. Once this digital communication channel is connected, the technician can observe and modify the operating characteristics of the device. Some smart devices operate by generating both analog (4–20 milliamperes direct current, or DC) and digital (1 or 0) signals over that same loop, while other smart devices operate by only sending a digital signal.

The advantages of using smart instruments have resulted in their increased use in industry. Remote device configuration, self-diagnosis of errors, and ease of calibration are some advantages plant engineers consider when selecting smart devices for an instrumentation project.

TYPES AND METHODS OF COMMUNICATION

The term "smart instrument" defines a category of devices. Every smart device is based on a *protocol*, or a formal definition that describes how data are to be exchanged. The protocol serves as a foundation for the exchange of information. Several protocols have been developed to create smart devices.

Two of the most common protocols are the highway addressable remote transducer (HART) protocol and the Foundation Fieldbus protocol. While other smart device protocols exist (Profibus, Modbus, etc.), HART and Foundation Fieldbus command most smart instrument installations. A defining characteristic of both protocols is that they are open, nonproprietary foundations for communication. A *proprietary protocol* is a nonstandard communication format and language owned by a single organization or individual. The open nature of both HART and Foundation Fieldbus permits any manufacturer to develop devices that are capable of communication. This open protocol environment has led to an installed base of HART and Foundation Fieldbus devices in the tens of millions.

HART

The HART protocol was initially developed by Rosemount in the 1980s. The ownership of the HART standard was transferred from Rosemount to the HART Communication Foundation in 1986. The foundation is a global nonprofit that oversees the development of standards and development for the HART protocol.

The digital communication used by the HART protocol transfers information through a series of discrete signals, often represented as 1s and 0s. In contrast, analog communication is a continuously variable value over the channel. The HART protocol works by simultaneously transferring digital information using the same channel as the analog signal. The Bell 202 frequency shift keying (FSK) standard is employed for the digital signal. This protocol is the same one used for caller identification on landline phones.

The Bell 202 protocol represents digital values by shifting the frequency of a

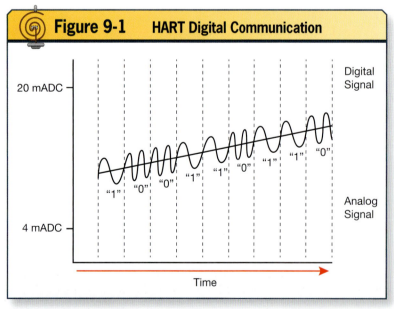

Figure 9-1. The digital signal is represented by a series of 1s and 0s over the analog signal channel.

continuous signal that is superimposed on the analog signal loop wires. The transfer of data occurs at 1,200 bits per second, quite slow by modern standards. A frequency of 1,200 hertz represents a 1, and a frequency of 2,200 hertz represents a 0. Because the average value of the superimposed signal is zero, the 4- to 20-milliampere DC signal is not changed. **See Figure 9-1.**

The HART protocol supports both point-to-point and fully digital multidrop installations. A point-to-point installation uses the 4- to 20-milliampere DC signal on a traditional instrument loop. In this installation, the digital communication is used for configuration and calibration of the HART device. The process variable is not transmitted digitally over the loop. Instead, the device converts the process variable into a 4- to 20-milliampere DC signal and transmits this information over the signal loop.

The multidrop installation allows up to 15 separate HART devices to reside on the same loop. Instead of transmitting information with an analog 4- to 20-milliampere DC signal, the process variable information is transmitted using digital communication. The multidrop configuration requires the technician to set a unique address for each instrument. Even though it lacks robust digital communication and speed, the HART platform is installed and maintained at a large number of facilities.

Foundation Fieldbus

Foundation Fieldbus is an all-digital standard for the transfer of process information on a plant network. Like HART, it is an open protocol that allows many manufacturers to create interoperable devices. Unlike traditional point-to-point wiring using 4- to 20-milliampere DC loops, Foundation Fieldbus operates on a common network, similar to a local area network (LAN). To understand the network layout, imagine an office building with several hundred computers, printers, and servers. Each of these devices is able to share a common network to transfer information. This network *topology*, or physical layout of a data network, represents the evolution of control networks toward a distributed control model.

In a traditional 4- to 20-milliampere DC analog loop, dedicated wiring carries information in an analog format to a controller to process the signal and generate an analog output (AO). In contrast, the fully digital Foundation Fieldbus standard allows information to be transmitted digitally from the sensor directly to the actuator. The digital devices implement the control strategy through the use of *functional blocks*, a software element that defines a specific characteristic of the process control function. Functional blocks include analog input (AI), proportional–integral–derivative (PID) control, and AO. The incorporation of these functional blocks into field devices distributes control toward the process and away from a centralized control system. **See Figure 9-2.**

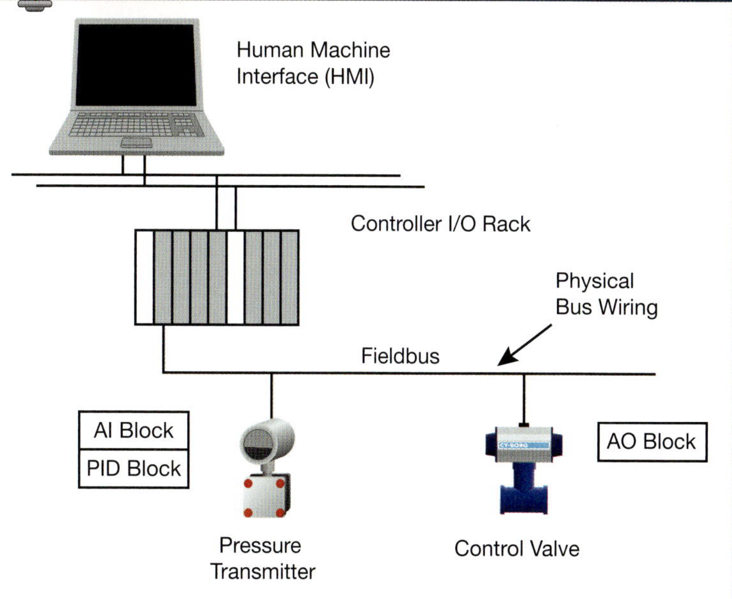

Figure 9-2. The pressure transmitter contains the AI block and the PID block. Digital information travels along the physical bus wiring to the control valve. An AO block converts the digital signal to an analog signal to drive the control valve. The I/O rack provides a connection for the input and output wires.

The installation of a Foundation Fieldbus system requires the technician to understand some specific requirements of the system. Unlike HART, which is not polarity sensitive, a Foundation Fieldbus network must observe polarity. Each device must have a unique identifier, commonly called a tag number. Devices on the network can be powered without additional power supplies, but care must be taken to ensure a minimum of 9 volts DC to each device for proper operation. Network terminators are devices used on a Foundation Fieldbus network to reduce electrical noise and ensure information is not corrupted on the bus wiring. Each bus must have two terminators located on either end of the trunk. The Foundation Fieldbus standard limits the overall length of the network to ensure the reliable and timely transfer of information. The topology of the wiring can be point to point, bus, daisy chain, or tree. **See Figure 9-3.**

As control networks continue to evolve toward distributed intelligence and all digital communication, standards such as Foundation Fieldbus provide real benefits to end users. However, the process control industry still relies on the standard 4- to 20-milliampere DC and 3- to 15-pound per square inch analog loops for most process instrumentation.

SMART INSTRUMENT COMMUNICATORS

A smart communicator is used to provide a working interface to a smart device. This smart communicator is the tool used to adjust the device's internal software configuration. Traditional devices are configured by turning screws or setting switches. The smart device is configured by changing the settings in the software. Because the smart device is microprocessor driven and controlled, it offers greater flexibility than legacy devices. Smart

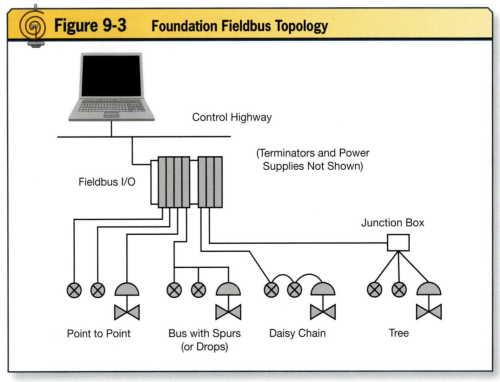

Figure 9-3. The topology of the Foundation Fieldbus wiring is flexible to suit the requirements of the process control design.

device communication can be achieved by a handheld device or a modem and terminal, such as a smartphone, tablet, or PC. **See Figure 9-4.**

Several manufacturers have products suitable for smart device communication.

Safety must be considered when communicating with smart devices. The digital communication signals are electrical impulses, so all safety hazards must be considered for hazardous areas. **See Figure 9-5.**

Technicians can observe the actions or functions of the smart device through a liquid crystal display (LCD). When connected to a device, the LCD provides an interface to the smart instrument's parameters and configuration. Most handheld communicators have alphanumeric keys for entering data. Many modern communicators also have touch-screen displays.

A HART communicator, when turned on, searches the connection for a HART device. If no device is found, the communicator will show that no device is connected. The technician can create an *offline configuration*, or a configuration created without a connection to a device. This configuration can be stored for later download to a smart device. In the offline configuration menu, a list of manufacturer's devices is available to select and configure. When a device is connected to a communicator, the technician can perform an *online configuration*, or a configuration that is performed with a real-time communication channel established. If a device is not found, the technician should troubleshoot the connections or wiring.

The communicator must have the appropriate device descriptions (DDs) to access all capabilities of a HART device. A DD is similar to a driver used by a computer to connect to a device such as a printer. If the appropriate DD files are not installed on the communicator, the technician will not be able to fully configure the device. For example, if a technician has an older communicator and a newer device, the communicator likely will not know that such a device exists and will not be able to recognize it. The closest model to the device will have to be

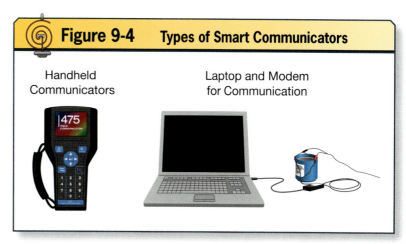

Figure 9-4. Smart communicators provide a working interface to a smart device.

ThinkSafe!

Never expose the HART communicator to voltages above the system's signal voltage rating (typically 24 to 48 volts DC). Connecting a HART communicator to voltages such as 120 volts alternating current can cause serious damage to the communicator and injury to the technician.

Figure 9-5 Typical Warning Labels for Communicators

⚠ **WARNING**
Explosions can result in death or serious injury. Do not make connections to the serial port or NiCad recharger jack in an explosive atmosphere.

⚠ **WARNING**
Explosions can result in death or serious injury. Before connecting the HART Communicator in an explosive atmosphere, make sure the instruments in the loop are installed in accordance with intrinsically safe or nonincendive field wiring practices.

Figure 9-5. All safety hazards must be evaluated when using smart communicators.

chosen for data exchange, or a generic menu may be available. The device can accept the information, because it should be compatible with older communicators, but it may not have the same parameters as the older model, such as the upper-range value (URV), lower-range value (LRV), and sensor type. Each manufacturer publishes their DD files in a file library, which is available from the HART Communication Foundation.

Navigating the series of menus to configure a device is documented by the manufacturer in the form of a menu tree. **See Figure 9-6.**

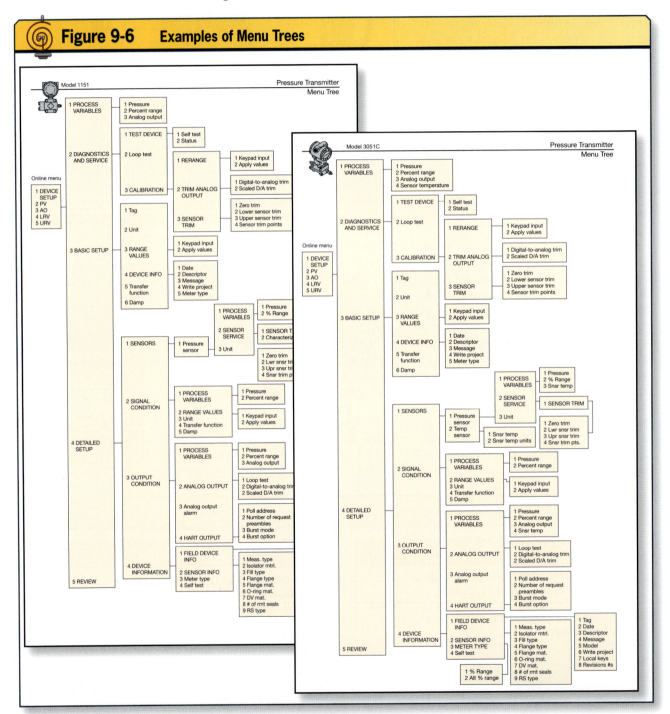

Figure 9-6. *The menu tree shows the structural layout of the series of menus used to configure a device.*

The actions of entering data are menu driven and follow a predefined sequence of steps. Device manufacturers also publish fast key sequences used to rapidly access specific options in the menu tree. **See Figure 9-7.**

Several menus are available for advanced diagnostics and troubleshooting techniques. These menus should not be activated unless the technician clearly understands the consequences of making changes in these areas.

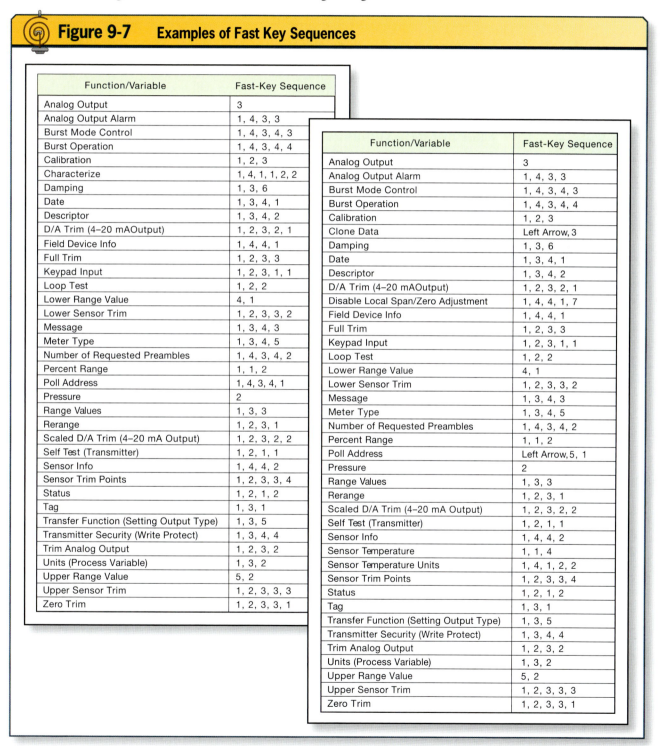

Figure 9-7. Examples of Fast Key Sequences

Function/Variable	Fast-Key Sequence
Analog Output	3
Analog Output Alarm	1, 4, 3, 3
Burst Mode Control	1, 4, 3, 4, 3
Burst Operation	1, 4, 3, 4, 4
Calibration	1, 2, 3
Characterize	1, 4, 1, 1, 2, 2
Damping	1, 3, 6
Date	1, 3, 4, 1
Descriptor	1, 3, 4, 2
D/A Trim (4–20 mAOutput)	1, 2, 3, 2, 1
Field Device Info	1, 4, 4, 1
Full Trim	1, 2, 3, 3
Keypad Input	1, 2, 3, 1, 1
Loop Test	1, 2, 2
Lower Range Value	4, 1
Lower Sensor Trim	1, 2, 3, 3, 2
Message	1, 3, 4, 3
Meter Type	1, 3, 4, 5
Number of Requested Preambles	1, 4, 3, 4, 2
Percent Range	1, 1, 2
Poll Address	1, 4, 3, 4, 1
Pressure	2
Range Values	1, 3, 3
Rerange	1, 2, 3, 1
Scaled D/A Trim (4–20 mA Output)	1, 2, 3, 2, 2
Self Test (Transmitter)	1, 2, 1, 1
Sensor Info	1, 4, 4, 2
Sensor Trim Points	1, 2, 3, 3, 4
Status	1, 2, 1, 2
Tag	1, 3, 1
Transfer Function (Setting Output Type)	1, 3, 5
Transmitter Security (Write Protect)	1, 3, 4, 4
Trim Analog Output	1, 2, 3, 2
Units (Process Variable)	1, 3, 2
Upper Range Value	5, 2
Upper Sensor Trim	1, 2, 3, 3, 3
Zero Trim	1, 2, 3, 3, 1

Function/Variable	Fast-Key Sequence
Analog Output	3
Analog Output Alarm	1, 4, 3, 3
Burst Mode Control	1, 4, 3, 4, 3
Burst Operation	1, 4, 3, 4, 4
Calibration	1, 2, 3
Clone Data	Left Arrow, 3
Damping	1, 3, 6
Date	1, 3, 4, 1
Descriptor	1, 3, 4, 2
D/A Trim (4–20 mAOutput)	1, 2, 3, 2, 1
Disable Local Span/Zero Adjustment	1, 4, 4, 1, 7
Field Device Info	1, 4, 4, 1
Full Trim	1, 2, 3, 3
Keypad Input	1, 2, 3, 1, 1
Loop Test	1, 2, 2
Lower Range Value	4, 1
Lower Sensor Trim	1, 2, 3, 3, 2
Message	1, 3, 4, 3
Meter Type	1, 3, 4, 5
Number of Requested Preambles	1, 4, 3, 4, 2
Percent Range	1, 1, 2
Poll Address	Left Arrow, 5, 1
Pressure	2
Range Values	1, 3, 3
Rerange	1, 2, 3, 1
Scaled D/A Trim (4–20 mA Output)	1, 2, 3, 2, 2
Self Test (Transmitter)	1, 2, 1, 1
Sensor Info	1, 4, 4, 2
Sensor Temperature	1, 1, 4
Sensor Temperature Units	1, 4, 1, 2, 2
Sensor Trim Points	1, 2, 3, 3, 4
Status	1, 2, 1, 2
Tag	1, 3, 1
Transfer Function (Setting Output Type)	1, 3, 5
Transmitter Security (Write Protect)	1, 3, 4, 4
Trim Analog Output	1, 2, 3, 2
Units (Process Variable)	1, 3, 2
Upper Range Value	5, 2
Upper Sensor Trim	1, 2, 3, 3, 3
Zero Trim	1, 2, 3, 3, 1

Figure 9-7. Fast key sequences are used to access specific commands quickly.

ThinkSafe!

Never connect a smart communicator to any loop that is in automatic control. A loop must always be in manual control before the smart communicator is connected.

ThinkSafe!

Never connect a smart communicator to any field device that is running in automatic control mode. This can cause erratic changes in the device and can create serious danger for the technician and the process. Always put the control loop in manual mode before connecting the communicator.

Technicians do not need to verify the calibration of smart communicators, as is necessary with other calibration equipment. The communicator does not contain measuring circuitry and does not measure analog or discrete values directly. A smart communicator is a configuration device that provides an interface to a smart device and is not subject to periodic calibration. When the communicator is displaying milliampere values, process variables, or other information, it is not measuring the variables; it is merely displaying the values that the smart device is reporting. In other words, the HART communicator does not measure the loop current directly; instead, it reads the digital signal generated by the device. Certified test equipment must be used to verify input and output values of an instrument.

The loop should be placed in manual control before communication is attempted over a field-installed device. This means that the plant operator must place the loop into manual control from an operation's standpoint. Placing the loop in manual control allows communication to a device without causing changes to the process. A prompt appears on most communicators to place the loop in manual control, but this is a prompt only. By selecting OK when the prompt appears, technicians are not placing the loop in manual control; they are acknowledging that they have asked plant operations to place the loop in manual control. **See Figure 9-8.**

A smart communicator should be accompanied by documentation to familiarize its user with its specific keystroke functions. In time, the technician will become familiar with the various menu trees used to navigate through the different functions for calibrating and checking the device.

SYSTEMS AND APPLICATIONS

A HART device connects to a process much like a legacy device does. However, the operations of measuring and transmitting the appropriate signal are quite different. A legacy transmitter operates by measuring the process variable as an analog value and generating an analog signal. When a technician calibrates a legacy transmitter, the zero and span screws are used to modify both the calibration and the range simultaneously. This concept, when applied to smart devices, can cause some confusion for the technician.

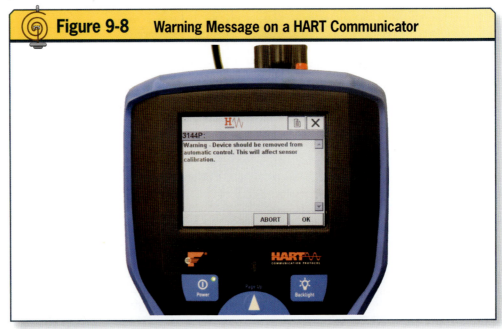

Figure 9-8. Always observe warning messages displayed on the HART communicator.

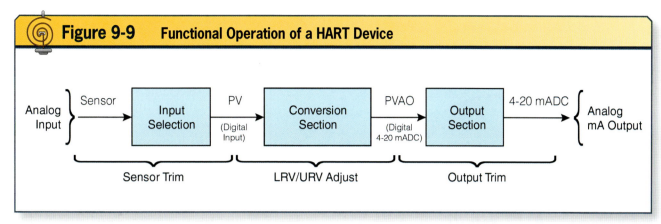

Figure 9-9. *A smart device requires the conversion of information from analog to digital and back to analog. The process variable (PV) and process variable analog output (PVAO) are represented digitally.*

The microprocessor-based smart instrument must convert the process variable signal from an analog signal to a digital representation. This is performed in the input section by the analog-to-digital converter (ADC). The output of this ADC is in digital form, which is required by the microprocessor. The conversion section of the smart device performs a mathematical conversion based on the configuration of the transmitter. This information is stored in the transmitter's *EPROM*, erasable, programmable, read-only memory that can be erased by ultraviolet light; a more up-to-date erasable, read-only memory may be electrically erasable and is usually called an EEPROM. When the transmitter is in operation, it looks at this stored information and uses it to produce an output signal in engineering units that corresponds to the input reference signal. After the mathematical conversion process is complete, the conversion section sends the information digitally to the output section. The output section consists of a digital-to-analog converter (DAC). This portion of the device then generates the appropriate output signal, typically 4–20 milliamperes DC. **See Figure 9-9.**

HART devices may have zero and span buttons for the technician to use for ranging the device. They are located either inside the electronic housing or under a plate external to the housing. **See Figure 9-10.**

These buttons allow a technician to set the LRV and URV by applying the proper process variable input to the sensor. To activate the buttons, many devices require the technician to depress the buttons simultaneously. Once activated, the controls are active for a fixed amount of time. The technician applies a LRV and presses the zero button to set the lower range. Next, the technician applies the URV and presses the span button to set the upper range. The technician must be aware that this operation only changes the device range and is not a substitute for

For additional information, visit qr.njatcdb.org Item #2120

ThinkSafe!
Always set the fail-mode switch immediately before making transmitter calibrations. Failure to do so could result in injury to personnel and equipment.

Figure 9-10. *The span and zero buttons are located inside the transmitter housing. These are used to re-range the device.*

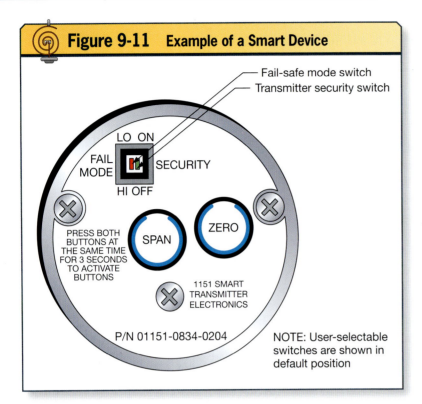

Figure 9-11. The device contains both a fail-mode switch and a security switch.

Most smart devices contain two types of user-selectable switches; a fail-mode switch and a security switch. **See Figure 9-11.**

The fail-mode switch allows the technician to configure the transmitter output to fail high (hi) or fail low (lo). **See Figure 9-12.**

If a device is set to fail high, when the internal diagnostic program of the transmitter detects an unrecoverable error, it will drive the AO signal to a value greater than the maximum output under normal conditions. Likewise, if fail low is selected, the transmitter will output a signal of less than the minimum output under normal conditions. **See Figure 9-13.**

The security switch must also be set by the technician according to the job requirements. The security switch must be in the off position before the technician can change, enter, or transmit configuration data. With the security switch in the on position, the communicator can typically connect to the device but is unable to make changes to the configuration. After the configuration switches are in their proper positions, transmitter configuration and calibration. If the zero and span buttons do not appear to work properly, they may have been disabled in the device configuration. This is commonly done to prevent tampering with the device. To enable them, the technician must use a communicator.

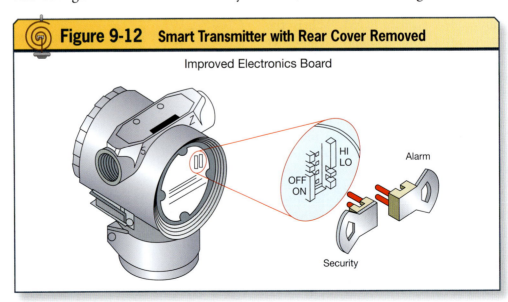

Figure 9-12. The technician must set the position of the security and fail-mode alarm switches as part of the instrument calibration procedure.

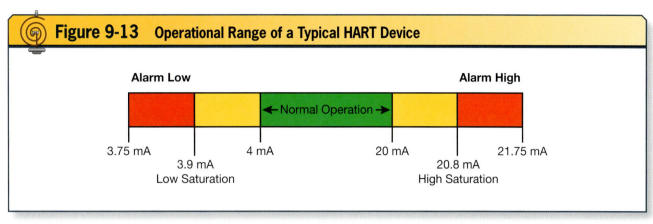

Figure 9-13. The normal operation range of a HART device is between 4 and 20 milliamperes DC. When the sensor has an input either lower than the lower range or higher than the upper range, the output saturates. If the device fails, the milliamperes DC value goes to either the Alarm Low value or the Alarm High value based on the position of the fail-mode switch.

testing can proceed. Smart devices arrive with the switches in their default position. These switches are easily overlooked during installation.

USING THE COMMUNICATOR TO CALIBRATE A SMART INSTRUMENT

Proper smart instrument calibration follows a series of steps, starting with establishing communication. After the technician ensures the communicator is connected to the proper smart device and all safety hazards are identified, the calibration procedure can begin. Following communicator connection, the steps of calibration are verification of the sensor input section, configuration of the conversion section, and verification of the AO section.

Connect the HART Communicator

The HART communicator can interface to a smart device at the site of the instrument, at a junction box, termination point, or in the control room. A HART communicator can establish communication with a smart device when attached in a signal loop. **See Figure 9-14.**

ThinkSafe!
The security switch is normally turned to the on position after calibration has been completed. This prevents inadvertent changes to the program while the unit is operating.

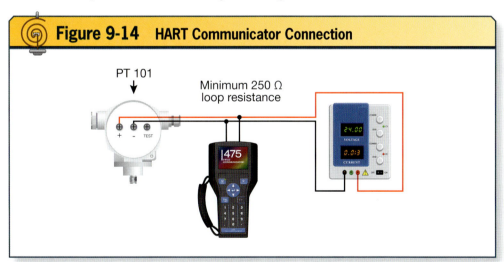

Figure 9-14. The HART communicator can be connected at multiple points in the loop.

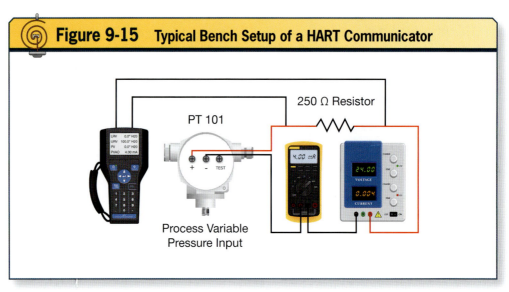

Figure 9-15. *The loop resistance is a minimum of 250 ohms of resistance. The communicator is connected across the 250 Ohm resistor.*

TechTip!
It is advisable to adjust the instrument's damping value to zero before performing tests or adjustments.

The communication leads are not polarity sensitive. When communicating between a HART device and a HART communicator, there must be a minimum of 250 ohms of resistance in the loop. **See Figure 9-15.**

Verify the Sensor Input Section

The HART device must convert the analog process variable value, such as pressure or temperature, to an equivalent digital representation. This digital representation of the process variable enables the device microprocessor to process the information. The sensor input device consists of an ADC. The technician must check to make certain that the smart device produces a digital signal that accurately represents the analog signal. If the digital signal does not accurately represent the analog value, the ADC must undergo a trim procedure.

The technician can verify the sensor input calibration by connecting a HART communicator and a suitable input calibration standard. For example, assume the device under test is a pressure transmitter measuring inches of water. The technician would connect a calibrated pressure gauge in parallel with the source pressure to the transmitter. If 60 inches of water is measured by the gauge, the communicator should show the device reports a process variable of 60 inches of water. This is not always the case. For example, if 60 inches of water is applied and the gauge indicates 60 inches of water but the communicator shows a value of 58 inches of water, this is a sensor error. The values displayed on the gauge are verified as accurate, and the values on the communicator screen are the digital representation as generated by the ADC. If discrepancies exist, the technician must navigate to the proper area of the communicator menu and perform a sensor trim procedure. **See Figure 9-16.**

The technician can perform full trim or zero-only trim. Full trim is a two-point calibration procedure in which first the LRV for pressure is applied and the device is trimmed. Next, the technician applies the URV for pressure and completes the trim procedure. This two-point trim procedure ensures the sensor input is calibrated throughout the range. Zero-only trim is a simpler, one-point process comparison. The technician applies a zero pressure input and trims the device. After the procedure, the values on both the pressure gauge and the communicator should be the same. Once this is verified, the sensor input is calibrated.

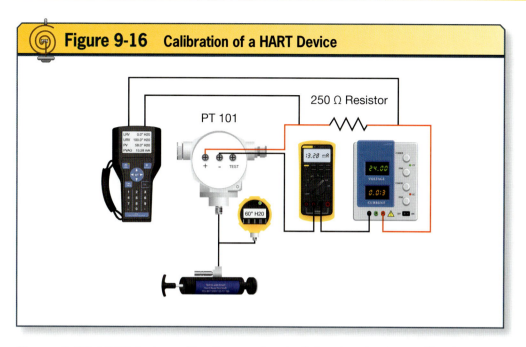

Figure 9-16. HART device calibration requires both input and output calibration standards. The HART communicator provides a working interface to the device.

Configure the Conversion Section

The conversion section of a smart instrument performs a mathematical conversion to the signal received by the sensor input section. No true calibration can be performed in this section of the device. Instead, the technician can change values such as LRV and URV, signal characterization such as linear to square root, and engineering units. In addition, the technician can alter process variable *damping*, defined as the progressive reduction in the cycling amplitude of a system. High damping values filter out process noise and unwanted fluctuations, whereas low damping values increase the response time of the instrument but can introduce rapid fluctuations of output values.

During the calibration process, the technician must enter the correct range values for the instrument in the conversion section. This is done by navigating to the proper area of the communicator menu and entering the values using the keypad. Re-ranging a smart instrument is not equivalent to calibration. Calibration can only be done by comparing the analog values on a calibration standard to the digital values shown on a communicator. The zero and span buttons on a smart device affect the way the smart device converts the input signal, but they do not replace calibration with a communicator.

Verify the AO Section

The third function of the HART device is responsible for converting the output calculated by the conversion section back into an analog signal. The conversion section of a HART device generates a digital value that represents the proper AO signal. Only after the digital value gets converted into an analog value by the digital-to-analog converter (DAC) can the 4- to 20-milliampere DC signal be transmitted on the instrument loop. The technician must check to make certain that the smart device produces an analog signal that accurately represents the digital signal. If the analog signal is not accurately generated, the device must undergo a trim procedure.

The technician must verify the AO by comparing the analog signal, as measured by a calibration standard, to

ThinkSafe!
Always verify that the engineering units of the device and those of the related process are the same (e.g., inches, pounds per square inch, or liters).

TechTip!
It is normal to have to perform digital trim for a new transmitter. If possible, digital trim should be performed after the device is installed and mounted. If this is not possible, perform digital trim on the bench before installation.

the device value, as read by the communicator. For example, a device under test could be a temperature transmitter with an AO of 4–20 milliamperes DC. With the calibrations standard and communicator connected, the values displayed on the calibration standard read 12.35 milliamperes DC. However, in this example, the communicator indicates the device is transmitting a 12.00-milliampere DC signal. This error of 0.35 milliamperes DC is significant, and it must be corrected by performing AO trim. To perform this trim procedure, the technician should navigate to the proper area of the communicator menu and select analog trim. AO trim procedures are commonly done at the LRV and the URV, typically 4 and 20 milliamperes DC, respectively.

Example of a Typical Installation Sequence for a Smart Device

The installation procedure for a HART device is similar to the installation of a conventional device. The following example illustrates a step-by-step approach to bench calibrating and installing a HART pressure transmitter. **See Figure 9-17.**

The example assumes a new device is to be configured and installed. If the device has been previously configured, the technician must perform an as-found test before making changes to the device.

The technician begins by connecting the pressure transmitter to an appropriate power supply and then connecting the HART communicator. The loop should have a minimum of 250 ohms of resistance; the technician may need to place a resistor in line with the loop to achieve this value. After turning on the communicator, the device appears on the communicator screen. Configuration of the device requires the process variable units to be set—in this example, pounds per square inch. Next, the input LRV and URV are set. The output type is set to linear output for this example. Finally, the damping value is set to the value required by the procedure. The device configuration is now complete, and verification can begin.

Device verification involves both the sensor input section and the AO section. The technician connects a pressure source from a calibration standard and verifies the input section on the communicator. Next, an AO calibration standard is connected. The milliamperes DC value shown on the standard should read the same as that on the communicator. Small errors may be acceptable within the accuracy tolerance specified by the calibration procedure.

Once the verification is complete, the device must be field installed. The fail-mode and security switches are set to the appropriate settings. Then, the transmitter is ready to be mounted per the manufacturer's specifications.

Next, the process wiring can be completed to the device. During loop checking of the process, the instrument is powered. In addition, the process connections are checked to verify correct operation. The device is checked to ensure the configuration is correct. The device may require trimming if the mounting position affects the accuracy of the transmitter. Finally, the transmitter is ready to be put into service during process start-up.

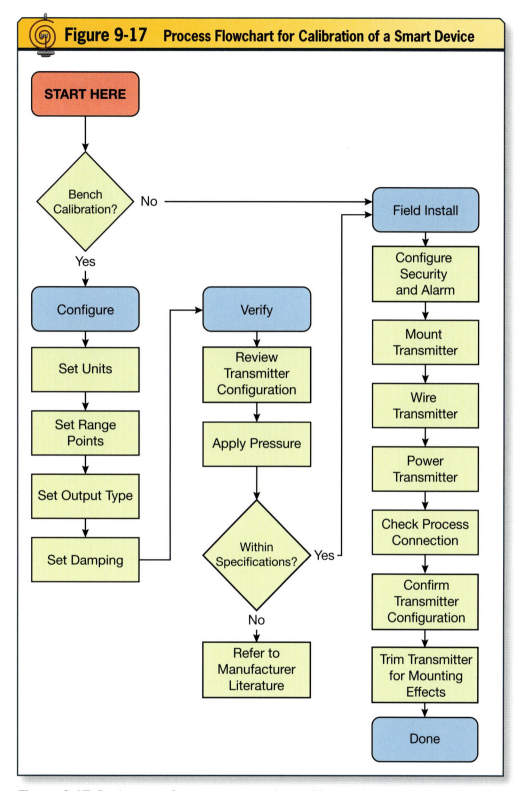

Figure 9-17. Device manufacturers commonly provide step-by-step device calibration instructions.

SUMMARY

Smart devices have several advantages when compared to conventional devices. The technician must have a working knowledge of the device and how it operates. The use of smart communicators for calibration and troubleshooting is common in an industrial environment. Too often, calibration procedures are not performed correctly because the technician lacks understanding of the calibration procedure and of the communicator's operation. A technician who can describe the connection, functions, and operation of the communicator is able to assist in the calibration and troubleshooting process. Knowing how to enter, retrieve, and interpret the data using a smart communicator is key to determining whether a device is calibrated correctly and performing correctly.

REVIEW QUESTIONS

1. A smart device operates based on a(n) __?__.
   a. continuous communicator signal
   b. internet connection
   c. microprocessor
   d. wireless signal

2. The __?__ protocol is used for smart devices.
   a. Foundation Fieldbus
   b. HART
   c. Profibus
   d. All of the above and more (many protocols exist)

3. When does a smart communicator require calibration?
   a. Never
   b. On a 2-year cycle
   c. Only when the batteries are replaced
   d. Yearly

4. The HART protocol operates using the __?__ method.
   a. Digital-only communication
   b. Duplex signaling
   c. Frequency shift keying
   d. Superposition

5. What is the minimum loop resistance for communication with a HART device?
   a. 4–20 Ω
   b. 20 Ω
   c. 100 Ω
   d. 250 Ω

6. Which section of a HART device converts an analog value to a digital value?
   a. 250-ohm resistor
   b. ADC
   c. Conversion section
   d. DAC

7. What is the purpose of switches on a HART device?
   a. Dampen process variations.
   b. Determine range values.
   c. Set security and alarm settings.
   d. Set the device address.

8. A pressure transmitter under test has 235 inches of water applied by a traceable pressure standard. The HART communicator indicates a pressure of 243 inches of water. What should the technician do in this situation?
   a. Adjust the pressure until the communicator reads 235 inches of water.
   b. Send the HART communicator out for recalibration.
   c. Trim the AO of the transmitter.
   d. Trim the sensor input of the transmitter.

9. A temperature transmitter under test has an output of 4–20 milliamperes DC. A traceable digital multimeter is connected and reads 7.42 milliamperes DC. The HART communicator indicates a value of 7.21 milliamperes DC. What should the technician do in this situation?
   a. Adjust the power supply until the milliamperes DC values match.
   b. Change the transmitter range until the milliamperes DC values match.
   c. Trim the AO of the transmitter.
   d. Trim the sensor input of the transmitter.

10. When sending changes to a HART transmitter with a communicator in an active process, what should the technician do to ensure the process is not affected?
    a. Isolate the transmitter using the three-valve manifold.
    b. Put the control loop in manual mode.
    c. Remove the 250-ohm resistor temporarily.
    d. Verify the process variable specific gravity has not changed.

Control Valves, Actuators, and Accessories

Many components work together in a control valve assembly to provide accurate flow control. The components that make up a control valve assembly may not be physically mounted on or with one another, but each component plays a crucial role in the function of the control valve. The principles of pressure and flow are essential to describing how a control valve regulates flow in a process line. Understanding the principles used to provide actuation and indication enables proper installation and maintenance of control valve actuators, pneumatic supplies, valve position indicators, and general control loop devices associated with actuators and indicators. Control valves, actuators, and several of their accessories are the foundation for control of processes in safe, productive, and profitable modes of operation.

Objectives

- » Explain the operation of a sliding-stem valve and a rotary valve.
- » Identify and give the purpose for sliding-stem and cage-guided valve assemblies.
- » Describe the functions of an actuator, a regulator, a transducer, and a positioner.
- » Determine the proper operation of a linear and a rotary actuator.
- » Describe the operations of direct- and reverse-acting actuators.
- » Explain the operation of a valve positioner.
- » Explain the operation of the nozzle-flapper system.
- » Explain valve failure modes and how they affect safety.
- » Describe the purpose and use of pneumatic volume boosters.
- » Describe the purpose and operation of valve position-indicating devices.

Chapter 10

Table of Contents

Control Valve Assembly Components 210
Principles of Pneumatics 210
Types Of Control Valves 211
    Sliding-Stem Valves 213
        Globe Valves 214
        Gate Valves 216
    Rotary Valves 216
        Butterfly Valves 217
        Ball Valves 217
Types of Actuators 218
    Pneumatic Actuators 219
        Diaphragms 219
        Direct- and Reverse-Acting Actuators 221
        Piston-Type Actuators 221
        Rack and Pinion Operators 222
    Electric (Motor-Operated Valve) Actuators 223
    Hydraulic Actuators 224

Types of Accessories 225
    Solenoids 226
    Pneumatic Regulators 226
    Nozzle-Flapper Mechanism 227
    I/P Transducers 228
    Pneumatic Boosters and Amplifiers 230
    Position-Indicating Devices 231
        Limit Switches 231
        Reed Switches 231
        Potentiometer Position Indicators 232
        Linear Variable Differential Transformers 232
    Valve Positioners 233
        Pneumatic Positioners 235
        Electropneumatic Positioners 236
        Smart Positioners 236
        Calibration 237
Summary 238
Review Questions 238

For additional information, visit qr.njatcdb.org
Item #2690

CONTROL VALVE ASSEMBLY COMPONENTS

A *control valve* is a device, other than a common, hand-actuated on/off valve or self-actuated check valve that directly manipulates the flow of one or more fluid process streams. Put more simply, the control valve regulates the flow in a process line. As a change in flow occurs, pressure, temperature, and level may be affected in the process. The control valve is normally used as the final control element in a process control loop.

A control valve is positioned so that a process flow is regulated to a desired rate. A control valve is not a device that can typically operate under its own guidance. The position of the control valve is usually determined by a controller. An *actuator* is a part of the final control element that converts a signal into a forced action. It is used to provide the necessary force to provide movement to the valve. The addition of a component called a *positioner*, a controller that is mechanically connected to a moving part of a final control element or its actuator and that automatically adjusts its output to the actuator to maintain a desired position in proportion to the input signal, provides a feedback loop, enabling the valve actuator to provide precise valve movements in response to the input signal. **See Figure 10-1.**

The interacting components of a pneumatic control valve arrangement can include the regulator, valve actuator, transducer, positioner, and valve body. The supply pressure to the valve assembly is controlled by a regulator. The valve *diaphragm* is the sensing element consisting of a membrane that is deformed by the pressure differential formed across it. The valve actuator is used to provide movement to the control valves. The transducer converts the controller's position signal, usually 4–20 milliamperes direct current (DC), to a force used by the valve diaphragm, usually 3–15 pounds per square inch of pressure.

PRINCIPLES OF PNEUMATICS

Pneumatics is the application of a compressed gas as fluid power to provide motion. In mechanics, *force* is the push or pull along a straight line. For fluid systems, force can be determined by the equation:

$$\text{Force} = \text{Pressure} \times \text{Area}$$

Force units include the pound (English system) and the newton (international system of units [SI]).

Perhaps the most common application of pneumatics is to use the force of a gas (e.g., air) to move a valve diaphragm or piston. In addition to providing movement to valve diaphragms or pistons, pneumatics can be used as a signaling medium.

In a pneumatic signaling system, data are carried by the pressure of a gas in a pipe. The pressure in a closed system is the same at all points, according to Pascal's law. If a length of pipe is used to transmit a pressure signal, increasing the pressure at one end causes the pressure

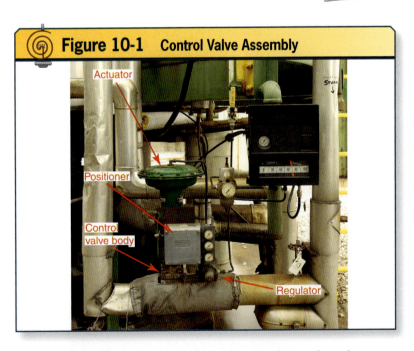

Figure 10-1. The valve actuator provides motion to the valve stem. The valve positioner measures valve movement and makes adjustments to ensure the valve position is correct based on the input signal. The regulator adjusts incoming air pressure to a stable, preset value.

to change down the length to the other end, where it also raises the pressure. This method of conveying information is slower than using electrical signals. The change of pressure travels down the length of the pipe close to the speed of sound (about 1,082 feet per second). For comparison, electric current propagates along a conductor close to the speed of light, which exceeds 983 million feet per second. For some process applications, this time delay is acceptable. However, more rapid responses are desirable in many cases. For these applications, electronic signaling is used to transmit long distances, and the signal is converted to a pneumatic signal close to the pneumatic valve actuator.

A pneumatic type of signal was used for many years before electronic devices began to emerge. Pneumatic systems are considered to be among the safest ways to control a process for some applications. They are still used when it has not been cost-effective to remove the system or when the threat of sparks from an electrical signal does not warrant a change. For most applications, the gas used is dry instrument air, nitrogen, or another inert gas for explosive hazardous areas. Signal information usually has been adjusted to a range of 3–15 pounds per square inch.

TYPES OF CONTROL VALVES

Control valves are used in industry to regulate a process by controlling the rate of flow or supply. By adjusting the opening in the valve assembly through which the controlled process fluid flows, the valve regulates the volume flow rate, which is calculated using the following equation:

$$Q = V \times A$$

The controller provides a control signal that the valve uses to achieve the desired flow rate. **See Figure 10-2.**

Several types of valves exist to suit the wide-ranging applications of different industries. Sliding-stem and rotary valves are probably the most common types. Each valve type has its own procedures for valve disassembly, assembly, and repair. Still, procedures vary among manufacturers for the same type of valve. The manufacturer's guidelines, along with specific requirements for the site where the technician works, should contain the necessary information to evaluate and repair the valve body. **See Figure 10-3.**

Figure 10-2. This sliding-stem control valve has a valve actuator, a regulator for incoming air, and a positioner to provide accurate positioning.

TechTip!
Always verify the control signal and control pressure first when troubleshooting a failed control valve. Typically, it is a failure of one of these two components that causes failure of valve operation.

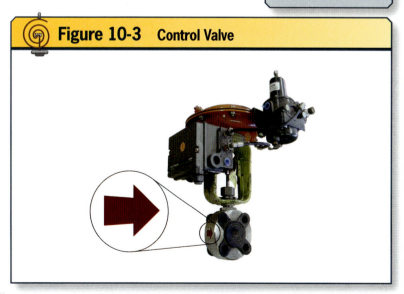

Figure 10-3. The red arrow shows the flow direction on the valve body. Many valves are designed to be installed in a single direction.

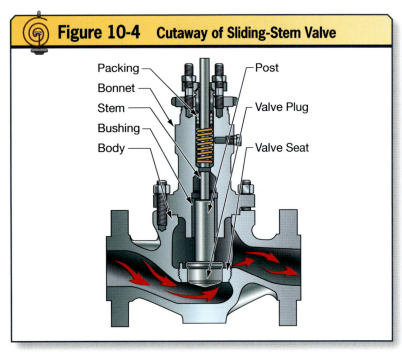

Figure 10-4. The valve body, seat ring (valve seat), valve plug (closure member), stem, packing, and valve bonnet are among the components indicated on this sliding-stem control valve.

All control valve types share some common terminology and components. **See Figure 10-4.** The *body* forms the framework that surrounds all internal components joined together inside of a valve. It is the largest structural component of a valve. The valve body provides a means of installing the valve to a piping system by use of the end connections. The control of internal fluid in a valve is provided by the various passages that are machined or forged within the valve body. The valve body also acts as one of the primary pressure boundaries used to contain and control the flow of fluids within a piping system.

The *seat* provides the sealing surface for shutting off or controlling the flow of a fluid within the valve itself. It is usually an immovable part of the valve body. The valve seat may be made of metal or a soft elastomer material such as Teflon or rubber. It may be an integral part of the valve body (permanently cast or welded), or it may be a removable and replaceable item. In some instances, the valve seat (or seat ring) may be hard-faced or made with a different material than the valve body casting to resist damage that can be caused by corrosion or erosion.

The *closure member* is the movable part of a valve that forms a seal against the valve seat to control the flow of fluid. Depending on the type of valve, the closure member may be a plug or a disc that uses linear (up and down) or rotational motion to modulate flow. The closure member can be made of a variety of materials similar to those used for the valve seat. However, if the seal is made of a soft material, the closure member is usually constructed of metal, and the opposite is typical if the valve seat is made of metal.

The *stem* is the device that transmits the motion of the valve actuator to the disc or plug, thereby raising, lowering, or turning it. The stem is connected to the valve actuator (handwheel, motor, etc.) at one end and the closure member at the other. Normally, the stem is constructed of a relatively tough corrosion resistant metal such as stainless steel.

The valve *post* is a rigid section connected to or integral to the stem. The post is aligned by the valve *bushing*, a precision machined guide that provides rigidity and aligns the stem throughout the stem travel. These two components work together to prevent lateral deflection and provide overall structural integrity.

The *stem sealing device*, commonly called the packing, provides leakproof closure around the stem of the valve while still allowing the stem to rotate or move linearly. This leakproof seal is located where the stem penetrates the bonnet of the valve. Although packing is the primary method of sealing around a valve stem, O-rings may be used on valves designed for use on low-temperature and low-pressure piping systems.

The *valve bonnet* provides the pressure boundary for the upper portion of the valve and allows leakproof closure of the valve body. It is attached to the upper portion of the valve body. The valve bonnet contains the stuffing box and packing, which together form a seal around the valve stem. The valve bonnet also provides support for the upper valve components, such as the valve actuator and the stem.

All valve types are required to provide leak-free flow control and ideally will provide years of service with minimal maintenance requirements. The selection of the proper control valve ensures the process will run at optimum efficiency and produce products to the specifications required.

Sliding-Stem Valves

A sliding-stem valve controls process fluid flow rates by the movement of a closure member in a linear motion. The closure member is typically a gate, plug, or diaphragm member that throttles fluid through the valve body. The linear motion can be provided manually through a handwheel and threaded stem or automatically by a valve actuator. Sliding-stem valves come in different body styles. The sliding-stem valve may have a conventional body, slant body, or angle body. The different valve body types have different applications. For example, the angle-body valve is used when space is limited, because the valve serves as both an elbow and a flow control device. Each type of sliding-stem valve has advantages and disadvantages. **See Figure 10-5.**

A common concern with all sliding-stem valves is seen by examining the stem. By its nature, the stem must travel through the sealing point of the valve body. A tight seal must be maintained, yet it must allow the stem to travel with minimal resistance. A packing material is typically used, though some valves use a metallic bellows to provide high levels of sealing. The packing material is a frequent source of valve issues and must be maintained regularly to prevent leaks or excessive friction.

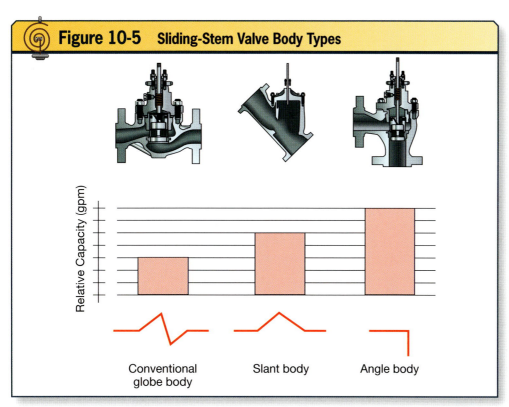

Figure 10-5. The globe valve on the left has the lowest capacity. This is because of the complicated flow path. The slant body and angle body provide higher flow rates for a given valve body size. The angle body also serves as a pipe elbow to reduce space requirements.

Figure 10-6 Single-Seated Globe Valve

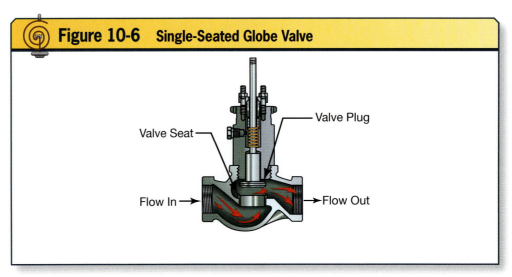

Figure 10-6. The single-seated globe valve has a single path for fluid to pass through the valve.

Globe Valves

The globe valve is a common style of sliding-stem valve that gets its name from the chamber of the valve body through which the process flows. A globe valve body can be either single seated or double seated. A single-seated globe valve is used when a tight shutoff is required. Single-seated globe valves are usually top guided. In a top-guided globe valve, the parts that guide the stem are located in the top of the valve. In a single-seated globe valve, the process fluid flows through a single valve opening. **See Figure 10-6.**

A double-seated globe valve is generally top and bottom guided. There are two paths for the divided fluid to flow. One part of the fluid flows through the upper valve opening, and the remainder flows through the lower valve opening. The advantage of this double-seated construction lies in the reduction of required actuator forces, but it is difficult to fully close a double-seated globe valve. As a result, these valves are not generally used for a tight shutoff. **See Figure 10-7.**

Stem-guided globe valves, where the sliding stem is supported by a bushing, take up less space inside the valve chamber and lead to less flow restriction. Stem-guided globe valves are relatively open and allow easy flow through them. For slurry and gritty flow media, stem-guided globe valves are good performers. Stem-guided types of valves are typically used at a lower operating pressure than the next type of valve: the cage-guided valve.

Figure 10-7 Double-Seated Globe Valve

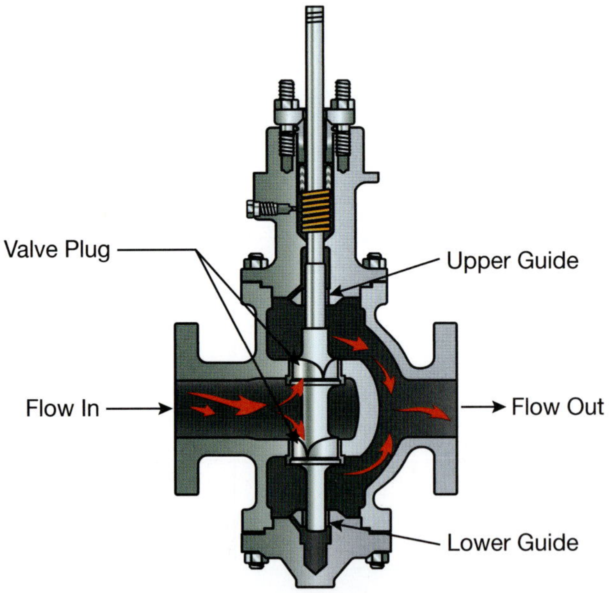

Figure 10-7. The double-seated globe valve has two paths for fluid to pass through the valve.

The cage-guided valve is another type of sliding-stem control valve. Most cage-guided valves have a rounded valve body, valve bonnet, stem, plug (closure member), valve seat, and packing assembly. The difference from the previously described designs is the cage-guided valve has an additional part, called the cage, which helps to guide the movement of the stem and plug. This part is usually a hollow metal cylinder. The valve stem and plug move through the cage as the valve opens and closes. There are ports in the walls of the cage to allow process fluid to flow through the valve opening. **See Figure 10-8.**

The cage design allows an even alignment throughout the entire range of the stroke of the sliding-stem valve. The stem-guided valves tend to be pushed to one side when subjected to higher pressures and volumes, but the cage assembly has the force spread out to cover the entire trim area of the valve, and valve seating is not endangered. The cage provides a means for rigidly seating the valve without the sideways push that is present to prevent chattering or noise.

Trim refers to the internal valve components that come into direct contact with process fluid. The cage-guided valve's trim can be changed without removing the valve body from the process piping. The cages are placed into the valve body through a top entry point. This makes it easy to inspect the valve and remove various components without removing the entire valve body from the process line. **See Figure 10-9.**

The cage openings through which flow takes place can be changed, which leads to a change in the valve's characterization. Valve characterization is the relationship between the valve stroke position and the equivalent flow. The sliding-stem control valve relies on differently shaped valve plugs to characterize flow. In contrast, the cage-guided valve relies on differently

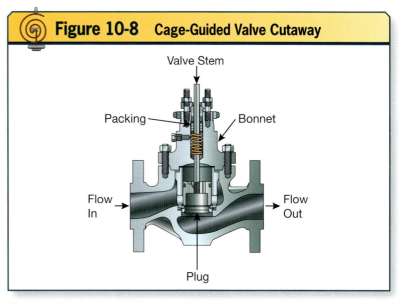

Figure 10-8. The addition of the cage to guide the valve plug leads to a more robust valve assembly.

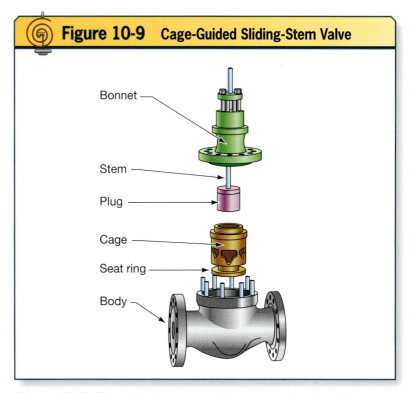

Figure 10-9. The valve bonnet may be removed and the trim may be changed on a cage-guided valve without removing the valve body from the process.

TechTip!

Stem-guided valves are typically used in low-pressure applications. Cage-guided valves are used in higher-pressure applications.

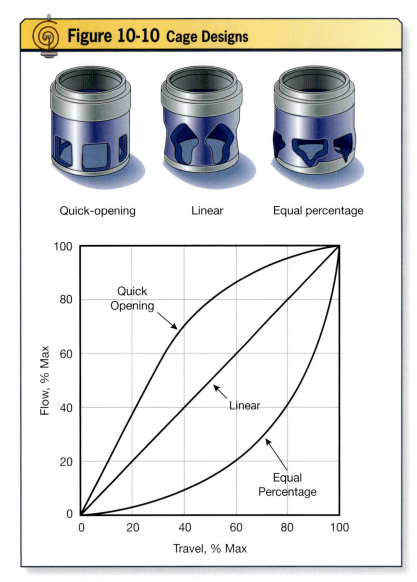

Figure 10-10. The cage designs have different-shaped cutouts. Each provides a different flow response based on the valve plug position.

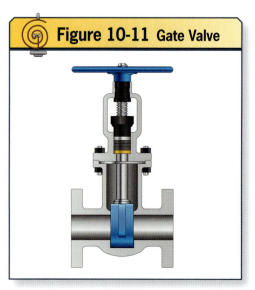

Figure 10-11. The gate valve provides tight sealing for on/off applications.

shaped cages to characterize flow. **See Figure 10-10.**

The quick-opening cage is used for on/off applications and results in high flow with relatively little valve plug motion. The linear cage causes flow to change linearly in relationship to plug travel. The equal percentage cage is used extensively in process control because of a unique characteristic: An equal increment of valve lift produces an equal percentage in flow change with an equal percentage trim cage. This response of the equal percentage trim helps to offset pressure losses in the system.

 TechTip!

Typically, valves are sized to run between 25% and 75% open at minimum and maximum normal process flow conditions.

Gate Valves
Gate valves are typically used to provide a positive, tight-sealing shutoff valve. The linear motion of the stem causes the gate assembly to lower into the flow profile and block the passage of material. Gate valves can be either manually operated or operated via a valve actuator. **See Figure 10-11.**

Rotary Valves
A rotary valve controls flow by moving a closure member in a circular plane. Many rotary valves have full travel of 90 degrees, leading to the alternate name of quarter-turn valve. **See Figure 10-12.**

The two primary types of rotary valves are the butterfly and the ball. Rotary valves can be manually operated or controlled with a valve actuator. Identifying a rotary valve can be done by observing the actuation of the valve. In contrast to a sliding-stem valve that uses a linear movement, the rotary valve requires a circular movement to operate. Many rotary valves use a linear actuator that is modified to convert a linear force to a rotational force. A gear box assembly is employed. Rotary valves have the distinct advantage of creating less pressure loss with greater flow in the system when compared to sliding-stem valves. Rotary valves also

Figure 10-12. This rotary valve has a butterfly closure member operated by a handwheel. The closure member has a maximum of 90 degrees of travel, leading to the alternate name of quarter-turn valve.

can be manufactured to control flow in large process lines, some in excess of 15 feet in diameter.

Butterfly Valves

Butterfly valves get their name from the operation of the closure member. A circular disc is used to rotate it across the opening so that it is parallel to fluid flow when fully open and perpendicular to fluid flow when shut. A butterfly valve can be used for both isolation and throttling purposes. The simplicity of the design, the variety of designs available, the ease of maintenance, and few moving parts are all reasons to use a butterfly valve when possible. The butterfly valve is also lightweight, space saving, and low cost, all with good flow control. The actuator of a butterfly valve is usually perpendicular to the shaft. In a butterfly valve, when the valve actuator receives a signal to open or close the valve, the actuator stem force is transferred to the closure member through a gear box or linkage mechanism. **See Figure 10-13.**

Ball Valves

Ball valves are commonly used for on/

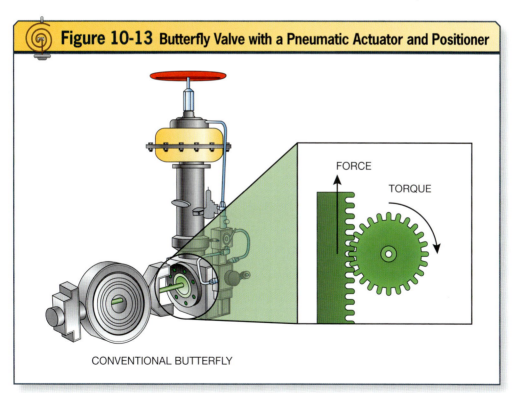

Figure 10-13. This butterfly valve has a pneumatic actuator that provides linear movement to a stem. Inside the gear box mechanism, the linear motion is converted to a circular motion that operates the closure member.

Figure 10-14. The full-flow ball valve is used for on/off service. The design provides a tight seal when actuated with a quarter turn of movement.

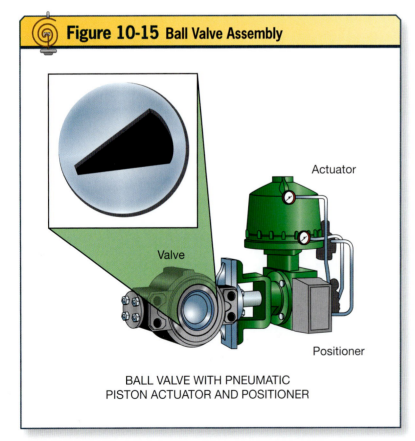

Figure 10-15. The ball valve assembly shows an actuator, a positioner, and the valve body. The ball has a V-shaped notch. This shape provides precise throttling control in a range of flow characteristics.

off applications, although some designs are used for throttling control. A full-flow ball valve is best suited for on/off service. The minimal flow disturbances that are introduced when a full-flow ball valve is fully open means the process can flow as though the valve were not present. These valve types provide tight-sealing shutoff capabilities. **See Figure 10-14.**

A second style of ball valve is the V-notch ball valve. This style is used for throttling control and can achieve precise flow rates across a range of positions. Instead of having a full-bore opening in the ball closure member, the ball has a V-shaped opening machined into the bore. These shapes can provide linear or equal percentage flow characteristics. A variety of V-notch styles can be suited to a range of process applications, making the V-notch ball valve a flexible valve type for throttling control in a process line. **See Figure 10-15.**

TYPES OF ACTUATORS

The valve actuator is the device that provides the force to open and close control valves. The most common valve actuator power source is pneumatic air pressure, though electric and hydraulic actuators are also used. The process conditions often dictate that a valve actuator produce large amounts of force to move that closure member. The selection of a valve actuator depends on several factors, such as the required force, failure mode, reliability, and availability of compressed air or electrical power. The failure mode of a valve actuator must be carefully considered to prevent catastrophic consequences in the event of power loss or valve malfunction. If the valve in question was controlling the supply of fuel to a burner, the valve should fail closed, preventing property damage or personal injury that could be caused by overheating. Conversely, if the valve was controlling the flow of cooling water to a exothermic process, , the valve should fail open, thus avoiding overheating. The failure mode can be fail open, fail closed, fail indeterminate, or fail locked. Fail-open and fail-closed valves typically use springs to cause the movement of the valve to the fail position. A fail-indeterminate valve will not fail in any specific position and may drift over time. A fail-locked valve will fail in the last position and will not move from that last position. The appendix contains a list of valve symbols including the standard symbols for valve fail states.

Pneumatic Actuators

The pneumatic actuator is a proven performer. This method of conversion of a signal to a force has been used for many years whenever mechanical force is needed to position a valve. In a pneumatic actuator, air pressure force is applied to a valve diaphragm or piston. The valve diaphragm or piston acts as a force multiplier, resulting in a large amount of force. By knowing the spring force (F_S) opposing the valve diaphragm or piston and knowing the supply pressure applied, the technician can calculate a valve position. Force equals pressure times area. The force of a pneumatic actuator (F_A) is determined by taking the surface area of the valve diaphragm or piston and multiplying it by the force applied.

For example, imagine a technician has a 10-inch circular diaphragm with 11 pounds per square inch applied and a 150-pound spring force. The technician needs to find the surface area (A) of the valve diaphragm:

$A = \pi R^2$

$A = \pi 5^2$

$A = 78.54 \text{ in}^2$

Next, the technician needs to find the force generated, or the actuator force:

$F_A = 78.54 \text{ in}^2 \times 11 \text{ psi}$

$F_A = 863.94 \text{ lbs}$

TechTip!
The actuator nameplate can be used to determine the safe operating pressure.

The technician then needs to find the force applied to valve stem (F_{Stem}):

$F_A = 863.94 \text{ lbs}$

$F_S = 150 \text{ lbs}$

$F_{Stem} = 863.94 - 150 = 713.94 \text{ lbs}$

As shown, the relatively low standard air pressure signal of 3–15 pounds per square inch can achieve a great force by using a valve diaphragm.

Diaphragms

The valve diaphragm is a flexible disc of natural or synthetic rubber containing one or more fabric inserts to provide additional strength. **See Figure 10-16.**

The valve diaphragm is mounted between the flanges of the cases. Its purpose is to convert the pressure from the output of a current-to-pressure (I/P) or pneumatic controller into a force for moving the stem of the valve. By applying a set pressure on one side of the valve diaphragm, the force becomes stronger than the opposing spring and valve stem movement occurs. It then moves an attached actuator arm that is connected to a valve. The compression

ThinkSafe!
Always verify that a valve actuator is supplied with air before venting. In some instances, nitrogen is used to actuate. In sufficient quantity, nitrogen can cause suffocation.

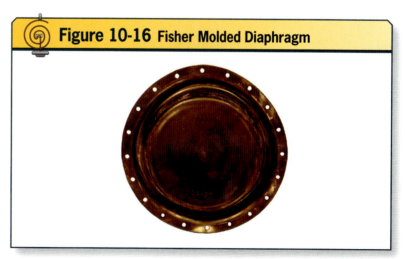

Figure 10-16. A valve diaphragm is a flexible membrane used to amplify force to move a pneumatic actuator.

ThinkSafe!

Many pneumatic devices are spring loaded. Always use extreme caution when removing mounting bolts, because great force can be exerted as the spring tension is relieved. This force can cause serious injury and equipment damage if not planned for and alleviated.

Figure 10-17 Components of a Diaphragm-Type Actuator

- Loading Pressure Connection
- Diaphragm Casing
- Diaphragm and Stem Shown in up Position
- Diaphragm Plate
- Actuator Spring
- Actuator Stem
- Spring Seat
- Spring Adjustor
- Stem Connector
- Yoke
- Travel Indicator
- Indicator Scale

Figure 10-17. *The valve diaphragm is placed between the top and the bottom cases. Air is applied through the 1/4-inch national pipe thread (NPT) fitting to apply pressure to the top side of the valve diaphragm. This pressure opposes the spring and causes the valve stem to extend.*

of the spring is set to restrict the amount of diaphragm movement for a specific input pressure. **See Figure 10-17.**

Valve actuator stem movement is set so that a full range of motion is obtained. A process called *bench set*, the initial compression placed on the actuator spring with a spring adjuster with the actuator uncoupled from the valve, is performed to ensure the valve actuator can travel the full distance in relation to the input signal. By altering spring compression, the initial linear motion of the valve actuator is determined. Actuators may have a travel stop device that physically limits stem movement in one or both directions. These travel stops must be adjusted to allow full movement of the valve assembly. The bench set should be rechecked once the valve–actuator assembly is mounted, because mounting may alter the previous adjustments. Valve manufacturers generally provide step-by-step literature to ensure the bench set procedure is performed correctly.

The device supplying the pneumatic signal to the valve actuator is typically calibrated for a range of 3–15 pounds per square inch, but the valve can be bench set to operate at different ranges of pressure. The amount of force stored in the spring is a direct linear function of its compression in inches, providing a linear relationship between input pressure change and valve travel. The total limit of travel can be limited by adjusting the stops located on the valve stem. A spring adjuster on the stem establishes the initial amount of input pressure necessary to initiate movement of the valve stem. This is done by compressing, or preloading, the spring with a certain amount of force, which requires an initial amount of input pressure before the valve stem moves.

For the common range of 3–15 pounds per square inch, the adjuster fixes the point of 3 pounds per square inch and the spring design fixes the desired travel for the output of 15 pounds per square inch. Changing the adjuster does not change the spring

TechTip!

Technicians should always break the bolts opposite them first when removing a valve from a process line. This allows trapped pressure or process fluid to be discharged away from the technician.

TechTip!

If the technician is asked to troubleshoot a diaphragm valve that is unresponsive when the proper air signal is applied, the diaphragm may have ruptured (developed a leak or hole) inside the case.

characteristics. The travel indicator is a metal disc mounted between the lock nuts, which lock the valve stem to the stem extension. The stems are joined by means of a machine screw thread. In conjunction with the travel indicator scale, it indicates the position of the valve plug (closure member) through the full range.

A diaphragm actuator may be a direct- or a reverse-acting type. The decision of which type to use must take into consideration the required valve movement and the required failure mode.

Direct- and Reverse-Acting Actuators

The direct-acting actuator causes the valve stem to extend from the valve actuator in response to an increase in pressure. A direct-acting actuator has springs that cause the valve to fail in the retracted stem position. In contrast, a reverse-acting actuator causes the valve stem to retract when air pressure is increased. Therefore, the failure mode of a reverse-acting actuator is a fully extended stem. Four distinct configurations use a combination of direct- and reverse-acting actuators and direct- and reverse-acting valve bodies. Direct-acting valve bodies cause the closure member to close when the stem is extended. In contrast, reverse-acting valve bodies cause the closure member to open when the stem is extended. **See Figure 10-18.**

Piston-Type Actuators

A piston-type actuator is another type of valve actuator. Theses valve actuators use a piston that is contained within a cylinder housing. A sealing ring prevents pneumatic pressure from leaking by the piston along the cylinder walls. A piston-type actuator normally is used when very large forces must be generated to overcome the process fluid forces. Because of its construction, a piston-type actuator can deliver force beyond what a diaphragm-type actuator can produce. A piston-type actuator typically is used

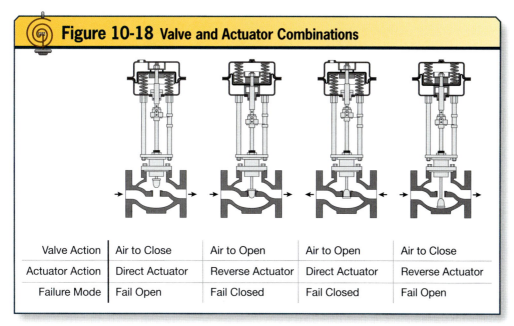

Figure 10-18 Valve and Actuator Combinations

Valve Action	Air to Close	Air to Open	Air to Open	Air to Close
Actuator Action	Direct Actuator	Reverse Actuator	Direct Actuator	Reverse Actuator
Failure Mode	Fail Open	Fail Closed	Fail Closed	Fail Open

Figure 10-18. The chart shows four combinations for actuators and valves. Notice the direction of stem travel and the spring opposition to the stem movement.

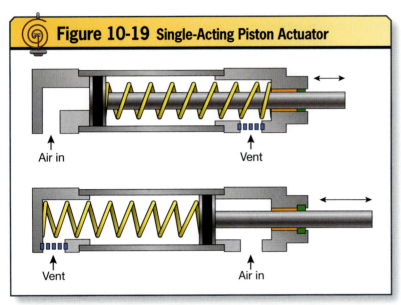

Figure 10-19. *The single-acting piston actuator on top is an air-to-extend version. The one on bottom is an air-to-retract version. Both have return springs installed.*

Single Acting. A single-acting piston actuator uses compressed air or other gas to drive the piston, causing the actuator stem to stroke in one direction. The single-acting piston actuator is typically fitted with a return spring to return the stem to the starting point. This valve actuator can be an air-to-extend or air-to-retract configuration. While the force exerted under the pressurized stroke can be high, the return stroke force is limited to the spring strength. This limitation, in addition to the limited stroke length because of the required spring return mechanism, leads to single-acting piston actuators being used primarily for light-duty and small-valve applications. The failure state of a single-acting piston actuator is typically controlled by the spring if air pressure is lost. **See Figure 10-19.**

under high-operating-pressure conditions, and extra care must be taken to ensure that the valve actuator is locked out before work is performed. These valve actuators can provide longer stroke lengths than those provided by diaphragm-type actuators, which makes them useful when actuating large valves. The use of a valve positioner is generally required when using piston-type actuators, because an increase in pressure does not always correspond to a linear change in valve stroke.

Double Acting. Double-acting piston actuators use compressed air or other gas to drive the piston in both directions of travel. In contrast to the single-acting piston actuator, which is under pneumatic power in only one direction, the double-acting piston actuator is pneumatically powered in both directions. No return spring is needed, though the manufacturer may install one to ensure the valve actuator fails in a specific position. Most commonly, the double-acting piston actuator fails in an indeterminate state, meaning the failure state is not consistent. As air bleeds from the system, the valve actuator may drift from one position to another. Double-acting piston actuators have two pressure ports. When the pressure is equal on both sides, equilibrium is reached and the piston does not move. As pressure is increased on one pressure port, the piston moves away until a new state of equilibrium is achieved. **See Figure 10-20.**

Rack and Pinion Operators

Rack and pinion operators use an assembly consisting of one or more racks and a pinion gear. Depending on the

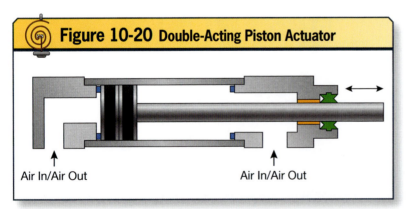

Figure 10-20. *The double-acting piston actuator has two pressure ports and is capable of traveling under pneumatic pressure in two directions.*

valve actuator design, the rack and pinion can convert a linear motion to a rotary motion or a rotary motion to a linear motion. **See Figure 10-21.**

The rack and pinion actuator is used to provide linear action to an application that requires the valve stem to be stroked in a linear manner. A motor is often the source of power to rotate the pinion gear, resulting in the rack movement.

When a rack and pinion actuator is used to provide rotary motion, such as to operate a butterfly valve, the pinion gear is affixed to a shaft that is connected to the butterfly closure member. A linear actuator is often coupled to the rack mechanism, providing the initial linear motion. Through the transfer of motion from linear to rotary, the closure member is opened or closed in accordance with the control signal. **See Figure 10-22.**

Electric (Motor-Operated Valve) Actuators

While pneumatic actuators are the predominant style used in industry today, electric motor-operated valves (MOVs) are gaining popularity because of their inherent accuracy and reduced operating costs. To operate, pneumatic

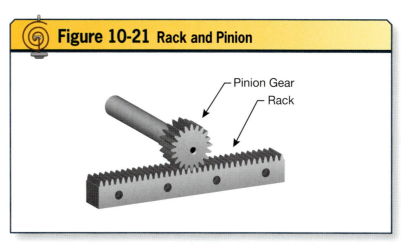

Figure 10-21. The pinion gear engages the rack, resulting in a transfer of movement to either the rack or the pinion, depending on the type of valve actuator used.

actuator requires compressed air or other gas, an expensive and maintenance-demanding resource. Pneumatic actuators also require careful attention to tubing fittings to avoid leaks that lead to inaccurate valve position. In contrast, MOVs are typically powered by a power source separate from the instrument signal loop, because the loop power has insufficient available power. The technician must exercise caution when servicing MOVs, because voltage and current that exceed standard signal

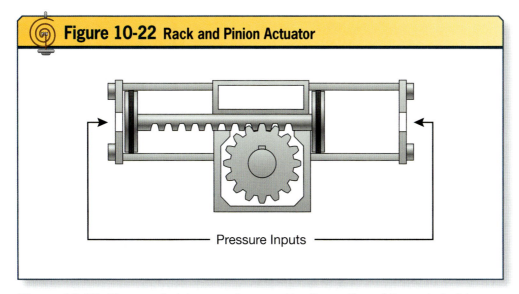

Figure 10-22. The round pinion gear is moved by the motion of the rack. In this example, the rack is connected to a double-acting pneumatic actuator capable of rotating the pinion gear in two directions under pneumatic pressure.

Figure 10-23. This MOV actuator provides linear motion to a globe valve.

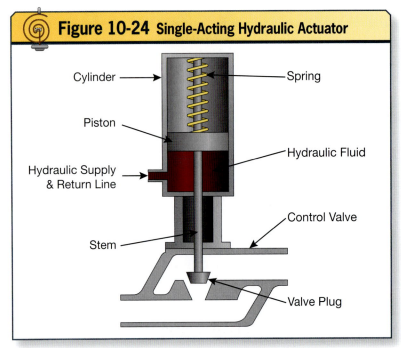

Figure 10-24. This single-acting hydraulic actuator is of the reverse-acting style and is coupled to a pressure-to-open valve. A loss of hydraulic pressure would result in the spring forcing the piston down. This means the valve assembly is a fail-closed design.

loops are commonly present. **See Figure 10-23.**

MOV actuators are used for both on/off and throttling applications. Two-position on/off valves typically use limit switches to provide *feedback*, or information about the status of the controlled variable that may be compared with the information that is desired in the interest of making them coincide, when the valve is fully open or fully closed. This prevents the motor from continuing to run and damage the gear mechanism or the motor. On throttling applications, two basic types of motors are used to ensure the closure member is properly positioned through the range of travel. These are the stepper and servo motors.

A stepper motor is a style of DC motor that rotates in a series of predefined and repeatable increments. Coupled to a gear reduction mechanism, the shaft movement can result in high resolution control. For instance, a stepper motor shaft may rotate 90 degrees for every command signal that is sent from the actuator electronics. A gear reduction of 1,000:1 would result in the butterfly closure member moving 0.09 degrees of a revolution. Because the increments are the same every time, there is no need for a feedback mechanism to ensure the valve is properly positioned.

The second style of electric motor used for a MOV is the servo motor. A servo motor, unlike the stepper motor, requires a feedback mechanism to determine the position of the motor shaft. The position of the motor shaft directly relates to the position of the closure member. The feedback mechanism sends information back to the servo motor drive circuitry, which in turn makes adjustments based on the control signal to accurately position the valve.

Hydraulic Actuators

Hydraulic actuators operate similarly to pneumatic piston actuators, with the main difference being the fluid used to apply pressure to the piston is a liquid instead of a gas. The benefit of using a hydraulic liquid is the relative

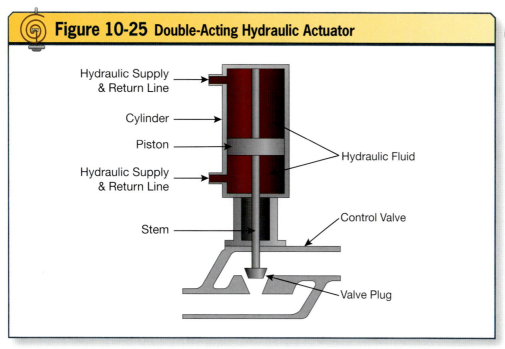

Figure 10-25. *This double-acting hydraulic actuator strokes the valve stem in two directions. A loss of hydraulic pressure would result an indeterminate fail state.*

> **ThinkSafe!**
> Verify the fail-safe position of the device before beginning work. On loss of signal or air, will the device close or open? Failure to address this issue may result in injury to the technician by pinching and crushing hazards caused by the device changing states unexpectedly. It can also disrupt the process if flow conditions are unexpectedly changed.

incompressibility. Whereas a gas shrinks in volume when subject to pressure, the hydraulic liquid maintains its original volume. This phenomenon leads to more force generated and a more linear relationship between an increase in pressure and an increase in piston stroke. A hydraulic actuator is typically used for large valves or valves that are subject to extreme process pressures.

Hydraulic actuators can be of the single-acting or double-acting stroke style. A single-acting hydraulic actuator travels under pressure in a single direction. A spring is commonly used to return the hydraulic ram to its original position. **See Figure 10-24.**

A double-acting hydraulic actuator travels under hydraulic pressure in both directions. All hydraulic actuators require a reservoir of hydraulic oil and a pump to pressurize the valve actuator. Failure modes in hydraulic actuators vary by design. A spring return may determine whether a valve actuator is a fail-extended or fail-retracted style. In addition to spring returns, some hydraulic actuators have locally mounted hand pumps to supply hydraulic pressure for emergency valve positioning. These are used in the case of an electric pump malfunction or electrical power loss. Similar to pneumatic systems, hydraulic oils leaks can cause the valve actuator to become nonresponsive. **See Figure 10-25.**

TYPES OF ACCESSORIES

The accessories attached to a control valve provide an efficient means for the control valve to maintain position. The valve body and valve actuator make up only two components of a control system. Several additional devices are required to make the system operate properly. For example, insufficient air pressure to a pneumatic valve prevents the valve from stroking its full travel. Likewise, many installations require a valve positioner to offset process fluid forces and valve friction. The technician must understand the relationships among the various components of the valve control system to install, maintain, and troubleshoot the system.

Figure 10-26 Electric Solenoid

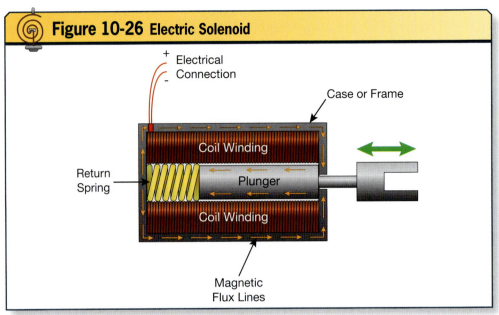

Figure 10-26. *The coil windings are energized with an electric current, creating a temporary magnet. The plunger is then pulled inward by the magnetic field, resulting in retraction of the plunger. When deenergized, the spring causes the plunger to extend to the original position.*

TechTip!

Always verify the solenoid nameplate coil voltage and current ratings. Solenoids are available in various voltages for both AC and DC sources.

Solenoids

A solenoid is a device that provides linear motion. The common electric solenoid consists of a coil of wire wrapped around a plunger made of a ferromagnetic material. When a current is applied to the coil of wire, a temporary magnetic field is produced. This magnetic field develops a north and a south magnetic pole. The plunger located inside the coil of wire is either repelled or attracted by the magnetic field and moves. When the electric current is removed from the coil of wire, the plunger returns to its original position because of spring force. **See Figure 10-26.**

Using the simple concept of a temporary electromagnet, a solenoid spool valve has been developed to direct the flow gases or liquids by the position of a spool within a chamber.

A solenoid air valve is typically used for full-open or full-closed positioning. A common application is an emergency overflow prevention system. For example, if an air-to-close control valve controls the outflow of a process tank and a high-level alarm is activated, the safety instrument system may cause the solenoid valve to vent all pressure to the valve. This will cause the valve to completely open and allow the tank level to reduce quickly. **See Figure 10-27.**

Figure 10-27 Solenoid Air Valve

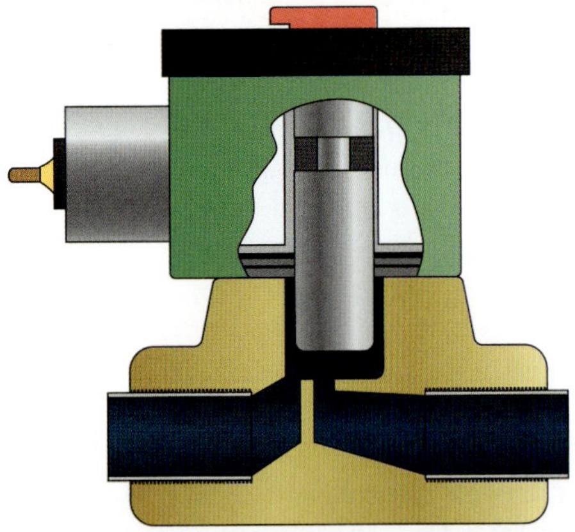

Figure 10-27. *The green housing on top of the solenoid valve contains the coil and electrical connections. The brass valve body contains the spool and inlet/outlet ports.*

Pneumatic Regulators

A regulator is used to maintain a

constant working pressure. Regulators for gas flow can be thought of as self-contained flow systems, consisting of a sensor, controller, and valve in a single unit. Most regulators use springs, cams, and diaphragms to operate with no outside power supplied. The necessity of regulators to maintain constant pressure downstream is shown by the nature of compressed air systems. Plant air is often supplied at more than 100 pounds per square inch to distribute it throughout a building. Most control valve positioners or I/P devices operate at a setting of 20 pounds per square inch. The pressure regulator provides a method of reducing the air pressure.

Several pieces make up the inner workings of a regulator. The adjustment screw sets the spring compression to a desired level. The diaphragm is deformed by the spring, allowing the poppet valve to open. As the poppet valve opens, gas connected to the inlet flows through the regulator body. As the outlet pressure reaches the desired value, internal pressure rises and the diaphragm is deformed upward, drawing the poppet valve closed and stopping gas flow. **See Figure 10-28.**

A regulator that is subjected to extreme temperatures and pressures can sustain damage to its internal workings. A damaged regulator may vent the working gas to the atmosphere, and care must be taken to ensure that the gases are not vented in a dangerous way or in a dangerous location, causing injury to personnel and contaminating the work environment.

A regulator is often ordered and received with factory-calibrated working pressures. Most regulators have a point of adjustment of working (outlet) pressure. Care should be taken not to

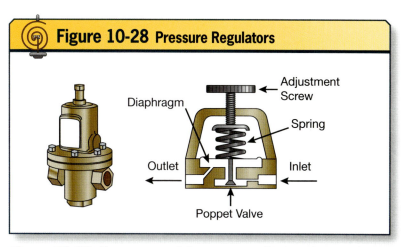

Figure 10-28. Pressure regulators use an adjustment screw to change the outlet pressure.

induce inlet pressure higher than the upper range specified on the regulator, which can cause damage. Adjusting the output pressure of a regulator requires proper ventilation of the pneumatic gas. The adjustments should be made under the same conditions as the process operating conditions. These include temperature, pressure, and by using the same gas such as clean dry air or nitrogen. The vent port of a regulator must remain open.

Nozzle-Flapper Mechanism

The nozzle-flapper mechanism is a simple yet effective method of controlling pressures inside a pneumatic system. The operation of the mechanism is easily understood by thinking of a garden hose with flowing water. Imagine a pressure gauge is installed in the run of the garden hose to show the amount of pressure developed as the fluid flows through the hose. If the gardener were to place a finger over the end of the hose, the pressure would increase inside the hose until it matches the supply

TechTip!
Most regulators are configured with one-directional control. If installed backward, the pressure will fail to regulate. The regulator body typically has a flow arrow indicating the correct flow direction.

TechTip!
One of the most common failure points of pneumatic devices is blockage of air orifices. Always verify that the airways are unobstructed when an error is encountered in these devices.

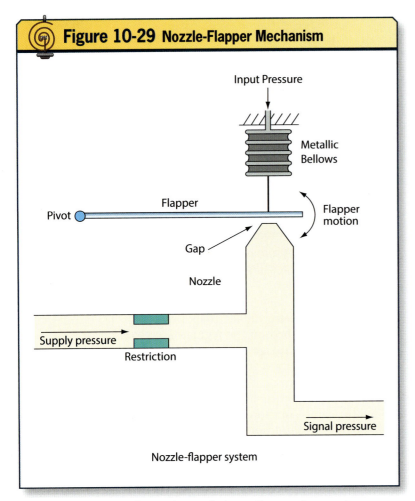

Figure 10-29. The flapper's proximity to the nozzle opening determines the amount of pressure that is developed in the system.

Figure 10-30. This I/P transducer converts a 4- to 20-milliampere DC input signal into a directly proportional output signal of 3–15 pounds per square inch.

pressure. As the gardener released the finger from the hose, the pressure inside the hose would decrease. Instead of using a finger, the nozzle-flapper mechanism uses a component known as a flapper. As the flapper approaches the nozzle, more pressure is built up inside the line leading to the nozzle. Various electrical and mechanical movements have been designed to move the flapper either closer or farther from the nozzle to control pressure. The nozzle-flapper method may be used to directly control pneumatic pressure to a valve. **See Figure 10-29.**

I/P Transducers

A common signal conversion technique used in industrial processes is the current to pressure conversion using an I/P transducer. Electrical signals travel quickly, whereas pneumatic signals have significant delays over long distances. Early process control systems were strictly pneumatic, with pneumatic tubing running from the field to the control rooms. In the control rooms, pneumatic controllers read the incoming pneumatic signals and generated a pneumatic signal based on the set point. These systems are generally considered outdated and are often replaced with electronic controllers that use electrical signals traveling throughout the plant. This presents a problem however, because most valve actuators are pneumatically controlled. To solve this problem, a transducer is used to convert the electrical signal to a pneumatic signal near the valve actuator. Transducers with the designation E/P convert a voltage signal to a pneumatic signal. Transducers with an I/P designation convert a current signal to a pneumatic signal. **See Figure 10-30.**

The industry standard for a current signal is 4–20 milliamperes DC, and the standard for output pressure is 3–15 pounds per square inch. Transducers can be installed in both direct- and reverse-acting configurations. For a direct-acting transducer, a 4-milliampere DC signal would result in a signal pressure that is 0%, a 12-milliampere

> **ThinkSafe!**
> Electric current loss to an I/P converter can cause a control device to change states or positions suddenly. Technicians should always determine the current and fail state of any device they are repairing to prevent an unplanned and unsafe condition.

DC signal would translate into a signal pressure of 50%, and a 20-milliampere DC signal would result in a signal pressure of 100%. For example, a direct-acting converter could have an adjustment so that 4 milliamperes DC would equal 3 pounds per square inch and 20 milliamperes DC would correspond to 15 pounds per square inch. A reverse-acting transducer contains the same relationship, but with the 0% pressure signal established at 20 milliamperes DC, not 4 milliamperes DC.

The transducer is a device that gives a method of converting energy of one form to another. Therefore, it provides a reliable and cost-effective method to regulate industrial process control valves.

Transducers require calibration, and the same characteristics of zero and span can be applied. Most transducers allow field adjustments to be made. It is considered good practice to verify device calibration on the bench before field installation to correct any error before installation into the process.

The working components of a device are subject to wear and deformation and therefore need periodic adjustment. These adjustments typically are made as a part of a well-managed preventative maintenance program. **See Figure 10-31.**

The same observations should be made for the transducer as for other pneumatic devices. The proper venting of bleed pressures, the venting of exhaust pressures, and general working order should be verified. Field technicians who troubleshoot or repair a transducer must have an adequate understanding of how it operates to keep the device operating at optimal efficiency.

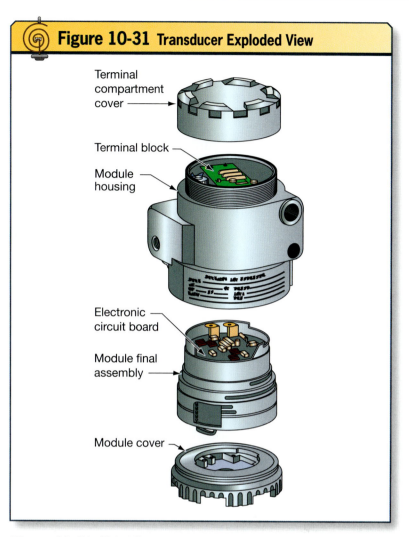

Figure 10-31. This I/P transducer assembly has a field-replaceable components and seals.

The technician can perform a simple self-test using working pressures and currents to determine an I/P transducer's operational design. The technician connects a current supply to the transducer, runs through the range of the input signal (usually 4–20 milliamperes DC), and records the output pressures of the transducer. The technician

> **TechTip!**
> Some applications require two split range I/P's, such as the case in temperature mixing valves. One I/P is commonly ranged at 4-12 mADC and the second I/P is ranged at 12-20 mADC.

should be able to determine whether the transducer is direct acting, with 3–15 pounds per square inch from 4–20 milliamperes DC, or reverse acting, with 15–3 pounds per square inch from 4–20 milliamperes DC.

Pneumatic Boosters and Amplifiers

Many control valves are actuated by a pneumatic signal that is 3–15 pounds per square inch. Pneumatic volume boosters and pressure amplifier are used in transducers and as standalone devices to allow increases in pneumatic pressure and flow. A transducer that flows a relatively small volume of air can be used with a volume booster to provide a high volume signal to the valve actuator. This increased volume capability results in much faster valve positioning. **See Figure 10-32.**

A simple analogy is to compare a volume booster to an electric relay with a 12-volt alternating current (AC) coil voltage that requires 1 ampere AC to energize. When the relay coil receives a current flow of 1 ampere AC, a set of electric contacts changes state from open to closed. Now the contacts can allow significantly more amperage to flow, based on the rating of the contacts. Instead of depending on the $1/4$-inch tubing coming from the I/P transducer to provide sufficient air volume to the valve actuator, a volume booster may have a 1-inch air line that can provide more volume in a shorter time frame.

A pressure *amplifier* is a device whose output by design is an enlarged reproduction of the input signal and is energized from a source other than the input. This device converts input signals to a higher pressure. A pressure amplifier is often used to raise or lower a specific area's controlling pneumatic pressure or to control the position of process control devices. Pressure amplifiers can increase or decrease the output signal in relation to the input signal. This relationship is determined by the device's gain number. Assume a pressure amplifier has a gain of 10, which means it increases the input signal

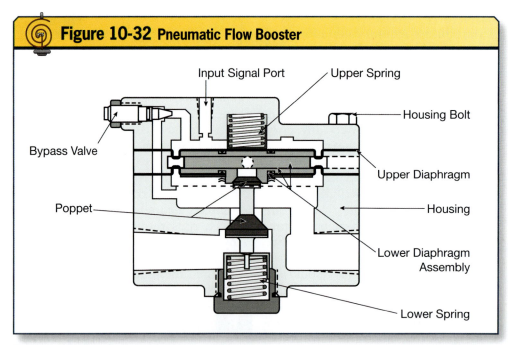

Figure 10-32. High-pressure air is connected to the supply port, and a signal that is 3–15 pounds per square inch is connected to the input signal port. As the pressure increases at the input signal port, the diaphragm assembly moves downward and the poppet valve opens. This allows additional air to leave the output port to provide a boost in airflow to a valve actuator.

10-fold. Applied to the standard operating pressure of 3–15 pounds per square inch, the result would be an output of 30–150 pounds per square inch.

Position-Indicating Devices

The internal components of a valve inside the valve body are not visible during operation. Position indicating devices provide a method to determine the position of the valve closure member. These devices can be a local indicator that a technician can visually inspect or the devices can provide remote indication of the valve status. The category of valve position devices ranges from simple discrete switches to elaborate closed-loop feedback electronic control systems. In all cases, the control loop is only as capable of maintaining the process variable at the set point as it is of accurately positioning the valve.

Limit Switches

A limit switch is a mechanical device that can be used to determine the physical position of a valve. For example, an extension on a valve shaft mechanically trips a limit switch as it moves from open to closed or closed to open positions. The limit switch gives discrete output signals that correspond to the valve position. Normally, limit switches are used to provide full-open or full-closed indications. **See Figure 10-33.**

Limit switches are mechanical devices with a finite life span. Failures are normally mechanical. If the proper indication or control function is not achieved, the limit switch will need to be inspected to determine whether the wiring or switch is defective. Caution must be taken when working near online valve actuators, because serious injury can occur from pinch points. Various types of limits switches are suitable for different environments. The technician must always make sure that the switch meets the requirements of the installation, such as for areas that are subject to frequent wash-downs or areas where hazardous materials are present.

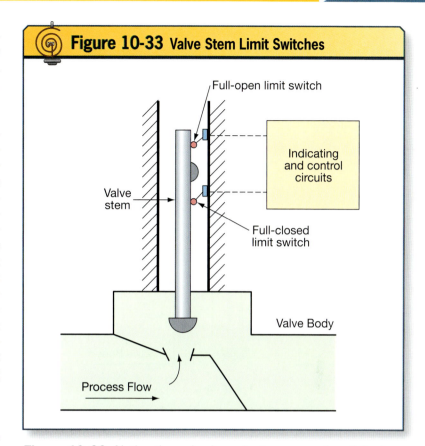

Figure 10-33. Notice the red valve actuators on each switch. As the valve stem travels to the fully closed or fully open position, the cam located on the shaft causes a discrete change of state in the contacts of the switch.

Reed Switches

Reed switches are constructed of flexible ferrous strips (i.e., reeds) that are located inside a housing. The housing is commonly a hermetically sealed chamber to prevent an arc from being exposed to the outside environment. Reed switches come in normally open or normally closed configurations. Similar to the limit switch, a reed switch can only provide a discrete signal. The reed strips are actuated when a magnet is brought near the switch. The reed switch is placed near the intended travel of the valve stem or control rod extension. A permanent magnet is mounted on the stem. As the magnet approaches the reed switch, the switch state changes. When the magnet moves away, the reed switch returns to the previous state. This on/off indicator is similar to mechanical limit switches.

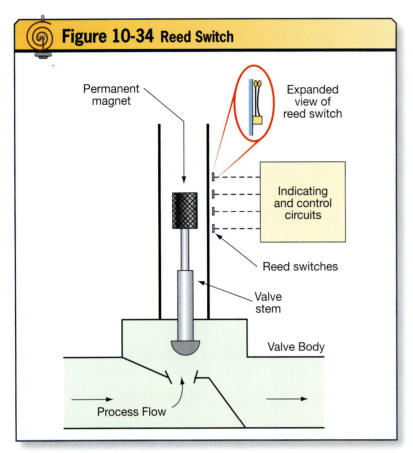

Figure 10-34. Notice several reed switches are used along the travel of this valve stem. As the permanent magnet approaches the switch, the reed contacts change state.

By using a large number of magnetic reed switches, incremental positions can be measured. This technique is sometimes used in monitoring a control valve's stem position. This approach allows greater valve position resolution and controllability. **See Figure 10-34.**

An important consideration for reed switch device is the relatively low current rating of the contacts. Failures common to reed switches are normally limited to a reed switch that is stuck open or closed. If a reed switch is stuck open or closed, a change in valve position will not elicit a change in switch indication.

Potentiometer Position Indicators

Potentiometer valve position indicators provide an accurate indication of position throughout the travel of a valve or control rod. Unlike the limit or reed switch, the potentiometer is only limited by resolution in its ability to indicate position across the valve's entire travel. *Potentiometers* provide a change in resistance proportional to the valve's travel. They come in both rotary and linear forms. In the linear form, the valve stem moves up or down and the resistance of the attached circuit changes. This change then alters the current flow in the circuit. The amount of current is proportional to the valve position. **See Figure 10-35.**

Potentiometer valve position indicator failures are normally electrical. Wear of the wiper and other electric components increase the valve movements, leading to a shorter life span. An electrical short or open circuit can cause the indication to fail at one extreme or the other. With an increase or decrease in the potentiometer resistance, an erratic indicated valve position occurs.

Linear Variable Differential Transformers

A linear variable differential transformer (LVDT) is a device that provides accurate position indication throughout the range of valve or control rod travel. Unlike the potentiometer position indicator, no physical connection to the stem is required. The valve shaft, or control rod, is made of a metal suitable for acting as the movable core of a transformer. Moving the extension between the primary and the secondary windings of a transformer causes the magnitude of induced voltage between the two windings to vary, thereby varying the output voltage proportional to the position of the valve or control rod extension. **See Figure 10-36.**

LVDTs are extremely reliable. As a rule, failures are limited to rare electrical faults that cause erratic or erroneous indications. An open primary winding can cause the indication to fail to some predetermined value. A failure of either secondary winding causes the output to indicate a full-open or a full-closed position. Other positive characteristics

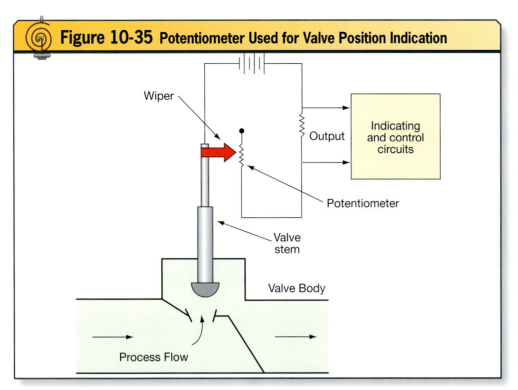

Figure 10-35. The red arrow represents the potentiometer wiper, which moves in relation to the valve stem. As the wiper moves, the circuit resistance changes proportionally.

TechTip!
Air quality can cause disruption in pneumatic equipment. Verify that the air supply is filtered and that moisture is controlled when possible.

of an LVDT are their inherent infinite resolution and friction-free design.

Valve Positioners

A valve positioner is a device used to increase the level of accuracy in relation to valve positioning. Pneumatic, analog I/P, and digital positioners are available and used in many industries. To understand how a positioner increases accuracy, the concepts of open- and closed-loop feedback must be investigated. In an open-loop system, information does not travel in a circular pattern. Instead, the controller generates an output, but there is no way of determining whether that output is achieved. In a closed-loop system, information travels in a closed loop with feedback to the controller. When the controller generates

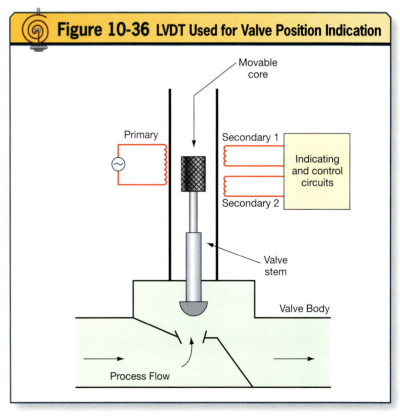

Figure 10-36. The primary winding induces an AC voltage to the two secondary coils by magnetic coupling through the movable core.

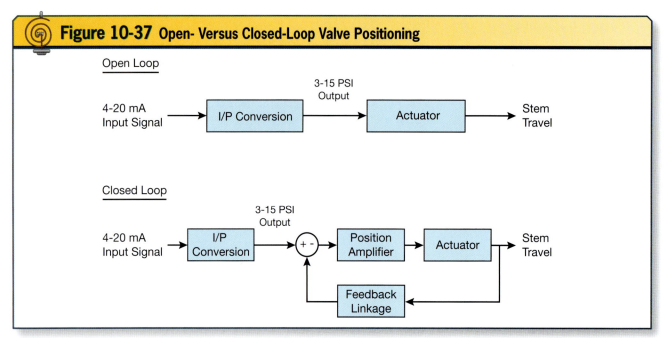

Figure 10-37. The open loop does not provide feedback to correct the valve stem travel. The closed loop incorporates a feedback linkage to ensure the valve stem travel is accurate.

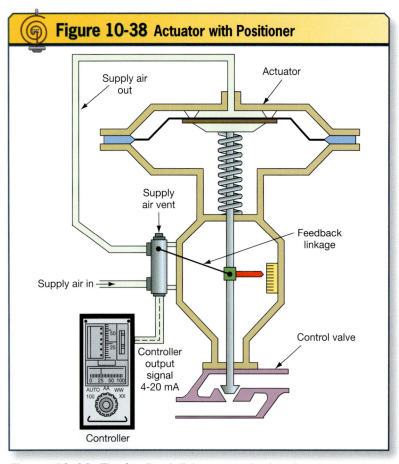

Figure 10-38. The feedback linkage attached to the stem provides information to the positioner for accurate valve travel.

an output, the closed-loop feedback verifies that output value is achieved. A valve positioner forms a complete, closed-loop feedback mechanism mounted directly on the valve actuator. The positioner ensures that the valve moves to the position called for by the controller. **See Figure 10-37.**

A positioner regulates the supply air pressure to a pneumatic actuator. It does this by comparing the valve actuator's demanded position with the control valve's actual position. The demanded position is transmitted by a pneumatic or electrical control signal from a controller to the positioner. The controller generates an output signal that represents the demanded position. This signal is then sent to the positioner. Externally, the positioner may contain an input connection for the control signal, a supply air input connection, a supply air output connection, a supply air vent connection, and a feedback linkage. Internally, it may contain electric transducers, air lines, valves, linkages, cams, and electrical circuitry. **See Figure 10-38.**

For a direct-acting actuator example, as the control signal increases, a valve inside the positioner admits more supply air to the valve actuator. As a result, the control valve moves downward. The linkage transmits the valve position information back to the positioner. This forms an internal feedback loop for the valve actuator. When the valve reaches the position that correlates to the control signal, the linkage stops supply airflow to the valve actuator. This causes the valve actuator to stop. If the control signal decreases, another valve inside the positioner opens and allows the supply air pressure to decrease by venting the supply air. This causes the valve to move upward and open. When the valve has opened to the proper position, the positioner stops venting air from the valve actuator and stops movement of the control valve. **See Figure 10-39.**

Pneumatic Positioners
Pneumatic positioners operate on a pneumatic input signal, typically 3–15 pounds per square inch. Common styles rely on components such as a beam level, interchangeable cams, and a movable flapper arm with a nozzle assembly. With mechanical parts, *ambient conditions*, or conditions around the device that are examined (e.g., pressure and temperature), play a large role in

> **ThinkSafe!**
> When determining the type of valve actuator needed for a given application, always ask this question: If the signal or air pressure is lost to the device, what should the process flow do? To stop the flow, a fail-closed actuator is needed. To continue the flow, a fail-open actuator is needed. The selection of the proper fail state is typically determined by the process system designer but must be verified by the technician.

determining whether a device is functioning correctly or will function correctly when installed. *Ambient compensation* is the design of an instrument such that changes in ambient temperature do not affect readings of the instrument or compensation for ambient conditions when mounting an instrument. It may be needed to adjust a transducer that is calibrated in a controlled environment but is installed in a working environment. The mounting locations of pneumatic devices also have to be considered, because ambient temperature changes affect their performance. A pneumatic device needs to be placed away from areas with large temperature swings, if possible.

A hazardous location for a pneumatic instrument may not be as great a

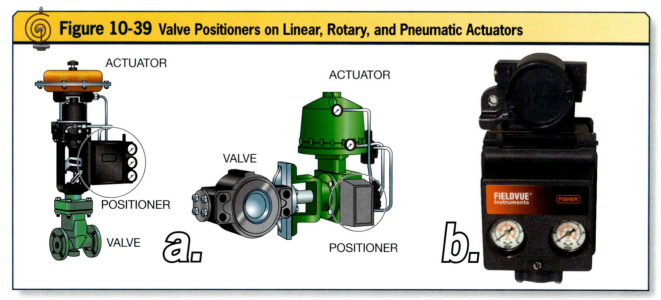

Figure 10-39. (a) Valve positioners are used on both linear and rotary actuators for accurate and repeatable valve positioning. (b) The gauges on the face of a valve positioner typically indicate supply and output pressures.

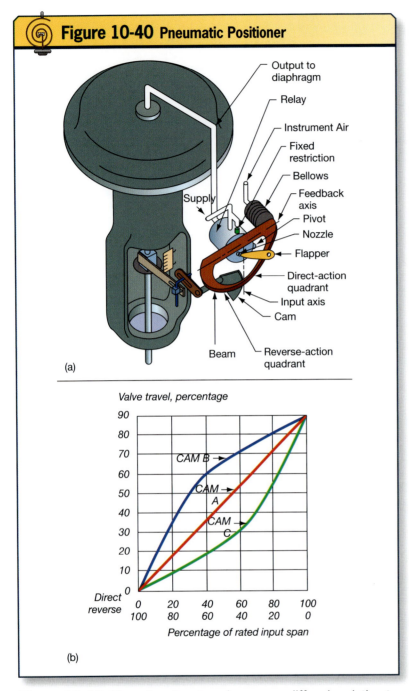

Figure 10-40. The valve stem travel response differs in relation to cams A, B, and C. These components, along with the nozzle-flapper mechanism, provide flexibility in control in this style of positioner.

calibrate the device in an environment simulated to be as close as possible to the environment of the location where it will be mounted. The mechanical parts in the device are subject to gravitational forces, and a device that is installed differently than calibrated has a tendency to drift toward the high or low limits of the range.

Pneumatic positioners are some of the oldest control devices in terms of technology, yet they are still installed on new control loops because of their proven effectiveness. A pneumatic positioner with interchangeable cams can provide both linear and nonlinear responses to the input signal. **See Figure 10-40.**

Electropneumatic Positioners

The electropneumatic positioner is generally viewed as an improvement over the pneumatic positioner. The electropneumatic positioner has an I/P transducer incorporated into the design. This allows the positioner to receive an analog signal, typically 4–20 milliamperes DC. The advantage of using a 4- to 20-milliampere DC signal is its speed and ability to travel long distances without being affected by temperature. **See Figure 10-41.**

Smart Positioners

The trend in valve positioners is to install smart positioners that incorporate microprocessor-based control and digital network communication. The term "smart" is a generic reference used by a number of manufacturers of digital valve positioners. Protocols such as HART, Foundation Fieldbus, and Profibus are used to communicate over a network to the smart positioner. There are several reasons a facility would want to replace legacy technology with smart positioners. For one, they have the capability to perform self-diagnostics on regular intervals. These tests can identify problems with valve packing, air leakage, excessive valve friction, and valve bench set. In addition, data gathered from the microprocessor can be used alongside valve software to tune a

concern as a hazardous location of an electronic instrument, but a dirty environment plays a large role in creating control errors, whether it is a calibration problem or fouling of a moving linkage. When calibrating a pneumatic instrument, care must be taken to

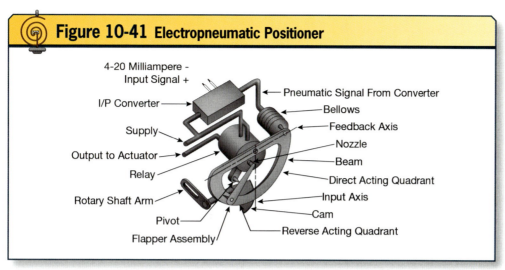

Figure 10-41. The major difference between a pneumatic positioner and an electropneumatic positioner is that the latter has an I/P converter. This allows the positioner to receive a milliampere DC signal.

loop for maximum performance. Some styles of positioners have digital readouts in the form of liquid crystal display (LCD) screens located at the valve, a benefit to maintenance personnel. Position indication is typically performed by the positioner, so no additional position-indicating devices are required. A smart valve positioner requires the use of a communicator compatible with the protocol used by the positioner. **See Figure 10-42.**

Calibration

To calibrate a positioner, the concepts of zero and span must be observed. The calibration of a positioner has to assume that the valve actuator has been adjusted as it should be within its proper working range (i.e., bench set). The complex nature of a valve positioner mandates a thorough understanding of the working principles and close adherence to the manufacturer's calibration procedures. The methods in which the zero and span adjustments are made vary by the design of the positioner. The objective of positioner calibration is to establish an accurate starting point called zero. The controller is manipulated to send an equivalent zero signal (4 milliamperes DC for an electronic signal and 3 pounds per square inch for a pneumatic signal), and the positioner compares the controller's signal with the valve position. The positioner is adjusted such that at the zero signal, the positioner is calling for a zero position: full open for a direct-acting, air-to-close valve. As the signal is increased, the positioner should be adjusted (if needed) to call for an increase in the valve actuator's working pressure. The two variables that must be monitored are the controller's signal and the valve stem position.

Figure 10-42. This smart valve positioner has an LCD screen to indicate valve travel and pressure values.

For additional information, visit qr.njatcdb.org Item #2417

The zero setting should be adjusted to allow the stem position to remain at zero and should be adjusted accordingly throughout the valve stem range. If the valve has a linear response, the valve stem position should be at 50% for a 50% controller signal. The span adjustment is adjusted until the span of the valve corresponds with the controller signal. The valve stem should indicate full travel when 100% of controller signal is received.

When setting the span for valve positions, do not make the span adjustment to the degree that it takes an exaggerated signal to move the valve from its furthermost point of travel. Some manufacturers suggest that the valve travel for 100% should be set at a value slightly less than full range. Likewise, the valve should be adjusted fully closed when a signal slightly greater than the zero position is received. This approach ensures that the valve's full range of travel is set given the full signal ranges.

SUMMARY

All components of a control valve assembly are essential to its function. The determining factors for correct operation depend on the proper installation, calibration, and maintenance of valve actuators, transducers, positioners, and regulators. Schematics, wiring, and configuration parameters are needed for the proper operation of these devices. The proper operation of the final control element assembly enables correct interpretation of control loop performance.

REVIEW QUESTIONS

1. **Which type of valve requires linear movement to change valve position?**
   a. Ball
   b. Butterfly
   c. Sliding stem
   d. V-notch ball

2. **A valve __?__ is a flexible membrane considered to be a sensing element.**
   a. actuator
   b. diaphragm
   c. piston
   d. positioner

3. **Force is equal to __?__ times __?__.**
   a. current / resistance
   b. kinetic energy / potential energy
   c. pressure / area
   d. pressure / diameter

4. **__?__ material is used to prevent leakage at the valve stem.**
   a. Closure member
   b. Packing
   c. Trim
   d. Valve seat

REVIEW QUESTIONS

5. A valve actuator that responds to increasing pressure by extending the attached stem is a __?__.
   a. direct-acting actuator
   b. fail-closed actuator
   c. fail-open actuator
   d. reverse-acting actuator

6. The internal component of a globe valve that comes into direct contact with the process fluid is called the __?__.
   a. actuator
   b. diaphragm
   c. trim
   d. valve bonnet

7. A __?__ valve is often used for large pipe diameters.
   a. ball
   b. butterfly
   c. globe
   d. v-notch ball

8. A pneumatic actuator is typically supplied with pressure in the range of __?__ pounds per square inch.
   a. 0–10
   b. 3–15
   c. 4–20
   d. 10–50

9. The process of verifying that the valve actuator can travel in a full range is called __?__.
   a. bench set
   b. characterization
   c. lapping
   d. packing

10. A __?__ actuator has a cylinder and piston assembly.
    a. diaphragm
    b. electric (MOV)
    c. hydraulic
    d. smart

11. The __?__ of a valve actuator is critical to ensure the process safety in the event of pneumatic pressure or electrical power loss.
    a. actuator force
    b. fail state
    c. limit switch
    d. mounting location

12. Which type of mechanism can convert linear motion to rotary motion?
    a. Handwheel
    b. Nozzle-flapper
    c. Pneumatic regulator
    d. Rack and pinion

13. A __?__ converts electrical energy into linear motion.
    a. I/P transducer
    b. LVDT
    c. potentiometer
    d. solenoid

14. A linear variable differential transformer is used for __?__.
    a. converting rotary to linear force
    b. determining valve stem position
    c. generating linear force
    d. measuring applied force

15. A valve positioner is used to __?__ for final control elements.
    a. decrease pressure
    b. increase accuracy
    c. increase pressure
    d. provide a connection for calibration equipment

16. An electropneumatic positioner uses a(n) __?__ that allows the positioner to receive a current signal.
    a. electric motor
    b. I/P transducer
    c. LVDT
    d. P/I transducer

Analytical Measurement

Various types of analytical measuring equipment are found in a process environment. These techniques and equipment are commonplace in industrial process facilities. Analytical measurements cover a broad category in the field of instrumentation. Accurate and reliable measurements are essential for safety, regulatory compliance, process equipment maintenance, and consistency in production facilities.

Objectives

- » Explain the need for analytical measurements.
- » Determine whether a solution is acidic or alkaline from pH measurements.
- » Describe conductivity measurement techniques.
- » Discuss oxidation reduction potential chemistry and measurement.
- » Explain how combustion efficiency and opacity are monitored.
- » Recognize the hazards of oxygen and flammable gases.
- » Describe various density measurement methods.

Chapter 11

Table of Contents

Analytical Measurement 242
Conductivity 242
pH Measurement 243
Oxidation Reduction Potential 246
Silica Analyzers 247
Continuous Emission Monitoring Systems 247
Combustion Analysis 248
    Opacity 249
    Measurement of Combustion 250
Gas Chromatography 251
Oxygen 253
    Life Safety Oxygen Measurement 253
    Dissolved Oxygen 253

Density 254
    Vibronic Density 254
    Oscillating Tube Coriolis Density 255
    Hydrostatic Density 255
    Buoyancy Density 256
Flammable Gas Detectors 257
    Catalytic Bead Sensors 258
    Flame Ionization Detectors 258
    Infrared Gas Detectors 259
Hydrogen Sulfide Monitors 260
Summary 262
Review Questions 263

For additional
information, visit
qr.njatcdb.org
Item #2690

ANALYTICAL MEASUREMENT

The branch of process instrumentation dedicated to analytical measurements is likely the broadest in the sense that a large number of measurement techniques exist. *Analytical measurements* are techniques to identify the presence or concentration of specific matter within a sample. The starting point for the measurement is acquiring a sample of the substance under test. Small samples may be tested at random from a large batch of product to serve as representations of the batch composition. In contrast, continuous sampling of a production building may be required for monitoring of hazardous or flammable gas leaks. The methods for sampling vary from manual preparation to constant and autonomous sampling systems. Once the sample is acquired, it is imperative the sample is not altered in any way, which would render the outcome of the analysis meaningless.

The sample is fed into an *analyzer*, an automatically operating measuring device that monitors a process for one or more chemical compositions or physical properties. Analyzers can be used for control purposes or for monitoring purposes. In a controlling function example, an analyzer measuring the pH in a cooling water system transmits a signal to a chemical pump feed system, indicating it should inject sulfuric acid to lower the pH. A monitoring function example would be the use of continuous infrared gas detection for the presence of hydrocarbons within a plant building, sounding an alarm if a leak is detected. No single analyzer technology exists to measure the variety of substances present in the process control field. Instead, many technologies exist, each with a list of positive and negative attributes for specific applications.

CONDUCTIVITY

Solution *conductivity* is the measurement of the relative ability to conduct electric currents. Conductivity measurements do not measure a quantity of a single substance. Instead, the conductivity can be used to infer the amount of dissolved electrolytes, such as salts, acids, and bases. Common applications for conductivity measurement are evaporative cooling water systems, known as cooling towers. **See Figure 11-1.** These systems use the principle of water evaporation to remove heat. As the water evaporates from the cooling tower to the atmosphere, the dissolved minerals concentrate. This concentration of minerals such as calcium, magnesium, chloride, and silica can cause scaling and corrosion to process piping and heat exchangers. A conductivity measurement system is used to detect the increased concentration of dissolved minerals and perform a controlling function to discharge, or "blow down," some of the cooling water and replace it with new makeup water to lower the concentration.

Figure 11-1 Cooling Tower Instrumentation

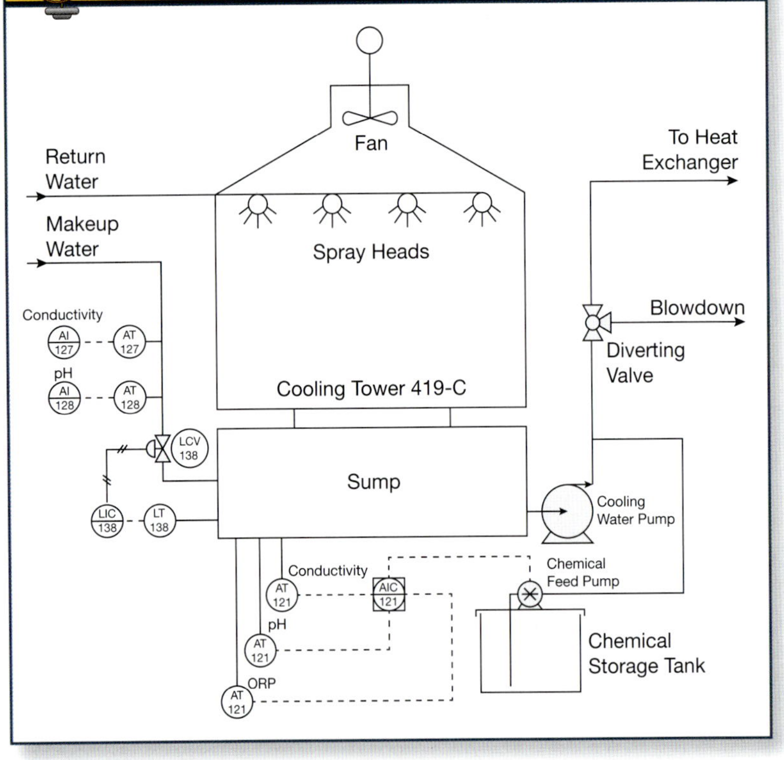

Figure 11-1. Analytical measurement for conductivity, pH, and oxidation reduction potential (ORP) is performed on the sump water of this cooling tower. The analytical controller (AIC) sends a signal to the chemical feed pump when chemicals are required to modify the water chemistry.

Conductivity, the reciprocal of resistivity, is measured in the unit siemens per centimeter. Because 1 siemen per centimeter indicates very high conductivity, most measurements are made in microsiemens per centimeter or millisiemens per centimeter. For example, pure water is a poor conductor (about 0.01 microsiemens per centimeter) in contrast to saltwater (about 30,000 microsiemens per centimeter). Temperature has a directly proportional effect on the conductivity of a solution; if the temperature increases the conductivity increases. Conductivity probes typically have thermal elements that measure the sample temperature and automatically compensate for temperature variations.

The measurement of conductivity can be made by contact probes or electrodeless inductive sensors. **See Figure 11-2.** The electrode style uses two to four metal electrodes that come in direct contact with the solution sample. The analyzer applies an alternating current (AC) to the electrodes and measures the resistance of the solution sample. By taking the reciprocal of the resistance, the conductivity is calculated. These contact-style probes can be affected by process materials that form resistive coatings or buildup. To eliminate these issues, noncontacting inductive sensors are available. This technology uses two wire-wound toroid coils encased in a nonconductive outer housing and submersed in the solution sample. The two coils act as a transformer that induces a voltage on the surrounding process material. The drive coil acts as the primary winding, and the receive coil acts as the secondary coil. The current measured between the two coils is representative of the conductance of the material.

Conductivity systems can be calibrated by using NIST traceable solutions with a known conductance as the input standards. Sensors can also be compared to other known accurate conductivity sensors to gauge the level of consistency between the two instruments, this is known as a one point check. Sensors may require cleaning to remove buildup that interferes with ability to accurately measure the conductivity of the sample.

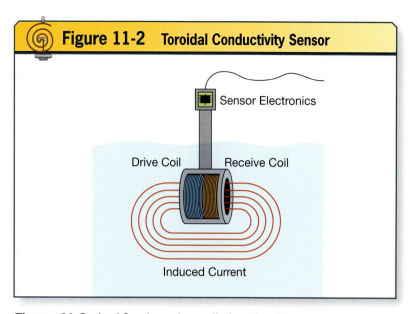

Figure 11-2. An AC voltage is applied to the drive coil on a noncontact conductivity sensor. Due to induced current in the process fluid, a voltage is induced in the receive coil. The amount of voltage induced in the receive coil is then used to determine conductivity.

pH MEASUREMENT

The analytical measurement pH is used to describe the acidity or alkalinity of a solution. Living cells are sensitive to pH levels. The human body maintains the pH of blood between 7.35 and 7.45. Any significant deviation can lead to coma or death. In some process applications, pH control is implemented to intentionally destroy bacterial growth. In contrast, regulatory compliance for wastewater discharge requires pH to be closely monitored and controlled to prevent damage to surrounding ecosystems.

The "p" in pH is a mathematical operator calling for the negative logarithm. The technician should remember that a logarithm is the inverse of exponentiation. For example, the statement

$$125 = 5^3$$

can be rewritten as follows:

$$\log_5(125) = 3$$

The "H" in pH is the chemical symbol for hydrogen. The pH is therefore the negative logarithm (–log) of the positively charged hydrogen ion concentration (H+) and is given by the following equation:

$$pH = -\log[H^+]$$

A pH of 7, or neutral, means the concentration of positively charged hydrogen ions are in balance with the number of negatively charged hydroxide ions (OH-). Think of pH like this: instead of saying there are 0.0000001 hydrogen ion moles in a liter of solution, the term "pH" allows the technician to relate the exact same information by saying the pH value is 7. **See Figure 11-3.**

For a sample with a pH containing a concentration of 1×10^{-7} positively charged hydrogen ion moles per liter, the equivalent pH value is found by the following method:

$$pH = -\log(1 \times 10^{-7})$$
$$pH = -(\log 1 + \log 10^{-7})$$
$$pH = -(0 + -7)$$
$$pH = 7$$

The product of positively charged hydrogen ions and negatively charged hydroxide ions is always $10 \times ^{-14}$. As the number of hydrogen ions increases, the number of hydroxide ions decreases. A solution with more hydrogen ions is acidic, and a solution with more hydroxide ions is basic (alkaline). Therefore, if the pH value is less than 7, the solution is acidic; if the pH value is greater than 7, the solution is basic. For an acidic sample with a pH containing a concentration of 1×10^{-3} positively charged hydrogen ion moles per liter, the solution is found by the following:

$$pH = -\log(1 \times 10^{-3})$$
$$pH = -(\log 1 + \log 10^{-3})$$
$$pH = -(0 + -3)$$
$$pH = 3$$

The measurement of a solution's pH is accomplished through several techniques. Perhaps the simplest pH measurement apparatus is pH-sensitive paper. A piece of the paper is dipped into the solution. The paper changes color and is compared to a chart to

Figure 11-3 Ionic Concentration and pH

[OH-] concentration (mol/l)	[OH-] concentration (mol/l)	pH	[H+] concentration (mol/l)	[H+] concentration (mol/l)	
1×10^{-14}	0.00000000000001	0	1	1×10^{0}	
1×10^{-13}	0.0000000000001	1	0.1	1×10^{-1}	
1×10^{-12}	0.000000000001	2	0.01	1×10^{-2}	Increasing Acidity
1×10^{-11}	0.00000000001	3	0.001	1×10^{-3}	
1×10^{-10}	0.0000000001	4	0.0001	1×10^{-4}	
1×10^{-9}	0.000000001	5	0.00001	1×10^{-5}	
1×10^{-8}	0.00000001	6	0.000001	1×10^{-6}	
1×10^{-7}	0.0000001	7	0.0000001	1×10^{-7}	Neutral
1×10^{-6}	0.000001	8	0.00000001	1×10^{-8}	
1×10^{-5}	0.00001	9	0.000000001	1×10^{-9}	
1×10^{-4}	0.0001	10	0.0000000001	1×10^{-10}	
1×10^{-3}	0.001	11	0.00000000001	1×10^{-11}	Increasing Basicity
1×10^{-2}	0.01	12	0.000000000001	1×10^{-12}	
1×10^{-1}	0.1	13	0.0000000000001	1×10^{-13}	
1×10^{0}	1	14	0.00000000000001	1×10^{-14}	

Figure 11-3. The concentration of hydrogen ions and hydroxide ions is used to determine the pH of a solution.

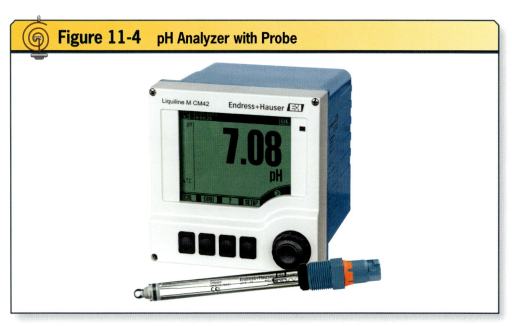

Figure 11-4. A pH analyzer converts the probe signal into a pH value and performs any necessary temperature compensation.

Courtesy of Endress+Hauser

determine the pH of the sample. While this method is quick and economical, it does not perform any controlling function or provide continuous measurement. Instead, pH paper is most useful for manual spot checks at regular intervals.

A more practical pH measurement technique for industrial process control is the pH measurement analyzer. **See Figure 11-4.** It contains three key components: the analyzer, the measuring electrode, and the reference electrode. The analyzer provides a method to convert the measurement reading into a signal that can be used by the controller or user. A measuring electrode is composed of a special type of glass that is filled with a potassium chloride solution that allows hydrogen ions to permeate the gel layer of the sensor. A silver-plated wire suspended inside the glass forms the measurement electrode. The reference electrode also has a silver-plated wire inside the housing and is used as a point of reference for the measurement.

The best way to understand pH measurement is to think about the voltage-measurement process on a battery. A reading cannot be obtained if only one lead is connected to the battery. If both leads are attached to the battery, a measurement can be taken. The reference electrode is like the reference lead (i.e., ground) connected to a battery. When the second lead (i.e., measurement probe) is attached, the circuit is completed and a measurement can be taken. In the case of pH, this potential difference, measured in millivolts direct current (DC), is converted to a signal by the analyzer and used by a controller.

Temperature changes in the sample can have an effect on the resistance of the measurement electrode's glass. As the temperature increase, the glass resistance decreases. Compensation is necessary, because the resistance of the

TechTip!

One of the most common causes of pH analyzer error is the buildup of contaminants on the probe (e.g., algae). All pH probes should have a routine preventative maintenance plan to clean the probes. Frequency of maintenance should be based on specific applications.

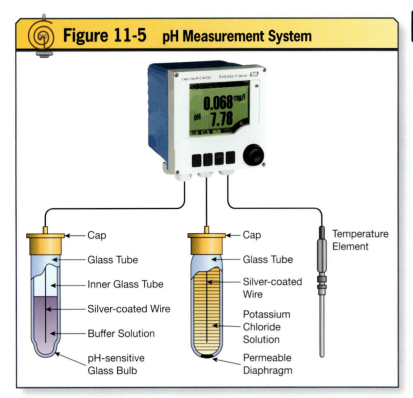

Figure 11-5. The three components of a pH probe are the measurement electrode, the reference electrode, and a temperature element.

TechTip!
Always check the expiration date on buffer solutions before using them for calibration.

OXIDATION REDUCTION POTENTIAL

The tendency for a substance to either gain or lose an electron is measured in many industrial processes. This measurement is defined as the ORP and is measured in millivolts DC. The millivolt DC value increases as the ORP value of a system increases. In an ORP reaction between chemicals, the molecule that loses the electron is the reducing agent and the molecule that pulls an electron is the oxidizing agent. This measurement is especially useful for controlling biological growth such as *Escherichia coli*, *Salmonella*, and *Legionella pneumophilia*. Research indicates a strong correlation between the ORP value and a reduction in biological growth. Common oxidizing agents are sodium hypochlorite (household bleach) and chlorine. These agents are added to cooling tower water, fresh-produce disinfectant solutions, and swimming pool water in metered amounts to reduce biological growth within the system.

ORP measuring probes are similar to pH probes with one exception: ORP probes rarely have temperature sensors integrated into the probe because of the miniscule effects of temperature on ORP values. The ORP probe's sensing electrode is typically made of platinum, and the reference electrode is made of silver chloride. An analyzer reads the millivolt DC potential between the two electrodes. ORP is frequently used, together with pH and conductivity control, to manage cooling tower water systems.

The ORP probe is calibrated using chemicals available commercially. One such product, quinhydrone, comes in powder form and is mixed with pH buffer solution at a specified concentration. The ORP probe is then immersed in the

glass directly affects the millivoltage reading made by the system. Most modern pH sampling probes contain a temperature-sensing element integrated with the body of the probe. The analyzer uses this temperature information to correct the signal and ensure an accurate representation of the pH value is measured. **See Figure 11-5.**

Calibration of pH systems require the technician to use specific pH input standards. These *buffer solutions* are solutions that have a constant pH value and the ability to resist changes in pH levels. Most pH meters require periodic calibration at several specific pH levels. A typical calibration procedure is performed at pH values of 4, 7, and 10. It is important to choose the correct buffer solution for the pH level desired for calibration. Care must be used to avoid contaminating the various buffer solutions during the test. Rinsing the pH probe with distilled water and drying them before and after immersing it in each buffer solution prevents contamination.

ThinkSafe!
Proper personal protective equipment should always be worn while working with pH devices and buffer solutions. These solutions are very acidic and very alkaline, and they can cause serious reactions if handled improperly.

prepared solution, which serves as the input test standard. As is the case with pH calibration, care should be taken to avoid contaminating the solution. Adjustment to the analyzer can be made to compensate for inaccuracies. ORP electrode cleaning is recommended when unstable or shifted readings are observed. Follow the manufacturer's guidance on proper cleaning procedures to prevent damage to the electrodes.

SILICA ANALYZERS

Silica is the second most common element on Earth, outnumbered in quantity only by oxygen. Silica is found in water in varying levels. Because ingesting water orally that contains silica has no known adverse health effects on humans, it is considered safe. However, silica presents a serious challenge for boilers, steam turbines, and piping. Given the right conditions, silica forms extremely hard scale deposits inside these pieces of equipment. Such a scale deposit, or silica glass, impedes the transfer of heat exchange and lowers the efficiency of the process. This translates to higher fuel costs for the operators of the system.

Silica can be measured via a silica analyzer that works on the photometric analysis principle. This device first takes a sample of the water in the system and then mixes the sample with a measured amount of *reagent*, a substance that is added to cause a chemical change to occur. The mixture is then fed through a chamber with a light-emitting diode on one end and a detector on the opposite end. The mixture absorbs a specific wavelength of light, whose absence is measured by a detector. **See Figure 11-6.** The amount of light absorbed is proportional to the concentration of silica in the sample. The silica analyzer must be restocked routinely with reagent bottles. The calibration of the analyzer is often an automatic process once initiated by the technician.

This value measured by the analyzer is typically transmitted in the form of a 4- to 20-milliampere DC signal. After the level of silica is measured, corrective actions often involve blowdown of water to reduce the concentration in the system. Prefiltering of makeup water using a reverse osmosis (RO) system or demineralization system can reduce the makeup water silica concentrations and may lower the demand on the water supply that frequent blowdown creates.

CONTINUOUS EMISSION MONITORING SYSTEMS

Continuous emissions monitoring systems (CEMS) is a measurement systems that determine the amount of air-polluting emissions and other by-products of industrial processes. CEMS is frequently required for regulatory compliance, such as with the U.S. Environmental Protection Agency's (EPA's) Clean Air Act and Acid Rain Program. These regulations are complex and provide many varying requirements on the design and operation of the plant. The amount of allowable plant pollution is capped at a calculated value; a plant that exceeds this pollution limit is penalized. To determine the amount of total pollutants the plant produces, a CEMS

TechTip!
ORP measurements are susceptible to electromagnetic interference and stray voltages. The probe conductors must be isolated from AC conductors to prevent interference.

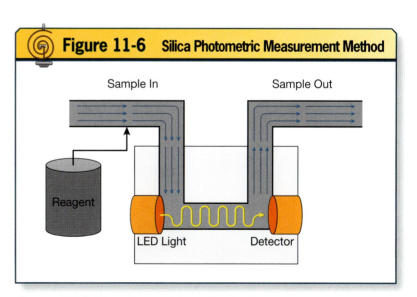

Figure 11-6. The photometric method mixes a reagent and sample fluid before entering a chamber with a light source and detector. The amount of light absorbed by the mixture is used to determine silica concentration.

unit collects and stores this information. The plant then must report total emissions based on these data. To ensure the data are accurate, rigorous commissioning and maintenance rules are needed.

CEMS are complex systems with many components. A typical extractive-style unit consists of sample probes, umbilical lines, sample pumps, pollutant concentration monitors, diluent gas monitors, stack gas flow rate monitors, and data acquisition and handling systems (DAHS). **See Figure 11-7.** This system typically has a redundant backup system for 100% uptime and a permanently installed calibration gas system. A CEMS measures variables such as sulfur dioxide (SO_2), carbon dioxide (CO_2), nitric oxide (NO), nitrogen dioxide (NO_2), oxygen, moisture, stack gas flow rate, and opacity.

Installation of a CEMS starts by mounting the components per the process design. Particular attention should be paid to mounting the components in a manner that permits future access for commissioning and maintenance. The system must pass a rigorous certification process once installed. This process requires significant documentation that must be completed to satisfy regulatory compliance. Both a 7-day calibration error test and a linearity test are completed to verify proper operation. The stack gas flowmeter and the DAHS also must show accurate and stable readings to be commissioned. After the system is installed and certified, any modifications or alterations of the system that could affect the operation of the unit require a complete recertification procedure. A relative accuracy test audit (RATA) is performed annually by an outside company. This test makes a head-to-head comparison of the plant's system to the readings of the test company's equipment. Daily calibration operations can be performed by plant personnel. These zero and span tests are typically automated by the control system after the operator initiates the test through a computer terminal.

COMBUSTION ANALYSIS

Fossil fuel–powered equipment, such as boilers and ovens, is widely used in industrial processes. The cost of fuel is significant and affects the profitability of the plant. In addition, compliance with regulations to limit pollution to acceptable levels mandates that plant owners optimize the efficiency of their fossil fuel–powered equipment. Combustion is the process of combining fuel, such as natural gas, fuel oil, or coal, with oxygen to produce heat. Fossil fuels are hydrocarbons made primarily of hydrogen and carbon. A common example of hydrocarbon combustion is natural gas, which is approximately 95% methane (CH_4). When methane reacts with oxygen (O_2), the result is the release of heat,

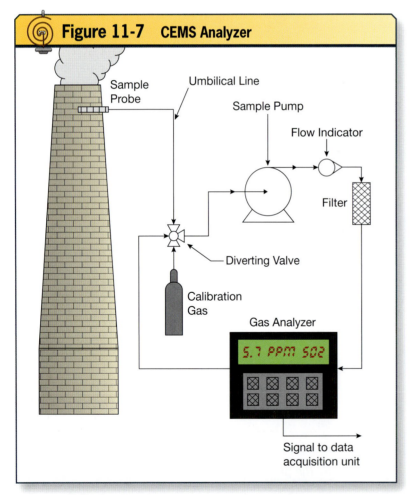

Figure 11-7. The CEMS analyzer constantly samples and measures the chemical makeup of exhaust gases.

through which two water (H_2O) molecules and one carbon dioxide (CO_2) molecule are formed. **See Figure 11-8.**

This reaction assumes a perfect ratio of fuel to air, called a stoichiometric condition. In contrast, a rich mixture has too much fuel, and a lean mixture has too little fuel. To achieve complete combustion, industrial processes typically are supplied with excess oxygen. This *excess air* is the percentage of air supplied above the calculated requirement for complete combustion. If the combustion has insufficient air, unburned carbon, called soot, and carbon monoxide may form. Supplying more air than is required ensures a sufficient amount of oxygen is available to react with the fuel. This excess air comes at a price: it carries heat away from the process and lowers the overall efficiency.

In addition to the air-to-fuel ratio, combustion analysis includes draft measurement and exhaust gas temperatures. The *draft* is the flow of gas through a heating system. The draft can be natural or may be powered by mechanical ventilation. In a natural-draft flue, heated gases inside the flue are less dense and are forced upward by cooler, higher-density gases entering the beginning of the system. The draft can be measured with a pressure sensor measuring pressure inside the flue. The flue temperature represents heat that was not turned into useful work, such as making steam. The flue temperature can reach high temperatures and is commonly measured with thermocouples rated for a flue environment. Combustion efficiency analysis can be assessed using the heat-loss method shown here:

$$\text{Combustion efficiency \%} = 100 - \frac{\text{Heat loss}}{\text{Fuel heating value}} \times 100$$

Opacity

Opacity is the measure of how much light can pass through a substance. Industrial processes that burn fossil fuels produce smoke in varying amounts. To

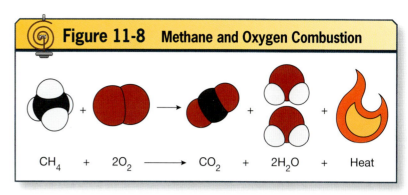

Figure 11-8. The products of methane and oxygen combustion are carbon dioxide, water, and heat.

comply with emissions guidelines, the opacity of the emissions must be monitored and controlled. Early opacity measurement methods were pioneered by Maximilien Ringelmann in the late 1800s. He observed thick black smoke billowing from coal-burning boiler stacks and correlated this with poor combustion conditions within the boiler. In an effort to standardize the opacity levels, he developed a five-step scale on which 0 is 0% opacity and 5 is 100% opacity. **See Figure 11-9.** The grid-patterned cards were held up and compared to the stack emissions. Surprisingly, this manual method is still taught and used today under the guidelines of EPA

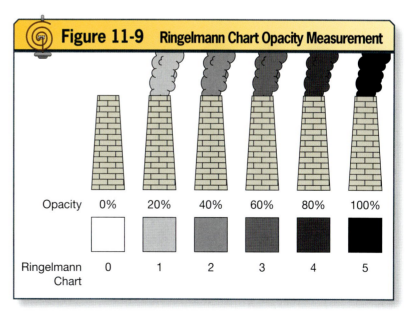

Figure 11-9. The Ringelmann chart uses a 0–5 scale to rate the opacity of stack emissions.

Figure 11-10 Transmissometer

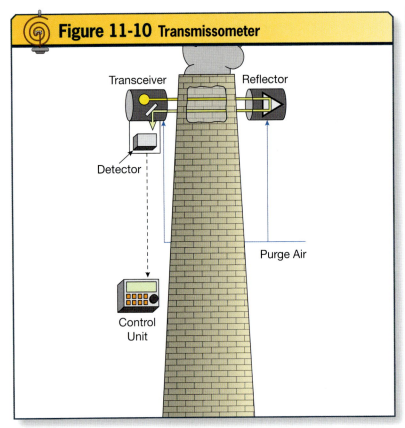

Figure 11-10. A transmissometer measures stack gas opacity by passing a beam of light through the gas and measuring the amount of refracted or absorbed light.

TechTip!
Proper alignment of the reflector and detector are critical for proper instrument operation.

A more sophisticated method of measuring opacity is to use a transmissometer. **See Figure 11-10.** This instrument relies on the principle of light transmission through a gas with suspended solid particles. These particles either absorb or scatter light as it attempts to pass through the sample. One such design uses a transceiver with a light source and detector mounted opposite a reflector. The light source emits a beam of light that travels through the stack, strikes the reflector, and returns to the detector. The control unit serves as the operator interface to the device and typically has a liquid crystal display screen to read current conditions. Clean purge air may be used to prevent particulate matter from entering the housing and fouling the lens or reflector. The mounting of a transmissometer should allow access for maintenance operations.

Method 9. A qualified observer is a person who has demonstrated the ability to accurately assess plumes of smoke within the previous 6 months.

Measurement of Combustion

The need to ensure a combustion system operates at maximum efficiency during commissioning and startup phases of a project is apparent. In addition, periodic combustion analysis checks are needed to ensure the system operates at maximum potential throughout its life cycle. As heating systems age, they frequently experience drift in efficiency. Seasonal changes can also affect the efficiency of a combustion system. To measure these changes, technicians use combustion analyzers to gather data such as gas

TechTip!
Flue gas samples must be collected as close to the equipment as possible and before any dampers or diverters, which allow additional air into the flue that dilutes the sample.

Figure 11-11 Portable Combustion Analyzer

Figure 11-11. Technicians use portable combustion analyzers to determine the efficiency of a combustion system.

Image courtesy of Testo, Inc.

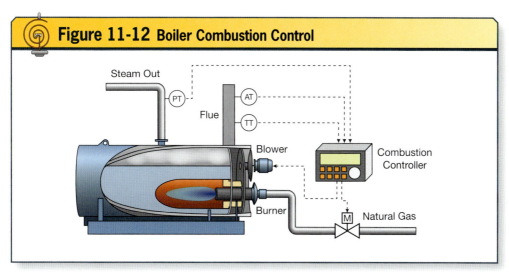

Figure 11-12. Variables such as flue temperature, oxygen content, and steam supply pressure are used by the combustion controller to optimize boiler efficiency and minimize emissions.

concentrations, flue temperatures, and system draft. A portable combustion analyzer is capable of performing all of these measurements. **See Figure 11-11.** The complex nature of combustion analyzers requires the technician to read and understand the specific manufacturer's testing procedures to ensure the date collected are valid.

A fossil fuel–powered boiler is commonly encountered in process plants and industrial settings. The production of steam involves heating water to the boiling point with heat from combustion. The inputs to this system, air and fuel, may be controlled by simple mechanical linkages or more sophisticated combustion control systems (CCSs). These systems are used in conjunction with burner management systems (BMSs), which are safety systems that provide safe start-up, monitoring, and shutdown of the burners. A natural gas boiler can be controlled by a continuous combustion controller. **See Figure 11-12.** The combustion controller monitors steam pressures to determine demand, flue temperature, and flue oxygen concentration. The outputs of the controller connect to the fuel gas valve, and the blower and damper assembly to control the flow of combustion air.

Performing a fossil fuel–powered boiler tune-up is similar to calibrating an instrument in that it requires a systematic approach. The procedure starts by measuring and recording the as-found data using a combustion analyzer. The flue gas temperature, flue gas concentration, and draft pressure are measured. Next, adjustments are made to the air control system to set the excess air at the desired value, commonly 15%. The amount of oxygen in the flue gas is an indicator of the amount of excess air that remained after the fuel completely burned. The concentration of emissions gases should be within the required values. Finally, as-left data are collected and documented. The data collected during this procedure serve as a record of maintenance and can be used for predicting the overall health of the combustion system.

GAS CHROMATOGRAPHY

Samples from industrial processes are rarely made up of a single component. Instead, samples contain a mixture of chemical components in varying concentrations. *Chromatography* is the method by which individual components of a substance are separated by the rate of travel through a medium, often a column or coil. The word

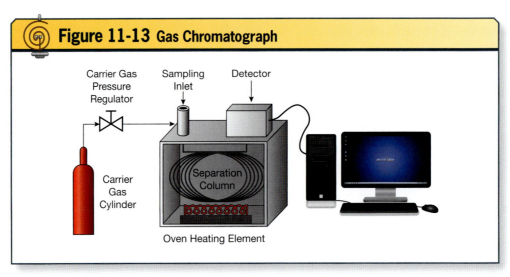

Figure 11-13. *Gas chromatography analyzers can separate samples of process fluids into individual substances.*

"chromatography" translates to "color writing." This technique was first used in the early 1900s to identify various color compounds in plants. Modern facilities use gas chromatography for measurements such as the presence of pollutants in a stack or the composition of petroleum distillates.

The components of a gas chromatograph include the carrier gas cylinder, pressure regulator, oven, separation column, detector, and sampling inlet. **See Figure 11-13.** To start the measurement process, the pressure regulator is adjusted for the appropriate amount of carrier gas to mix with the sample. High-purity helium, hydrogen, nitrogen, and argon are common gases that are used based on the sampling mixture and detector type. The sample is inserted either manually into the sampling inlet or automatically via sampling tubing to the process. Heat from the oven surrounding the separation column then vaporizes the mixture. As the sample mixture moves through the separation column, the individual components separate into homogenous groups and exit the separation column at various time intervals. The detector measures the presence of each substance as it exits the column. This signal is processed and sent to a computer for further analysis. A graphical representation a chromatogram shows the separation of the individual components of a mixture. **See Figure 11-14.**

Gas chromatography equipment is both some of the most complex and some of the most expensive instrumentation encountered in a facility. Factory installation guidelines must be followed to ensure proper operation. The electronic components inside the unit

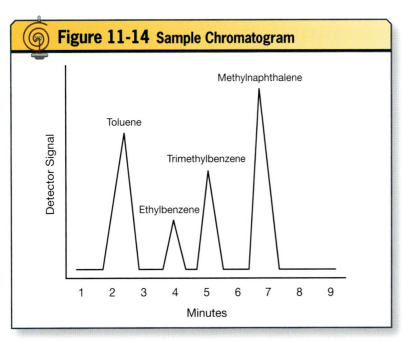

Figure 11-14. *A chromatograph shows the individual constituents of a sample.*

may be damaged if electronic static discharge precautions are not observed. Calibration procedures for these units vary; however, a two-point calibration using high- and low-calibration gases is common. These calibration gases are traceable to the National Institute of Standards and Technology (NIST) and serve as an input standard. The output signal of the detector can use either analog or digital signaling to transmit information.

OXYGEN

Oxygen is an essential element to sustain life. It is the most prevalent element on Earth, making up 20.8% of Earth's air. Oxygen also accounts for 89% of the mass of a water molecule. Though there are two hydrogen atoms in a water molecule, oxygen has a much higher atomic mass unit (AMU): 16 AMU compared to hydrogen's AMU of 1.

Life Safety Oxygen Measurement

Technicians who enter a confined space or work area that may have abnormal oxygen levels must monitor the atmosphere at all times to verify the oxygen levels are adequate to sustain life. The Occupational Safety and Health Administration (OSHA) defines the appropriate level of oxygen concentration for worker safety to be between 19.5% and 22%. Levels below this range are classified as oxygen deficient; levels above this range are classified as oxygen enriched. In addition to health risks encountered through breathing excess oxygen, an oxygen-enriched environment becomes a flammable hazard and can lead to spontaneous combustion of otherwise poor fuel sources. Serious physiological effects occur when the oxygen levels drop below 19.5%, including dizziness, disorientation, loss of physical strength, and unconsciousness.

Personal oxygen monitors are available for technicians to wear on their body. **See Figure 11-15.** These oxygen monitors sound an audible alarm when oxygen levels deviate from the normal range. The monitor should be calibrated regularly, bump tested daily, and charged adequately to provide monitoring during the entire work shift.

Permanently mounted oxygen analyzers are also used to determine oxygen concentration in industrial facilities. Consider a storage room in a process plant where cylinders, regulators, and distribution lines for inert gases such as nitrogen or argon are housed. A leak in this area could displace oxygen and become a hazard for nearby workers. To combat this possibility, an oxygen analyzer sensor can be mounted inside the room and connected to a remote mounted indicator. Before a technician enters the area, the monitor should be checked. In addition to the visual indication, the analyzer may have audible alarms to alert the worker to changing oxygen concentrations.

Dissolved Oxygen

Dissolved oxygen levels are essential for applications such as fish farming and wastewater treatment. Treatment of sewage commonly uses a combination of mechanical, chemical, and biological processes. Bacteria and protozoa are

> **ThinkSafe!**
> Never enter a known oxygen-deficient area in an attempt to rescue a worker. Call 911 and implement the emergency response plan. Tragically, many workers have lost their lives while attempting to save another worker.

For additional information, visit qr.njatcdb.org
Item #2572

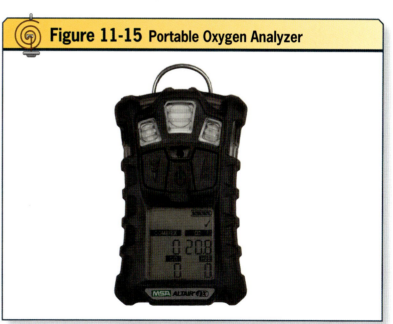

Figure 11-15. Portable oxygen analyzers provide continuous personal protection from oxygen-rich or oxygen-deficient atmospheres.

tiny microorganisms that thrive given the proper environment. These beneficial microorganisms play a vital role in the treatment process. The organic material in sewage is rich in nutrients for the microorganisms. This process is *aerobic*, or a process that requires oxygen. As the microorganisms digest the organic material, they release carbon dioxide. If the oxygen levels drop too low, the microorganisms die. Forced aeration replenishes the oxygen in the material and maintains the optimum levels of dissolved oxygen.

Dissolved oxygen measurement requires a sensor and analyzer installation. A membrane-covered amperometric sensor operates on the principle of current flow due to ionic activity in an electrolyte. The membrane is designed to allow oxygen permeation. Once inside the sensor, the oxygen is reduced to hydroxide ions at the cathode and a current that is proportional to the dissolved oxygen concentration flows through the electrolyte. This current flow is read by the analyzer and converted to a parts per million value. Finally, the information is transmitted using a robust signal, such as 4–20 milliamperes DC, or digital Foundation Fieldbus.

Sensor calibrations require the sensor be subjected to a calibration standard with a known value to which the sensor reading is compared and adjusted. For a dissolved oxygen sensor, this calibration standard is normally an environment that is 100% saturated with air and water, resulting in a 100% relative humidity environment. One method of creating a 100% relative humidity environment is to suspend the probe directly above a container of water. The proper distance between the water level and the probe is specified in the manufacturer's literature. A common distance is 1-2 inches. Next, the technician must follow a series of steps defined by the manufacturer's literature to complete the calibration process. This typically involves selecting a 100% relative humidity slope calibration from the analyzer's menu.

For additional information, visit qr.njatcdb.org Item #2573

 TechTip!
A common problem with dissolved oxygen sensors is fouling. Frequent cleaning or automated cleaning systems result in more consistent and reliable measurements.

DENSITY

Density measurements are used in a range of applications. For example, density is used to determine the concentration of an acid mixture, to measure the fat content of milk in a dairy, and to assess the heating potential of steam. Most substances exhibit an increase in density in relation to a decrease in temperature. Based on this, temperature compensation is frequently used for higher accuracy. Density is measured by taking the mass of the sample over the volume of the sample. An important consideration for density measurement is to ensure no gas bubbles are trapped in the sample, which this will lead to an erroneous reading. Methods to measure density include sensors that operate on the vibronic, oscillating tube, hydrostatic, and buoyancy principles.

Vibronic Density

The vibronic density sensor measures density by detecting a change in vibration frequency in relation to the mass of the substance surrounding the sensor. **See Figure 11-16.** Consider a tuning fork: it vibrates at a consistent and repeatable frequency in air when struck. If the tuning fork was immersed in water, the vibrating frequency would be dampened, or lowered, because water is denser than air. Vibronic sensors are not struck manually; instead, they are excited by piezoelements and a feedback amplifier circuit. These sensors can be mounted directly into the process fluid stream or tank to provide continuous measurement.

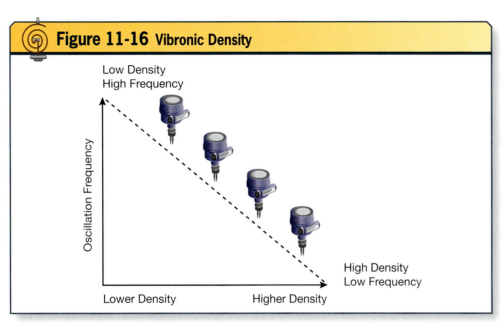

Figure 11-16. The oscillation frequency of a vibronic sensor is inversely proportional to the density of the sample.

Oscillating Tube Coriolis Density

The mass of an object determines the natural frequency of oscillation. The oscillating tube density measuring method operates on the principle that fluid inside a tube affects the oscillating frequency of the tube. For example, a tube with pure water at 4°C has a density of 62.43 pounds per cubic foot. If the same tube were filled with molasses at 4°C, which has a density of 89 pounds per cubic foot, the oscillation frequency would decrease. A common design uses a drive coil to provide motion to the tubes. The oscillation frequency of the tube is detected by pick-off sensors mounted on the tube. Temperature compensation via a resistance temperature detector is used in most designs and continuously monitors the tube temperature.

Calibration of a Coriolis density meter is typically performed by comparing the reading of the meter under test to a certified flow calibration rig. Multiple options are available to perform the necessary calibration. Many calibration labs offer on-site service by bringing a portable flow rig mounted in a pull behind trailer or mounted on the back of a flatbed truck. These portable rigs can calibrate smaller flowmeters (4" or smaller). Another option for calibration is to remove the meters from service and send them to an accredited calibration lab. While lab calibration results in better accuracy, turnaround time and cost can make this method undesirable.

A final option is to use a removable secondary meter, commonly called a proving meter. This meter is installed in line with the primary meter. This meter can be removed from the process on a periodic basis and sent to a calibration lab. This method of proving a density meter differs from calibration but provides a way to verify the primary meter is reading within the specified tolerance.

Hydrostatic Density

The liquid pressure exerted on a pressure sensor in a process tank is directly proportional to the density of that liquid. The versatile pressure transmitter may be used for density measurement in a vessel by measuring the amount of pressure and the height of the column. One installation method uses two transmitters with a fixed distance between the sensors. The specific gravity,

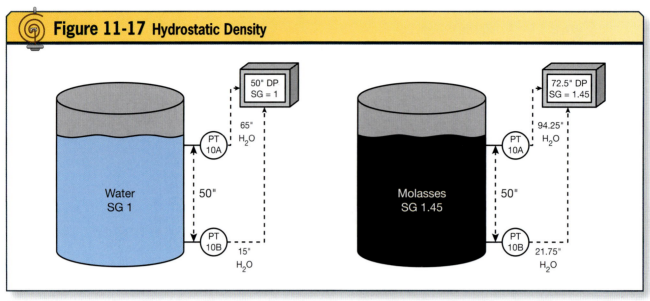

Figure 11-17. Hydrostatic pressure can be used to determine density of a process fluid.

and therefore the density of the liquid, is found by the following equation:

$$\frac{L-U}{D} = \text{Specific gravity}$$

where
L = Lower transmitter signal
U = Upper transmitter signal
D = Distance between transmitter sensors

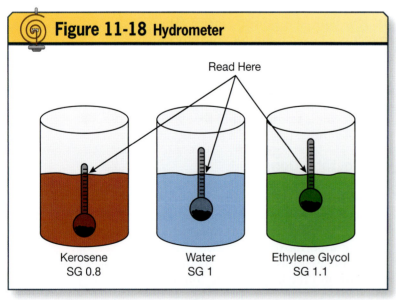

Figure 11-18. The density of a fluid is measured by observing the submersion depth using the hydrometer stem marks.

For example, suppose a tank containing water has a distance of 50 inches between the sensors. **See Figure 11-17.** The differential pressure readings are 65 inches of water for the lower transmitter and 15 inches of water for the upper transmitter. When inserted into the equation, the specific gravity equals 1. Now consider the second tank containing molasses. The differential pressures are 94.25 inches of water for the lower transmitter and 21.75 inches of water for the upper transmitter. These numbers result in a specific gravity of 1.45.

Buoyancy Density

The buoyant force exerted on a submerged object is directly related to the volume of displaced fluid and the density of the fluid. The Dead Sea is a naturally occurring example of this principle. The high salinity levels of the Dead Sea result in a density of more than 77 pounds per cubic foot, a specific gravity of approximately 1.25. This highly dense water permits bathers to float with minimal effort due to the increased buoyancy. Industrial processes such as beer brewing and maple syrup manufacturing are good applications for employing hydrometers to measure density.

The hydrometer is a glass or plastic apparatus similar in appearance to a thermometer. The bottom bulbous section houses weights to cause the hydrometer to partially sink. The hydrometer submerges until a state of equilibrium is reached between the downward gravitational forces and the oppositional buoyant force. Once this state is achieved, the graduated marking on the stem of the hydrometer indicates the density of the fluid. **See Figure 11-18.** The measurement range of a hydrometer is inherent to the design and cannot be modified. Calibration can be verified using a fluid of a known density.

FLAMMABLE GAS DETECTORS

Industrial processes both use and generate as by-products combustible gases and vapors. If these gases or vapors develop sufficient concentrations and an ignition source is present, the outcome results in both costly accidents to people and expensive damage to process equipment. The four elements necessary for a fire can be arranged on a fire tetrahedron: heat, fuel, an oxidizing agent (commonly oxygen), and a chemical chain reaction. **See Figure 11-19.** Oxygen in sufficient concentrations and heat in the form of ignition sources are present in most process environments. Therefore, the most reliable method to prevent fires and explosions caused by combustible gases or vapors is to limit their concentration to a safe level.

The concentration of these flammable substances directly affects their ability to combust. Two levels of concentrations are given and are expressed by volume concentration. The *lower explosive limit (LEL)* is the minimum concentration of a flammable substance that will support combustion and continue to burn after the ignition source is removed. The *upper explosive limit (UEL)* is the maximum concentration of a flammable substance that will support combustion and continue to burn after the ignition source is removed. Any amount of concentration between LEL and UEL is considered within the flammable range. **See Figure 11-20.** It is rarely practical to maintain a flammable substance above its UEL. Instead, limiting the concentration to well below the LEL is the accepted industry practice. For example, National Fire Protection Agency (NFPA) 86 states that when continuous monitoring is installed in a batch process oven, the vapor concentration shall not exceed 50% of the published LEL value. Published data tables show the LEL and UEL for each flammable

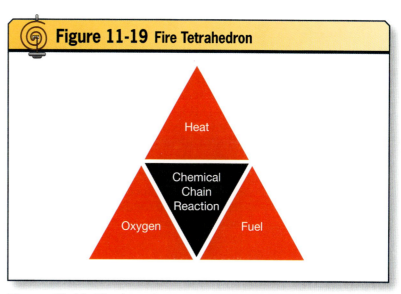

Figure 11-19. The four elements necessary for a fire are heat, fuel, oxygen, and a chemical chain reaction.

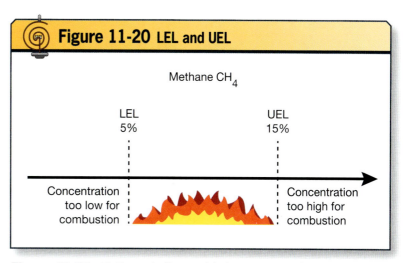

Figure 11-20. Methane is flammable in concentrations between 5% and 15%.

substance. However, the data are based on test conditions and may not apply to all process applications.

Three types of detectors are catalytic bead sensors, flame ionization detectors (FIDs), and infrared gas detectors. These detectors use different technologies to identify flammable substances. Each style of detector has limitations based on its operational principle, and each requires specific installation and calibration methods to ensure the measurement made by the instrument is accurate.

Catalytic Bead Sensors

The catalytic bead sensor relies on the principle of a *catalytic reaction*, an increase in the rate of a chemical reaction due to the addition of a catalyst. A common example of a catalytic reaction occurs in the aptly named catalytic converter of a gasoline-powered automobile. As exhaust gases pass through the converter, a chemical reaction known as reduction reduces the nitrogen oxide to nitrogen and oxygen. The addition of a catalyst, platinum, increases the rate of this reaction.

One design uses two catalytic beads in a flammable gas detector: the active element is exposed to the gas under test, and the reference element is sealed inside a reference chamber. **See Figure 11-21.** As electric current flows through the beads, the catalyst coating causes flammable substances to combust inside the detector. To prevent the detector from becoming an ignition source for the surrounding atmosphere, the active bead is typically protected by a sintered metal housing.

Flammable substances that reach the active element and combust produce heat, which causes a temperature differential between the reference element and the active element. This temperature differential is in direct proportion to the amount of flammable substance in the sample. Catalytic bead sensors should be calibrated for a specific substance by exposing the sensor to a NIST traceable calibration gas. The sensor should be recalibrated frequently thereafter. Two notable characteristics of this type of detector are the high likelihood of sensor drift over time and the susceptibility to element poisoning. Element poisoning occurs when certain compounds adhere to the element and reduce the element's sensitivity.

Flame Ionization Detectors

The FID is an analytical instrument that detects the presence of hydrocarbon content within a sample. Hydrocarbons are organic compounds composed of a mixture of hydrogen and carbon. Common hydrocarbons encountered in process environments include methane, acetylene, and propane. The FID contains a continuous flame fueled by hydrogen gas. **See Figure 11-22.** As the sample gas enters the chamber, any hydrocarbon compounds are ignited. Hydrocarbons produce ions as a by-product of combustion. The FID measures the ionic concentration by measuring the change in electric current through the FID's electrodes. This detector is commonly referred to as a "carbon counter" because of the direct proportionality between the amount of carbon in the sample and the measured electrical signal.

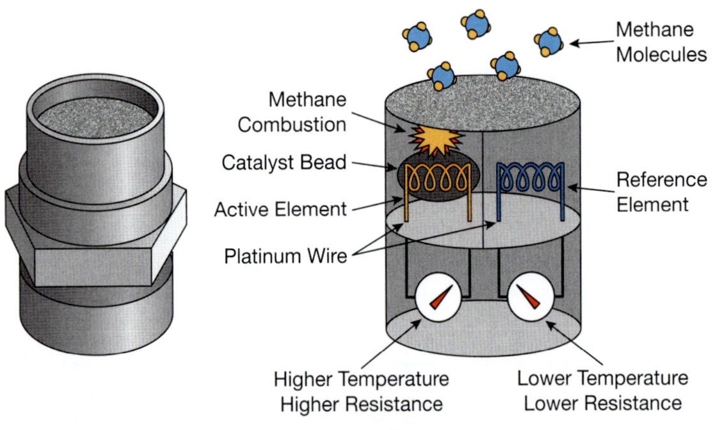

Figure 11-21. The active element is coated with a catalyst that promotes combustion at this active element in a catalytic bead sensor. A sintered metal housing prevents combustion from leaving the sensor.

The manufacturer's calibration procedure for a flame ionization detector must be followed to ensure the analyzer is functioning properly. A typical calibration procedure requires the technician to introduce NIST traceable gas into the sample inlet of the instrument. After navigating to the calibration menu, the technician would select the zero calibration setting. Next, the technician would introduce the NIST traceable zero gas into the sample inlet. After this step is complete, the technician would select the span calibration setting. Next, the technician would introduce the NIST traceable span gas into the sample inlet.

Infrared Gas Detectors

Infrared gas detectors detect combustible gases by measuring the absorption of specific wavelengths of infrared radiation. *Infrared radiation* is an electromagnetic wave with a wavelength longer than the color red and therefore cannot be seen by the human eye. However, the infrared radiation from a bed of hot coals can be felt as warmth on the skin.

The infrared detector typically consists of two primary components: an emitter and a detector. The emitter sends a beam of infrared radiation across the area to be measured. The detector then measures the amount of infrared energy that travels through the sampled area. **See Figure 11-23.** The underlying principle of operation is that specific wavelengths of infrared radiation are absorbed by hydrocarbons. The detector is capable of identifying this absorbed energy and determining the concentration of hydrocarbon gas present in the sample area.

The infrared method can be used as a single-point leak detection system, or it can be used to monitor large areas. Some versions of open-path infrared gas detectors can measure areas several hundred feet across. Installation in these applications requires the emitter and detector to be precisely aligned and the path to be free from obstructions. Routine calibration of an open path

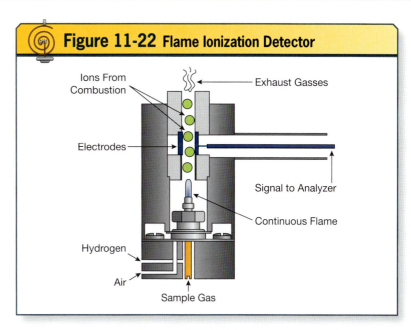

Figure 11-22. Hydrogen, air, and the sample gas are fed into the flame, and the ions that result from combustion indicate the amount of carbon present in the sample.

infrared gas detector may not be required by the manufacturer. However, installation and verification procedures must follow the manufacturer's specified guidelines. A typical procedure begins by ensuring no gasses are present and then performing a zeroing operation. Next, the detector must be checked to ensure the output signal changes based on the presence of the detected gas. This can be accomplished by holding up the plastic test sheets provided by the manufacturer to simulate infrared absorption by a gas. The technician should note that this test

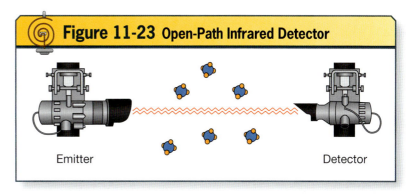

Figure 11-23. The detector measures the absence of infrared radiation that is absorbed by the methane molecules.

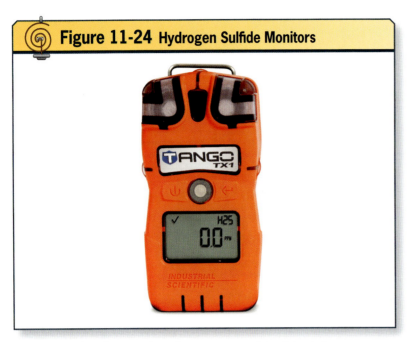

Figure 11-24. Personal analyzers are used to detect dangerous levels of hydrogen sulfide gas.

Courtesy of Industrial Scientific Corporation

verifies functionality but is not a true calibration procedure.

HYDROGEN SULFIDE MONITORS

Hydrogen sulfide (H_2S) is a flammable, colorless gas with a distinct odor, often compared to rotten eggs. Hydrogen sulfide exists naturally in natural gas and petroleum deposits. It is also a by-product of the breakdown of organic material. Because it is heavier than air, hydrogen sulfide finds low-lying areas and sumps and accumulates over time. The danger of hydrogen sulfide is twofold: it is both a toxic gas to breathe and an explosion hazard at elevated concentrations. Inhaling hydrogen sulfide causes lung and airway irritation, dizziness, vomiting, and unconsciousness. Just a few breaths of this toxic gas can lead to serious injury or death.

Natural gas and petroleum products that contain hydrogen sulfide are considered "sour" gas and typically go through a hydrodesulfurization process. This process removes the hydrogen sulfide and converts it into the element sulfur, a yellow powder used for fertilizer and pesticides. A relatively recent standard for diesel vehicle fuel in the United States is the requirement for fuel to be "ultra-low-sulfur diesel" that contains no more than 15 parts per million of sulfur. Previous standards allowed sulfur content of up to 500 parts per million.

Hydrogen sulfide monitors are available for permanent installation on the process equipment and in a portable form for individual worker protection. **See Figure 11-24.** A commonly used measuring method for these monitors is the electrochemical cell principle. In the electrochemical cell, the chemical energy of the target gas is converted to electrical energy. An oxidation reaction occurs inside the cell, resulting in free electrons that form a measureable current flow across the electrodes. This electric current is directly proportional to amount of gas, in this case hydrogen sulfide, present in the sample.

Any hydrogen sulfide monitor must be routinely serviced and verified to be certain it is operating within the specified guidelines. Calibration for personal monitors is performed on a site-specific interval and employs a calibrated gas. The calibration gas is a NIST traceable standard for verifying monitors. Daily bump tests are also performed at a bump test station to serve as a function check before the start of every work shift. Although not equivalent to calibration, bump tests confirm the unit is not clogged or inoperable. **See Figure 11-25.**

Workers who enter confined spaces where hydrogen sulfide may be present must test the space before entering. Hydrogen sulfide is denser than air and generally concentrates in low-lying areas. If gas is detected, continuous ventilation should be used to remove it. If this is not possible, the proper respirator apparatus to protect the worker is required, along with any other personal protective equipment.

Figure 11-25. A bump test station is used to verify that personal monitors sound an alarm when exposed to a test gas.

SUMMARY

The technician must understand the basic concepts involved with analytical measurement. The variety of measurement methods used for analytical instrumentation requires the technician to receive continuous training and consult the manufacturer when working with unfamiliar equipment. Life safety is the primary objective for all job tasks. Personal protective equipment, including gas monitors, must be used according to design and must be maintained properly. Many hazards in a process environment give no warning, such as smell or visible queues, for the technician to recognize.

REVIEW QUESTIONS

1. What is the measurement unit for conductance?
   a. Farad
   b. Henry
   c. Ohm
   d. Siemen

2. Which of the following is a substance that is added to a sample to cause a chemical change?
   a. Catalyst
   b. CEMS
   c. Hydrogen sulfide
   d. Reagent

3. Power plants report their total emission based on data from what system?
   a. CEMS
   b. Hydrogen sulfide analyzer
   c. Oscillating tube density
   d. pH analyzer

4. What term describes the perfect ratio of air to fuel?
   a. Excess
   b. Lean
   c. Rich
   d. Stoichiometric

5. Which description best describes chromatography?
   a. A color changing strip
   b. The absorption of infrared radiation as measured by the detector
   c. The separation of individual components by rate of travel through a medium
   d. The wavelength of light absorbed by a sample after exposure to a reagent

6. Calibration gases for analyzers must be traceable to which organization?
   a. EPA
   b. FDA
   c. NIST
   d. OSHA

7. An atmosphere containing 15.2% oxygen is classified as __?__ by OSHA.
   a. Acceptable
   b. Deficient
   c. Enriched
   d. Explosive

8. As __?__ increases, the __?__ decreases in most substances.
   a. ORP / natural oscillation frequency
   b. pH / temperature
   c. temperature / density
   d. temperature / pressure

9. The __?__ of an object determines the natural frequency of oscillations.
   a. mass
   b. percentage of hydrocarbons
   c. pH
   d. temperature

10. A hydrometer is read by observing the __?__.
    a. amount of deflection
    b. color
    c. digital readout
    d. graduated markings on the stem

11. A __?__ increases the rate of a chemical reaction.
    a. catalyst
    b. hydrocarbon
    c. ionic selective membrane
    d. reagent

12. Which detector is also known as a carbon counter?
    a. Catalytic bead sensor
    b. FID
    c. Gas chromatograph
    d. Infrared gas detector

13. Which highly dangerous gas present in many industrial processes smells like rotten eggs?
    a. Carbon dioxide
    b. Hydrogen sulfide
    c. Nitrogen
    d. Sodium hypochlorite

Process Controllers

Automatic controllers are used in a variety of applications. Batch control, process control, and emergency shutdown (ESD) systems are examples of automatic control systems. The technician needs to understand the purpose of automatic control, input and output (I/O) signal types, the components of a microprocessor, and the general functionality of automatic controllers. A technician who installs and calibrates field I/O should be familiar with the equipment used for automatic control.

Objectives

- » Compare the concepts of automatic control and manual control.
- » Understand the operational components of microprocessor controllers.
- » Describe basic pneumatic control devices.
- » Understand the various connection methods of I/O signals.
- » Describe types of programming.
- » Identify the purpose of ESD systems.

Chapter 12

Table of Contents

Fundamentals of Controllers 266
Automatic Control Concept 266
Field I/O Signals 267
    Binary Input 268
    Binary Output 270
    Sink and Source Devices 271
        Sinking Inputs 273
        Sinking Outputs 273
        Sourcing Inputs 273
        Sourcing Outputs 274
    Analog I/O 274
    Highway Addressable Remote Transducer Input Cards 275
    Resistance Temperature Detector or Thermocouple Input Cards 275

Systems and Applications 276
    Pneumatic Controllers 276
    Dedicated Loop Controllers 277
    Programmable Logic Controllers 278
    Instrument-Based Controllers 281
    Digital Communication Mapping 282
Control Program 282
    Ladder Logic 283
    Functional Block Programming 283
    Structured Text 284
Emergency Shutdown Systems 284
Human Machine Interfaces 285
Summary 286
Review Questions 287

For additional information, visit qr.njatcdb.org
Item #2690

FUNDAMENTALS OF CONTROLLERS

A controller is the key component to processing the information that is sensed throughout a control system. A controller resides at the midpoint of a control loop and performs in much the same way as the human brain: gathering information, performing complicated calculations, and ultimately orchestrating all movements necessary to perform a function. Information gathered by the various sensors is transmitted as binary, analog, and digital signals to the controller. Once this information is computed, a desired output signal is generated and transmitted to manipulate a final control element, such as a control valve.

Depending on a system's complexity, a process may require only one pneumatically operated valve controller, or it may require an assortment of grouped electronic controllers, computers, displays, and other devices for effective control. **See Figure 12-1.**

A complex process uses a combination of instrumentation devices and controllers, which allow a great deal of data to be collected, analyzed, stored, and manipulated to achieve the desired outcome, and at an ever-increasing speed.

Requiring an operator to manually take all required corrective actions is impractical and often impossible. An operator's attention can be focused on specific tasks, particularly startup and shutdown procedures, as well as maintenance procedures, while the controller or systems of controllers operate the process at a high rate of efficiency. No matter how sophisticated a system is, there is no substitute for the experience of a seasoned operator to monitor the control system and implement necessary manual methods of control when necessary, such as in an emergency situation. **See Figure 12-2.**

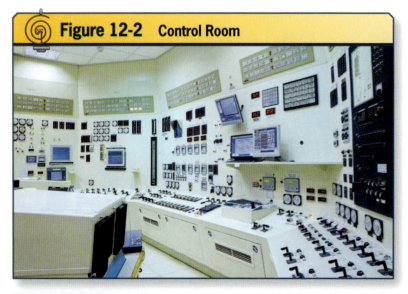

Figure 12-1. Pneumatic controls operate effectively on a clean, dry source of compressed gas, such as compressed air.

Figure 12-2. A centralized control room provides indication and status of an entire process, as well as a means of manipulating the process to make corrections.

AUTOMATIC CONTROL CONCEPT

Automatic control is the application of a control method for regulation of processes without direct human intervention. Imagine trying to consistently control a home heating system without automatic control. It would be significantly time consuming and inconvenient to constantly turn the heating system on and off to maintain a comfortable temperature within the house.

A basic home heating system thermostat operates on temperature and disregards the other atmospheric parameters of the house, such as humidity and continuous circulation of air. When the temperature drops below the value selected by the occupants of the house, the system activates to raise the interior temperature. **See Figure 12-3.**

When the temperature reaches the desired value, the system turns off. Automatic control systems neither replace nor relieve the operator of the responsibility for maintaining the process, because there is always a possibility for failure. Most importantly, if a control system fails, the operator must be able to take over and control the process manually. **See Figure 12-4.** In most cases, understanding how the control system works aids the operator in determining whether the system is operating properly and in choosing the actions required to maintain the system in a safe condition.

In older process environments, field operators relied on documents with historical data, such as logbooks, to help perform calculations to determine the process mixtures, valve alignments, and other factors. The modern process environment still calls for operators to predict what will happen to the process, but the control of the process is often dictated by a microprocessor-based controller such as a distributed or programmable logic control system. Remember a computer cannot perform functions that it has not been programmed to initiate. Computer systems can perform calculations extremely quickly; however, they can only perform calculations that were first designed and programmed by humans. As technology continues to evolve, controller software will take on artificial intelligence and will give the control system even more power to essentially think and reason through issues.

In today's *enterprise systems*, large-scale application software packages that support business processes, information flows, reporting, and data analytics in complex organizations, the interconnection of the factory floor and the front office is becoming more prevalent. Process management systems are becoming interconnected with inventory control systems, which are integrated with accounting systems, and so on. These systems work to make the process as efficient as possible.

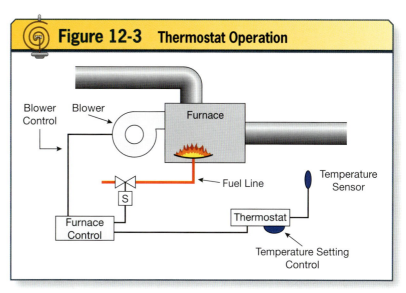

Figure 12-3. A thermostat-operated furnace maintains desired temperature conditions within a room.

FIELD I/O SIGNALS

Accurate transmission of signals is a key element of any process control setup. Typically, signals are thought of as being purely electrical. However, this

Figure 12-4. A remote control station is used for manual control of a process.

Figure 12-5. Wiring and tubing carry various signals among control devices.

Figure 12-6. Shielded wiring is often terminated by removing the foil shield, insulating the shield, and attaching the ground wire to an appropriate terminal.

TechTip!
Control wiring should not be mixed with power wiring.

is not always true. For example, clean, dry, compressed air is routinely used. In addition, a process medium is sometimes used as well, such as on a natural gas pipeline, where resources maybe limited. Regardless of the control signal medium, the piping or wiring must be completed in a neat and workmanlike manner. **See Figure 12-5.**

Electronic signaling offers a great deal of flexibility in terms of transmitted information, distance traveled, and signal accuracy. However, specific types of wiring and methods are required to guard against interference and sparks. Generally, shielded cables are used. The shields should only be grounded at one end, typically the controller or specified terminal block prior to the controller. This minimizes interference and eliminates ground loops. **See Figure 12-6.**

Binary Input

Binary inputs are typically high or low and on or off. When using pressure, the logic is set or reset much like a pressure switch. However, the action does not operate contacts; it merely controls air pressure. For instance, when air pressure builds to a specific amount, some type of action may occur. This action may initiate airflow to another device, or it may be self-contained such as an air pressure safety valve. The pressure relief valve spring tension is adjusted to a comparable air pressure so when air pressure builds above this pressure, the valve is forced open to atmosphere, relieving the pressure to a safe level. **See Figure 12-7.**

Microprocessor-based controllers accept an on/off signal in a variety of ways, depending on the manner in which they are used. The use of AC input cards allows an input to be connected much like a typical electric load

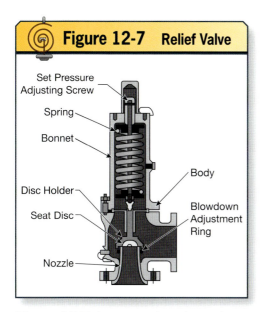

Figure 12-7. A pneumatic safety valve does not utilize any electricity. It is a purely mechanical device, which relieves pressure once a set point has been reached.

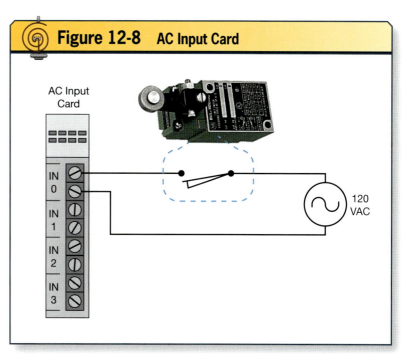

Figure 12-8. AC input cards can accept a range of voltage inputs to directly monitor the status of a control circuit.

For additional information, visit qr.njatcdb.org Item #2614

in a variety of voltages, such as 24, 120, and 240 volts AC. When the full voltage potential is applied across the input terminals, a signal is generated. Depending on the configuration of the input, the presence of voltage may generate a high or a low signal, similar to a closed or an open contact. In a conventional configuration, full potential across the input results in a high signal being generated. **See Figure 12-8.**

Another style of input card supplies its own voltage on the card, which must be switched by some type of contact closure. **See Figure 12-9.** These inputs can are generally meant for short distances, as well as applications in which low voltage control is in use. Because the voltage output by the card is generally low, a long wire could result in a voltage drop so severe that that the circuit could not be completed, regardless of contact position. In addition, the signal voltage could be susceptible to disturbances and dangerous conditions if the wiring is comingled with devices that use higher voltages. When signal cabling is comingled with power wiring, there is always the danger of high voltage making contact with the low-voltage conductors. However, it is more likely that voltages would be induced into the low-voltage wiring, which could create circuits that appear to be complete. These complete circuits would falsely trigger control components.

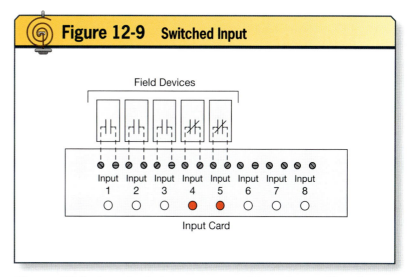

Figure 12-9. A contact closure across a binary input will register a high signal to the input.

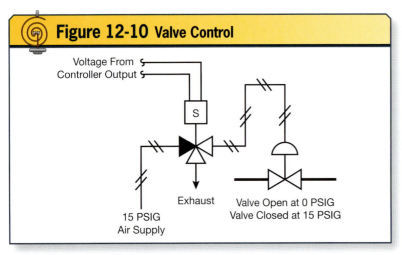

Figure 12-10. A solenoid valve is used to operate a larger process control valve, by controlling the air supply to the larger valve.

Binary Output

Binary outputs are the most basic output signals, because they have only two states: on and off. These outputs are excellent ways to control indicator lights, start and stop motors, open and close valves, or any of the possibilities that require something to be on or off.

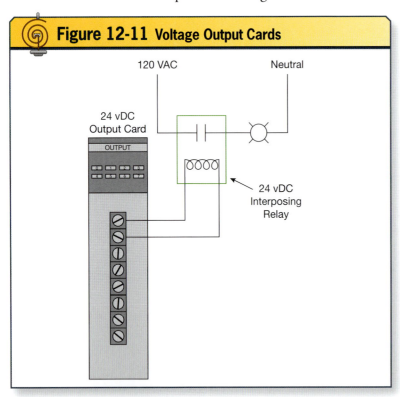

Figure 12-11. An interposing relay is used to isolate the controller output voltage from the control circuit.

Pneumatic signals outputs can be generated using some type of valve mechanism, known as a pilot device, to turn on the pressure. A variety of pressures can be controlled in this manner. The use of different pressure levels is much like the use of different voltages. Higher pressures can perform considerable work; however, lower pressures require less robust and expensive construction. Pressure levels, whether high, medium, or low, are defined by the design of the control system and facility standards. **See Figure 12-10.**

Programmable logic controllers (PLCs) can be used to perform a control function, which makes the output electrical. Voltage output cards control an output by applying a voltage potential to the device. For instance, a properly rated lamp may be connected directly to the output, without a subsequent power supply. These types of outputs, also known as sourcing outputs, can be convenient to use; however, they are highly susceptible to overloads and require properly sized fusing to protect the control equipment. To isolate and protect the electronic components of a controller, interposing relays are often used. *Interposing relays* are relays that provide a level of isolation between two electrical circuits. These devices, often conventional control relays, may appear unnecessary in some circuits until the need to isolate differing voltages is identified. **See Figure 12-11.**

Controller manufacturers have improved their product by incorporating the interposing relay on the controller circuit board; therefore, the field wiring terminates directly to the switching contact. These cards incorporate an on-board control relay so that the controller signal voltage is isolated from the field wiring. These types of outputs merely close a contact, so a voltage source must come from an auxiliary location. The use of relay outputs allows a variety of voltages, in both alternating current (AC) and direct current (DC) forms, to be switched by one type of card without auxiliary equipment, such as an interposing relay. **See Figure**

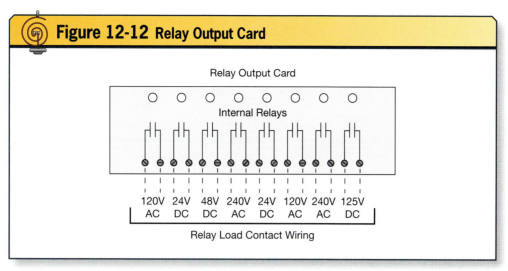

Figure 12-12. *Relay output cards provide a dry contact to perform the switching process.*

12-12. Whenever differing voltages are present, specific design and safety procedures must be followed, such as the *National Electrical Code® (NEC)* and *Underwriters Laboratories (UL) 508a* standard for industrial control panel construction.

Sink and Source Devices

The concepts of sinking and sourcing are the basis for electronic switching using transistors. With the increased use of solid-state switching devices in various industries, cards have been designed to sense the presence of DC voltage or ground by completing a sourcing or sinking input. These electronic inputs are capable of being used in a variety of applications, including sensing mechanical contact closure, but they shine when used in high-speed switching applications that are solid state driven. Mechanical switching processes rarely complete their switching action within a period that is adequate for the speed at which the process is operating. **See Figure 12-13.** In addition, because a current is being switched on and off, arcing occurs each time the contacts are opened and closed. This arcing results in wear of the contact surfaces.

Transistors are semiconductor devices made up of three distinct regions—the collector, the base and the emitter—capable of amplification, switching, and rectification. They offer a great solution to these issues. A transistor uses a small current flow to control a much larger current flow. Because of this unique property, the transistor makes an excellent choice for amplification and switching. Each region is constructed of either a P-type or an N-type semiconducting material in one of two combinations, either NPN or PNP. Properly biasing and applying a current to a

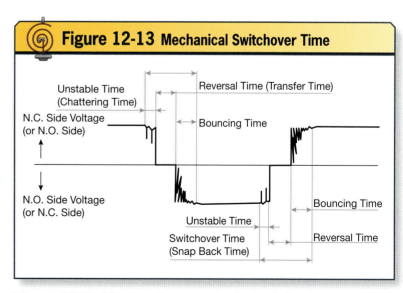

Figure 12-13. *The time that a mechanical switch takes to actuate can be unsuitable for high-speed applications.*

For additional information, visit qr.njatcdb.org Item #2615

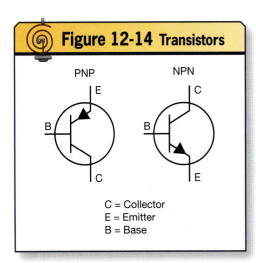

Figure 12-14. Transistors are semiconductor devices with three terminals and are available in two formats.

semiconductor to the correct regions of a transistor allows current flow in a specific direction. The application of a proper forward-biased current of these regions causes a larger current to flow, thereby amplifying the smaller current. However, the application of a reverse-biased current inhibits a current's flow. In other words, forward bias application of voltage or current turns a semiconductor on, and reverse bias application of voltage or current turns a semiconductor off. **See Figure 12-14.**

The sensitive nature of solid-state based inputs and outputs which make them ideal for solid-state switching applications creates special issues which must be addressed. These issues could lead to false triggering, erratic operation, and damage to microprocessor inputs and outputs.

Given the sensitivity of microprocessor-based inputs, it is possible for the input status to float in neither a fully on or off state, which will result in erratic operation. By using an appropriate sized resistor, an input may be pulled high or low to satisfy its normal state of operation, thereby eliminating any chance of a false trigger. For a digital input which references ground through a switched contact, a *pull-up resistor* is used to hold the input of the controller at a high value in the absence of an input signal, or closed contact to ground. Once the contact closes to ground, the circuit logic is set to the low value. Conversely, for a digital input which references a voltage source through a switched contact, a *pull-down resistor* is used to hold the input of the controller at low state in the absence of an input signal, or open contact to a voltage supply. Once the contact is closed to the voltage supply, the circuit logic is set to high.

Digital outputs of microprocessor-based controllers must be protected from damage when switching inductive loads, such as relays. The residual magnetic field present in an inductor (relay, contactor, solenoid coil), can result in dangerous voltage spikes which can destroy the solid-state outputs of microprocessor controllers. To combat this phenomena, some type of surge or spike suppression is necessary. Within AC or DC control circuits, a *snubber*, an electric circuit intended to suppress voltage spikes, is used. The snubber device is made up of different components such as capacitors, resistors, diodes, and varistors, depending on the intended application. The components are typically assembled in a package which makes connection to the control circuit relatively convenient. For DC control voltage applications, a flyback

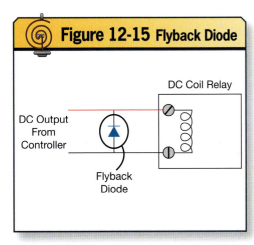

Figure 12-15. A flyback diode installed in parallel with the coil of a relay protects the output of a controller from harmful voltage spikes.

diode, may be all that is needed. A *flyback* or *snubber diode* is a diode used to eliminate a sudden voltage spike present in a control circuit when an inductive load, such as a relay or solenoid, is switched off. This diode is installed in parallel with the inductive coil, and absorbs the excess energy once the circuit is deenergized. **See Figure 12-15.**

As controller manufacturers continue to innovate, manufacturers may integrate these options within the cards or modules, thereby minimizing the need for additional components. Always consult the manufacturer's instructions and design documents for the proper use and calculation of values necessary for these devices.

Sinking Inputs

Sinking input modules or cards use a remote power supply in which the positive is connected to the field device (switching device) and the negative is connected to the input common. Once the switching device is closed, the circuit is completed and the input becomes high. Sinking inputs need to be used with PNP-style solid-state devices, because of the current path through the transistor within the device, to work properly. PNP-style devices are available in either normally open or normally closed styles. **See Figure 12-16.**

Sinking Outputs

Sinking output modules or cards use a remote power supply in which the positive is connected to the field device (the load) and the negative is switched high or low for on-or-off control of the load. **See Figure 12-17.**

Sourcing Inputs

Sourcing input modules or cards use a remote power supply in which the negative is connected to the field device (switching device) and the positive is connected to the input common. Once the switching device is closed, the circuit is completed and the input becomes high or turns on. Sourcing input modules or cards need to be used with NPN-style solid-state devices because

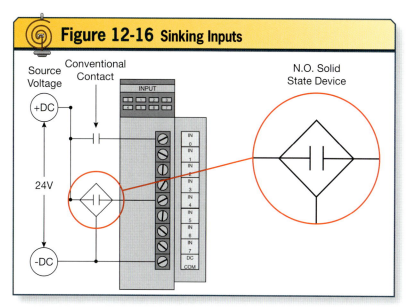

Figure 12-16. Sinking input connections can be used for both conventional and solid-state input devices. Solid-state input devices are identified by a diamond drawn around the conventional electrical symbol.

of the current path through the transistor within the device. To work properly, NPN-style devices are available in either normally open or normally closed styles.

A major disadvantage of this style input is that ground is used as the

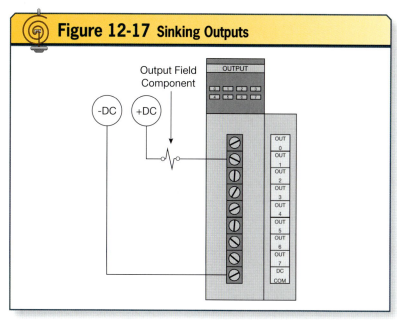

Figure 12-17. Sinking outputs switch the negative reference on and off to control a load.

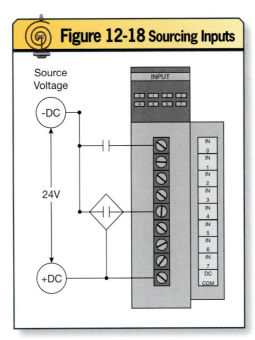

Figure 12-18. Sourcing input connections can be used for both conventional and solid-state input devices.

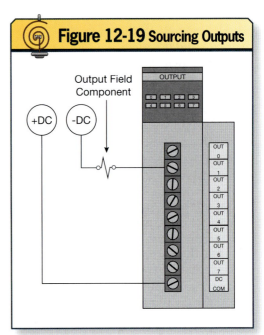

Figure 12-19. Sourcing outputs switch the positive reference on and off for a controlled load.

switch reference to determine whether the input is high or low. If the wiring becomes compromised or shorted to ground, it is possible to falsely trigger the input. **See Figure 12-18.**

Sourcing Outputs
Sourcing output modules or cards use a remote power supply in which the negative is connected to the field device (the load) and the positive is switched high or low for on-or-off control of the load. **See Figure 12-19.**

Analog I/O
Analog I/Os represent continuously variable signals. Whether these signals are a temperature transmitter sensing a rise in heat or the output of a controller commanding a cold water valve to open, they are constantly changing. When used in a pneumatic system, the variable pressure not only can operate a device smoothly but also can yield a powerful signal, depending on the size of the actuator. A signal pressure of 3 to 15 pounds per square inch gauge, when combined with an actuator area of 100 square inches, can result in a thrust force of 300 to 1,500 pounds.

Microprocessor controllers that accept or transmit analog signals can generally work with a variety of voltage and current spans. Depending on the manufacturer, cards are available as voltage, current, or both (user selectable). Field Instruments often need auxiliary power for operation. This voltage source most likely will be

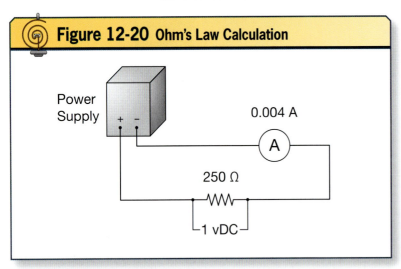

Figure 12-20. A voltage drop can be determined when the current and resistance are known.

separate from the controller power source; therefore, it is important to follow the manufacturer's directions for both pieces of equipment to ensure they are properly interconnected.

Because they are robust and reliable, 4- to 20-milliampere DC analog signals are present in an array of scenarios. However, in certain situations, only a voltage signal of 1 to 5 volts DC will suffice. The early electronic controllers would only accept such a voltage signal. Although *signal conditioners*, devices that convert one type of electronic signal into another type of signal, are available, they may not be necessary; a 250-ohm resistor and Ohm's law are often all that are needed. **See Figure 12-20.**

Placing the 250-ohm resistor in parallel with a 4- to 20-milliampere DC analog signal means the resulting voltage drop will be 1 to 5 volts DC. By using the resistor, it is possible to connect a current-based input or output to a voltage-based input or output. **See Figure 12-21.**

Highway Addressable Remote Transducer Input Cards

Some cards have the ability to accept not only a conventional analog signal but also a simultaneous communication signal on the same input. These cards, often called highway addressable remote transducer (HART)-enabled I/O cards, eliminate the need for specialized modems to access the additional signal information. This communication protocol has been adopted by many manufacturers to allow seamless communication between devices while still using the robust features of conventional 4- to 20-milliampere DC signal wiring. **See Figure 12-22.**

Resistance Temperature Detector or Thermocouple Input Cards

Resistance temperature detector (RTD) or thermocouple input cards allow direct connection of a temperature sensor without the need for a transmitter. This is especially useful if both devices are close to each other or if it is a critical

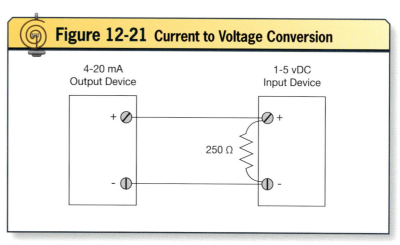

Figure 12-21. The connection of a fixed resistor generates a proportional voltage to the connected current.

Figure 12-22. HART input cards allow integration of the HART communication protocol without auxiliary modems directly to the controller, through existing wiring.

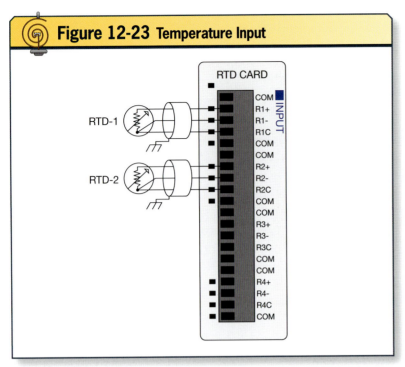

Figure 12-23. Temperature input controller cards allow direct connection of an RTD, eliminating the need for a transmitter.

the degree of accuracy required. Controllers are available for mechanical, pneumatic, electrical, hydraulic, and electronic methods of control. In each of these methods, the controller's job is to have the system give a timely response to any deviation of the process from the set point and thus to prevent an unsafe operating condition. Many of today's controllers are part of an overall Internet-connected enterprise. However, there are still pneumatic controllers in use that are purely mechanical, operating entirely on air pressure signals. The process plant is often so large that a *distributed control system (DCS)*, an assortment of communication, controlling, and measuring devices that the operations division uses to control the process, is implemented.

Pneumatic Controllers

In this technological age, it is easy to overlook that controllers began life as purely pneumatic devices. Using air pressure, springs, bellows, levers, and valves, air pressure can be precisely controlled. An example of a basic controller is the pneumatic timing relay. By varying the amount of air that can escape a compressed bellows, a delay can be instituted, thereby delaying the transition of a contact state. **See Figure 12-24.**

process for which added components could increase the risk of potential failure. **See Figure 12-23.**

SYSTEMS AND APPLICATIONS

Controllers use many techniques to initiate control actions, depending on

Figure 12-24. As air bleeds out of the nozzle, the delay action is created.

While a single timer works well under predictable and repeatable conditions, the need for controllers to react to ever-changing process conditions requires the addition of more components.

With today's smart instrumentation and programmable controllers, it is easy to insert a math function, such as square-root extraction for an orifice flowmeter signal to perform a square-root extraction. An older method is to use a pneumatic square-root extractor. By using a series of levers, cams, and other devices, an otherwise curved signal output of a differential pressure flow transmitter can be linearized.

Proportional–integral–derivative (PID) control may also seem like a control method reserved for computer-based control; however, pneumatic PID controllers have been operating reliably for years. **See Figure 12-25.**

Dedicated Loop Controllers

Dedicated loop controllers can operate as single-loop controllers or can be grouped as multiloop controllers. As the name implies, a *single-loop controller* is a device that performs a control function of one process variable only. Single-loop controllers are prominent throughout industry in both pneumatic- and electronic-based designs. In many cases, it is not only advantageous to control a loop separately but also required in a code safety situation, such as in gas-fired burner combustion systems. The early single-loop controllers included fixed instruction with minimal user configurations, but the controllers of today offer more flexibility for the single-loop application. Many of the available controllers are software-configurable systems, much like PLCs. In addition, these controllers can be monitored through communication networks just like other plant equipment. **See Figure 12-26.**

"Multiloop controller" is a broad term that can identify a grouping of single-loop controllers inside one enclosure or a computerized controller that can be configured to perform

Figure 12-25. A pneumatic PID controller is a purely mechanical controller, which requires zero electricity to operate.

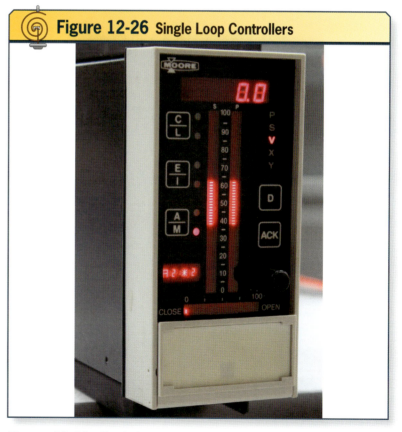

Figure 12-26. Single loop controllers are ideal for critical applications such as boiler burner control.

Figure 12-27. A multi-loop controller, such as a PLC, can perform a variety of control functions, often operating the multiple loops at once.

many control loops, such as a PLC. **See Figure 12-27.**

Programmable Logic Controllers

A microprocessor controller, commonly called a PLC, uses *digital control*, a control method based on mathematical calculations of numerical values of control signals known as bits. Just as a smart instrument must convert analog current signals into bits, a controller must do the same through the use of analog-to-digital converters (ADCs). The *central processing unit (CPU)* is the portion of a controller that decodes the instructions, performs the actual computations, and keeps order in the execution of programs.

There are several manufacturers of PLCs, but all models function generally the same way. Regardless of the type used, field technicians can interface with it and understand how it works with their field I/O. Advances by manufacturers are packing more processing power, speed, and communications into smaller packages. **See Figure 12-28.**

Figure 12-28 PLC Block Diagram

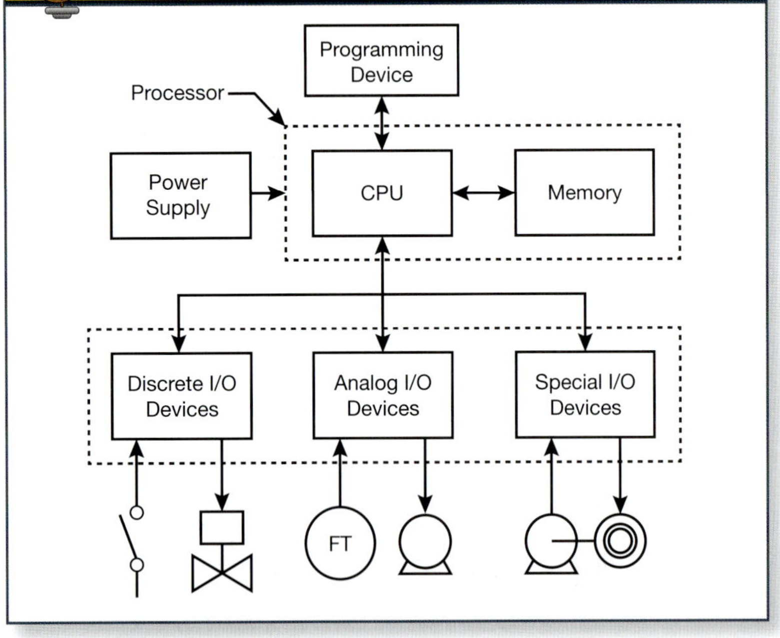

Figure 12-28. The basic components of a PLC include a power supply, a controller, input cards, and output cards.

> **TechTip!**
>
> Static electricity can pose a major threat when working with microprocessor-based devices. Always use electrostatic discharge protective measures when handling these devices. Failure to do so may cause immediate or latent damage to the microprocessor-based equipment.

Conventional slot or rack-mounted PLCs offer an array of functionality to operate modern processes. The modular design of rack-mounted PLCs allows a variety of configurations. Depending on a project's needs, the number of I/Os may vary greatly, so only the necessary cards need to be purchased and installed. Additional advantages include *controller redundancy*, with a backup controller in place that automatically assumes control if there is a failure of the primary controller, and the ability for *hot swapping*, or the ability to change controller components, such as I/O cards, without having to shut down the control system. These advantages minimize downtime and increase efficiency. **See Figure 12-29.**

The rack or back plane provides the communication and power links between the various plug-in modules. Racks have a finite number of *slots*, provisions for installing control cards such as communication, CPU, and I/O. By using the rack method, a card can be removed while maintaining power and communication to other cards. If all of a rack's available slots are in use, manufacturers typically have a system of auxiliary racks or a distributed I/O system. *Distributed I/O* consists of input or output modules that convert signal information to a communication protocol that can be communicated directly to a controller, simplifying the physical wiring layer. **See Figure 12-30.**

Typically, the power supply module is designed to handle the controller and auxiliary cards for the PLC. In some instances, they can also provide the loop power required for field-operable devices; however, it is often necessary and desirable to use a separate power supply for field devices. By separating the

TechTip!
Controller faults are often caused by faulty connections in the I/O devices. Always check wiring connections before focusing on the controller as the source of a faulty reading.

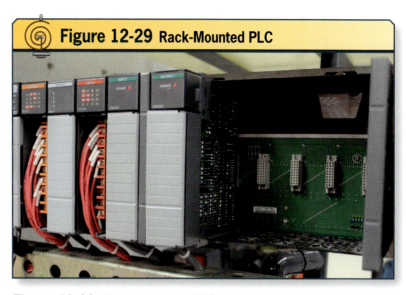

Figure 12-29. A rack-mounted PLC uses a chassis with plug-in slots to accept I/O cards.

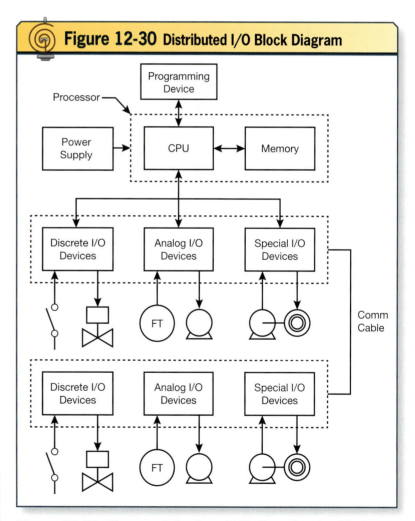

Figure 12-30. The use of distributed I/O allows for simplified wiring and ease of system expansion.

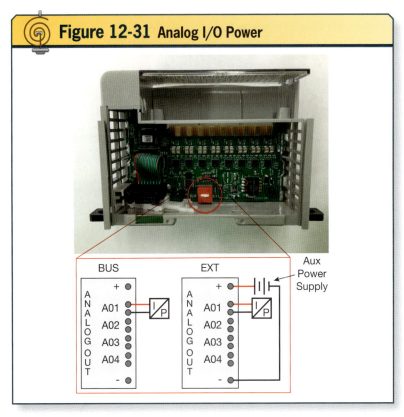

Figure 12-31. Analog output cards contain a switching method to select whether loop power is sourced from the PLC power bus or an external source.

Figure 12-32. Communication ports connect the controller to the automation network.

controller and field device power supplies, a failure of one will not affect the other. The PLC voltage supply typically is 120 or 240 volts AC input, with an output voltage of 24 volts DC. Once the 24 volts DC are wired to the controller, the voltage is supplied to all auxiliary cards using an internal bus link that connects all devices. Typically, analog current output cards require a switch setting to determine whether the PLC or the auxiliary power supply is used to generate the 4- to 20-milliampere DC output card. **See Figure 12-31.**

Communications are vital to a PLC's operation and monitoring characteristics. Although a program may continue to run if communication is lost, remote devices that are not hardwired to the controller will become inoperable. Additionally, remote notification systems such as email and text messaging will not function. Many controllers include native protocols within the controller, such as serial communication standards like Telecommunications Industry Association and Electronic Industries Alliance (TIA/EIA) *Recommended Standard (RS) 232* and *RS-485* or Ethernet-based communications. Native protocols do not require another piece of equipment, such as a converter or gateway, to communicate within the protocol. Direct connection of a shielded twisted pair or pairs, Ethernet cables, or even fiber-optic cables and some internal software configurations allow the controller to communicate with the connected world. **See Figure 12-32.**

In addition, many systems offer remote I/O points that assist in managing wiring. These I/O points use network communications to transmit I/O to a specific or group of controllers via programming. The use of remote I/O in a master-and-slave setup can significantly reduce infrastructure, because one Ethernet cable may carry the information that would be carried by multiple control cables.

PLCs are used with conventional operator interface controls, such as selector switches, potentiometers, and pushbuttons, as well as output devices

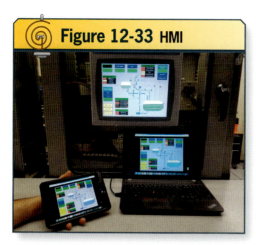

Figure 12-33. The use of graphics allows visualization of the process and places operator controls; such as buttons, switches and dials in a compact package that can be accessed in a variety of ways.

Figure 12-34. Compact PLCs offer some of the features of conventional controllers in a smaller, convenient package.

like relays, contactors, and solenoids. In addition, the advent of touch screen technology has made graphical user interfaces (GUIs) and human–machine interfaces (HMIs), the interaction between a human operator and a machine controller, increasingly prevalent. The systems of today are not confined to a typical control room or station. Remote interfaces are available through computers, tablets, and even cell phone apps. **See Figure 12-33.**

Compact PLCs offer a small package that encompasses the controller and I/O in one housing. These devices have a limited amount of expansion I/O and memory capability; however, they often support several communication protocols. **See Figure 12-34.**

Instrument-Based Controllers

Controllers that are integrated within either a primary or a final control device operate in a relatively stand-alone configuration. By providing control functions directly on the instrument, flexibility is gained. Instrument-based controllers, such as a differential pressure transmitter in combination with a flow computer, can independently control a valve without additional equipment, thereby maintaining required flow with less infrastructure. **See Figure 12-35.**

A flow computer, when used with a solar panel, battery, and cellular data connection, requires no other electrical connection. Additionally, a flow computer may include expandable I/O, which can be programmed to control additional equipment, such as a flow control valve, or safety apparatus.

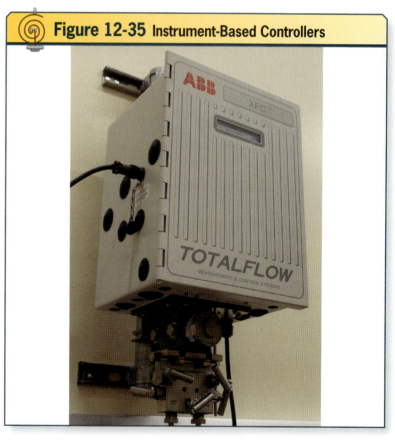

Figure 12-35. Instrument-based controllers offer stand-alone operation, especially in remote locations.

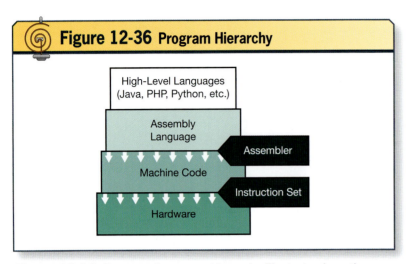

Figure 12-36. Program language hierarchy illustrates how the program that is written operates the hardware.

This type of communication, along with transmitter advances, means that more informational points are available from one sensor and one wire connection. A Coriolis meter, which once only transmitted a 4- to 20-milliampere DC, can now provide a variety of information to the control system. This device can now transmit the fluid temperature, density, housing temperature, power supply input, and onboard electronics temperature to monitor the overall device health. Each of these points is accessible network wide.

CONTROL PROGRAM

A controller, no matter how simple or complex, needs some type of routine to follow, known as a program. The International Electrotechnical Commission (IEC) has developed the standard *IEC 61131-3* identifying programming languages. Three common languages are ladder logic, functional block, and structured text.

The programmer develops the operation sequence using a high level language such as Python or Java. High level languages are used so humans can comprehend the instructions in the program. This language is then compiled by a *compiler*, a program that translates a higher-level language into a machine language that the CPU can execute. This process checks the program for execution errors and converts the information into the *machine language*, a computer programming language consisting of binary or hexadecimal instructions that a computer can respond to directly and the processor can follow. **See Figure 12-36.**

A program is read line by line, sequentially, via a process known as the *scan*, or periodically sampling in a predetermined manner each of a number of variables. The time it takes a controller to read through a program is known as the *scan time*, the time it takes for a controller to read the inputs, update the program, and apply the new output. This process happens in fractions of a second, which seems fast but in some situations

Digital Communication Mapping

With the further refinement of communication protocols, it is common to have primary input and final control output devices with onboard communication modules. These devices communicate using a protocol such as Modbus or Ethernet Industrial Protocol. When connected to the overall enterprise network, specific parameters can be transferred or mapped across the communication system through a direct connection or a gateway device. A gateway translates information from one communication standard to another.

For additional information, visit qr.njatcdb.org Item #2617

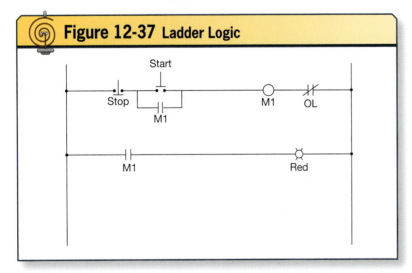

Figure 12-37. Each line of the drawing includes contacts and coils to operate a motor starter with a status light.

is not fast enough. Like a home computer, the processor and memory must be adequate to handle the written program. In many high-speed processes, if the scan time is too long, a desired change will be made after the process has been already changed by another factor. This is a particular problem with robotic-driven processes.

As the controller scans the program, it is simultaneously referencing each input, both physical and virtual. Each read value is calculated as required by the program, and the result is sent to either a physical or a virtual output.

Ladder Logic

Ladder logic programming appears to be simple, because it closely matches the ladder logic of motor control drawings. It is simple for technicians to learn as long as they have an understanding of motor control drawings.

Each line of the ladder logic program is built by dragging and dropping contacts and outputs to rungs of the "ladder." Outputs must be connected in parallel, but inputs can be connected in series or parallel, depending on the intended function. **See Figure 12-37.**

Functional Block Programming

Functional block programming is a programming method utilizing a graphical diagram that contains a preprogrammed control function. Functional

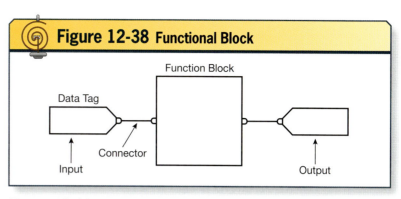

Figure 12-38. Function block input data connects on the left of the block, while output data exits on the right of a block. The interconnection of data tags and additional blocks is accomplished with connector lines.

blocks are set up to perform math, timing, counting, and PID functions, to name a few. In addition, some software programs allow custom blocks to be created to cover repetitive types of control logic, such as a motor starter holding circuit. This can simplify the building of a program. **See Figure 12-38.**

A typical functional block diagram shows several elements. **See Figure 12-39.** These elements include the following:
- Functional blocks—Blocks that contain the program functions
- Connecting lines, or wires—Lines that connect functions with inputs and with outputs
- References—Memory locations where information is stored

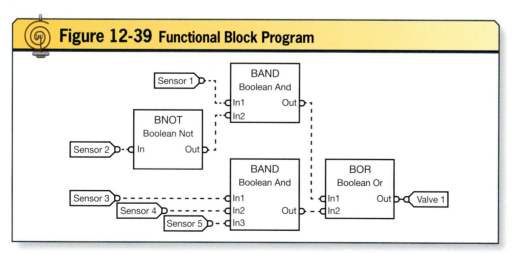

Figure 12-39. Functional block programming uses logic blocks connected together to perform an overall function.

Figure 12-40 Structured Text

```
1       IF bPressure_Switch_One AND bPressure_Switch_Two THEN
2           bShutdown_Solenoid   :=   TRUE ;
3
4       ELSIF bPressure_Switch_One AND bPressure_Switch_Three THEN
5           bShutdown_Solenoid   :=   TRUE ;
6
7       ELSIF b Pressure_Switch_Two AND bPressure_Switch_Two THEN
8           bShutdown_Solenoid   :=   TRUE ;
9
10      ELSE
11          bShutdown_Solenoid   :=   FALSE ;
12      END_IF
```

Figure 12-40. *Structured text uses textual statements to identify the logic of a controller.*

Structured Text

Structured text is a text-based control programming language, which uses statements such as "IF...ELSE...AND...THEN..." to accomplish a task. For example, if the tank level is high, then start the pump. Although seemingly simple, there is a *syntax*, which are the rules governing the structure of a language, to how the text must be input to the processor. **See Figure 12-40.**

As with ladder logic programming, the text is read left to right and top to bottom. Structured text is especially useful when the program must use an array of mathematical functions and complex equations.

EMERGENCY SHUTDOWN SYSTEMS

In areas that have hazardous substances or mechanical operations that pose a danger to people, technicians may find a PLC acting as a guard over the process. These PLCs are a part of the critical control systems often called the emergency shutdown (ESD) system. The DCS does not have extra shutdown actions written into the control program because the set points used for controlling are accessible to the operators who can change them. An ESD system is a stand-alone controller that receives inputs from field instruments, DCS control actions, valve positions, and more, but it does not have process variables that can be changed by human interaction. An ESD system monitors the process to observe that a dangerous condition does not exist. If a hazardous situation does arise, the ESD system assumes control of the process and executes an orderly shutdown. Sometimes, the field I/O assignments used for the DCS are the same as those for the ESD system, and sometimes, the ESD I/O assignments are separate. In extremely critical areas, a field technician often finds multiple sensors recording the same process variable, and all are part of the ESD system. These sensors are evaluated through the use of software to verify accuracy, deviation, and comparisons of like recordings. If a deviation occurs, procedures should be available for correction.

The Occupational Safety and Health Administration mandates that critical control I/O should be tested at intervals that ensure a failed component of the system will be detected before its failure causes a hazardous result. Dedicated procedures for testing ESD I/O should be available from the site where the technician works. Maintenance of critical I/O should be limited to the standard established for a particular device. With the development of increasingly smart I/O, the procedures are likely to cover a broader use of communication techniques for monitoring the process and for monitoring the accuracy, repeatability, and all other functions of instrumentation. ESD test results must be documented, and records should be retained for verification that the device was calibrated and maintained in a manner that meets all regulatory, environmental, and safety laws and standards.

Specialized safety PLCs and relays have been developed to assist in simplifying control systems. Although in many cases these individual controllers are more expensive than their counterparts, the wiring and infrastructure can be greatly simplified. The added cost for safety controllers is due to the built-in redundant processors, I/O monitoring, and communication equipment. Codes and standards dictate that these specialized controllers be identified separately from conventional controllers, typically by color coding them red.

For additional information, visit qr.njatcdb.org Item #2618

HUMAN MACHINE INTERFACES

As technology has progressed in terms of video and touch screen monitors, the image resolution and range of colors is endless. Unfortunately, it is easy to use too many colors, seemingly keeping things in order but in reality confusing the situation.

Situational awareness graphics are used to mitigate the confusion created by the instinct to use various colors to indicate statuses. Typically, these graphics are gray-scale images, with various shading of gray based on levels of concern. **See Figure 12-41.**

By using the gray-scale background, the operator's eyes do not tire from the range of colors and high-definition images. At this point, when a color is shown against the gray backdrop, it is discernible and draws attention to that status.

Although instrumentation technicians typically are not involved in process operation, they could view HMI screens as part of their work. Gray-scale imaging allows errors and issues to grab the attention of inexperienced operators as they become familiar with a system.

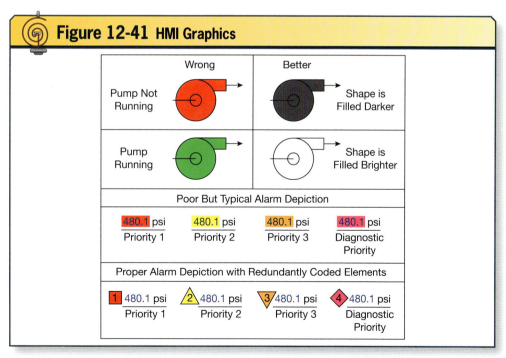

Figure 12-41. *HMI graphics utilize gray scale to indicate normal operating conditions, while the use of color is reserved for alarm conditions.*

SUMMARY

The many applications and types of controllers are a field of study that covers a broad range of disciplines. The technician must have a firm grasp of the objectives, components, and operation of an automatic controller. Many field instrument technicians, lacking understanding of its capabilities and limitations, see the controller as a source of faults. The technician must instead understand the necessity of providing adequate I/O and determining field I/O accuracy. Technicians should further investigate the languages, I/O configuration, settings, and control logic used in these systems.

REVIEW QUESTIONS

1. Which component of a microprocessor-based control system performs the calculations?
   a. Controller
   b. I/O Modules
   c. Power Supply
   d. Rack

2. When dealing with an automatic control system, the operator __?__ .
   a. actively starts and stops parts of the process
   b. changes set points continuously to make the process efficient
   c. does nothing
   d. observes the process and data provided by the system and makes minute adjustments if needed when an issue arises

3. A type of controller commonly used for boiler burner control is the __?__ .
   a. Compact PLC
   b. DCS
   c. Single Loop Controller
   d. PID Controller of any type

4. What common PLC programming language resembles a motor control diagram?
   a. Basic
   b. Functional Block
   c. Ladder Type
   d. Structured Text

5. Once a PLC program is written using one of the programming languages, it is __?__ .
   a. checked for conflicts and compiled into machine language by the software
   b. checked for conflicts and issues by another programmer
   c. compiled into machine language by the software
   d. ready for use

6. What does the acronym ESD stand for in terms of a process control system's safety design?
   a. Electrical Signal Dispatch
   b. Electric Static Discharge
   c. Emergency Shut Down
   d. Energized System Diagram

7. Safety relays and PLCs are typically identified by which color?
   a. Orange
   b. Purple
   c. Red
   d. Yellow

8. Solid state switching devices are indicated by a __?__ around the contact.
   a. circle
   b. diamond
   c. square
   d. triangle

9. Which type of control card turns positive voltage on and off to a device?
   a. Sinking Input
   b. Sinking Output
   c. Sourcing Input
   d. Sourcing Output

Fundamentals of Control

Industrial processes are controlled using different methods that are selected based on both the desired outcome and the nature of the process design. An automatic controller may use several control methods to allow an automatic control process to achieve the desired result. The instrument technician must understand how the controller implements the control method to properly maintain and troubleshoot a system. In addition to the understanding the control method, the instrument technician must understand the inherent characteristics of the process, such as process capacity and process lag.

Objectives

- » Identify and explain the operation of closed-loop and open-loop control.
- » Explain the operation of two-position control, such as on/off and time-cycle control.
- » Discuss continuous throttling and its elements of proportional, integral, and derivative control.
- » Explain advanced control schemes, such as cascade, ratio, and feedforward control.

Chapter 13

Table of Contents

Fundamentals of Control 290
Control Applications 290
    Open-Loop Control 291
    Closed-Loop Control 292
Control Modes 293
    Two-Position Control 293
        On/Off Control 293
        Time-Cycle Control 294
    Continuous Throttling Control 294
        Proportional Control 294
        Integral Control 295
        Derivative Control 296
        PID Control Action 297

Advanced Control 298
    Cascade Control 298
    Ratio Control 299
    Feed-Forward Control 300
Summary 300
Review Questions 301

For additional information, visit qr.njatcdb.org Item #2690

FUNDAMENTALS OF CONTROL

The controller is the portion of the process loop that makes a decision. This decision is made according to the design of the controller, either mechanical or electronic. Early process controllers relied on cams, levers, and springs to implement control. These methods have been replaced largely by electronic controllers that are capable of higher accuracy and speed.

An algorithm is a set of rules used to solve a problem. No matter how fast or accurate a controller may be, the controller will only ever be as good as the algorithm it executes. Humans use algorithms every day, from following a recipe to cook dinner to deciding whether to wear a warm coat. The human brain has a distinct advantage over the process controller in that it can continually learn and adapt an algorithm for optimum performance. While computer scientists are continuing to develop computers with learning capabilities, controllers such as programmable logic controllers and single-loop controllers only execute the algorithm programmed by a human.

The process controller algorithm must be suitable for the process under control. Take bathing as an example: simply turning the hot water valve fully open or fully closed would not achieve the desired output. Similarly, driving an automobile would be quite an adventure if the gas pedal only had two positions: fully depressed or fully released. Both of these situations require the ability to throttle the control element to achieve the desired result. In contrast, consider a sump pump in the basement of a home. The operation requires the pump to come on when the water level in the sump rises and to go off when the water level falls below a specific point. Simply turning the pump on or off is sufficient to achieve this outcome. Purchasing an expensive electronic controller that uses a sophisticated control algorithm would not provide a better outcome.

CONTROL APPLICATIONS

To select a proper method of control, the process must be understood thoroughly. Though the ongoing advancement of electronics continues to improve many processes, there is still a time and a place for nonelectronic methods of control that use pneumatic, hydraulic, or even water pressure to transmit signals.

A control loop typically consists of three elements: the sensor, the controller, and the final control element. The sensor measures the process variable, the controller executes the control algorithm, and the final control element implements the change as directed by the controller. The response of the system depends on several variables. Process *capacity* is the ability of the process to store energy or material. Consider a process tank with a volume of 10,000 gallons. If the inflow rate of a product is 7 gallons per minute, the tanks capacity will prevent the level in the tank from changing rapidly. Now consider a different process tank with a 20-gallon capacity. An inflow rate of 7 gallons per minute would lead to a rapid change in

TechTip!
Many process controllers have an auto-tune feature. This feature assists the technician in selecting the values for the control algorithm that lead to a stable process.

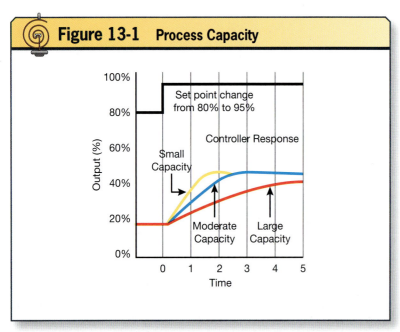

Figure 13-1. A process with a small capacity responds to a set point change faster than a process with a large capacity.

level. A temperature-controlled process in a large tank of fluid may take several hours to rise 10°C. In contrast, a heat exchanger fed by a steam supply may experience fast temperature changes if the inflow product temperature changes. Process capacity is a major factor in evaluating the response time of a control algorithm. **See Figure 13-1.**

Dead time is the time that elapses between when an instrument input changes and when the instrument responds to the change. Dead times are additive in a system. For example, the dead time of a temperature transmitter to respond adds to the dead time that the controller requires to run a program scan and set the output. This delayed output then travels to an I/P transducer, a device that requires some time to convert an electric current signal (I) to a pneumatic pressure signal (P) and adjust the pressure output. Finally, the pneumatic actuator diaphragm must deflect due to the change in pressure. All these actions require time to complete. Meanwhile, the change implemented by the control valve is based on information collected by the sensor some time ago. The value of the process variable may have changed since the measurement was taken, resulting in a correction that may not achieve the desired outcome. Depending on the overall speed of the process change, dead time can cause major trouble in controlling a process. The location of sensors can play a major role in the dead time of a process.

The combination of process capacity and dead time causes a lag of the process variable behind the controller output. This *process lag* is time it takes for a process to recover from a change in loading or set point. **See Figure 13-2.** The curve of this response can be described by a measurement of time called a *time constant*, the time required for a system to reach 63.2% of the final value. A system that has fully recovered is commonly assumed to have reached the final value after 5 time constants, even though the system has truly only reached 99.3%. **See Figure 13-3.**

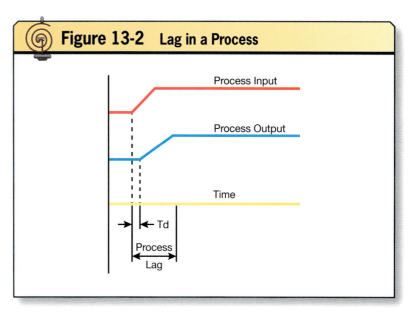

Figure 13-2. All processes exhibit some degree of both dead time (Td) and process lag.

Open-Loop Control

Open-loop control is control without feedback and is a relatively simple form of control. This makes it suitable for only certain types of applications. If the process is generally steady and produces consistent results based on historical data, an open-loop controller can give satisfactory results. If the process is not

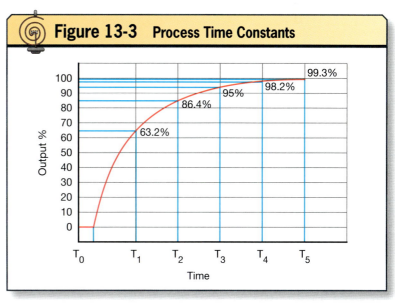

Figure 13-3. The time constants (Tx) of a process each consist of 63.2% of the distance between the current state and the final state of the process.

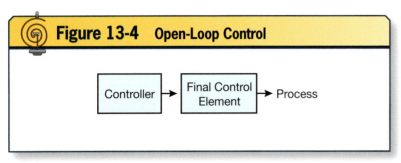

Figure 13-4. Open-loop control does not use feedback to achieve the desired outcome.

steady, inconsistent results may occur because open-loop control does not monitor the outcome of the process and any deviation in the process control will not be realized or corrected. **See Figure 13-4.**

A controller in an open-loop control scheme can be a simple as a valve and garden hose. When a plant needs to be watered, the valve is turned on and allowed to run. There is no measure of the amount of water or even whether water begins to flow through the hose to the plant. Often, once the water appears to build up enough around the plant or a specified amount of time has passed, the controller—through a manual, timed, or other command—closes the valve. In many cases, the expected outcome of an open-loop controller is based on prior experience or historical data. In this application, open-loop control relies on the process variable, the water source, always behaving in a predictable manner; i.e., water flows when the valve is opened, and it always does so at the same rate. **See Figure 13-5.**

Open-loop controllers serve the purpose of trying to remove error from a system to get desired results. In the previous example, the dry soil needed water and a final control element controlled a substantial source of water; therefore, the error was high in the system and was addressed by commanding the valve to open. The results of open-loop control can be unpredictable if the process deviates from the state for which the control method was designed.

Closed-Loop Control

Closed-loop control uses a feedback mechanism to collect information on the effectiveness of the current controller output and makes corrections based on the process variable. By measuring the actual results of a process following the initiation of a change to that process, the controller can make adjustments accordingly. Closed-loop control is also called negative feedback control. **See Figure 13-6.**

When open-loop control is used to water a plant based on a schedule, the outcome of the control loop is based purely on an expected condition. Unfortunately, there is no way to guard against any type of deviation occurring within the process. For example, what happens if the hose becomes partially kinked, limiting water flow, or a hot, dry day is causing the water to evaporate too quickly to saturate the soil? By adding feedback into the controller, the effectiveness of the control algorithm can be determined.

In the case of the plant-watering scenario, the control system could use feedback by measuring the soil moisture. A moisture sensor could be added

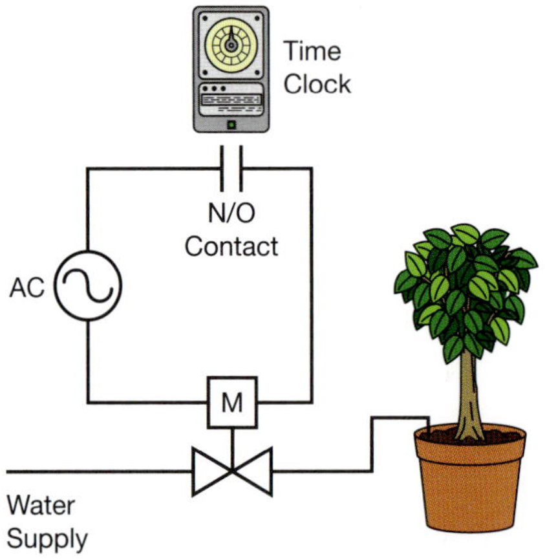

Figure 13-5. The time clock operates a set of normally open contacts. When closed, the motor-operated (M) valve opens and water flows to the plant for a specified amount of time.

to the soil surrounding the plant's root system to indicate how much moisture is reaching the roots. A closed-loop controller works by trying to minimize the error present in the system. To determine error in the system, a desired set point is required, in this case 40% moisture content. A measuring device first senses the process variable: 5% moisture in this example. An error comparator in the controller then evaluates the difference between the process variable and the set point:

Error = Process variable − set point

Error = 5% − 40%

Error = −35%

The resulting value, −35% moisture in this case, is used by the controller. The control algorithm in the controller exams the error input and determines whether the valve needs to be open or closed. In this case, the process variable is lower than the desired set point, so the final control element should cause the water valve to open. This operation runs continuously until the measured error is zero.

CONTROL MODES

The *control mode* is the type of control action used by a controller to perform a control function, such as on/off or proportional–integral–derivative (PID) control. Whether simple or complex, control modes are the basis of how a process maintains its stability. Some processes can be effectively controlled using simple on/off control of a single control loop, while others have multiple control loops that use complex control algorithms to throttle the final control element. Working knowledge of how control modes function, along with how a process is expected to operate, is essential to understanding whether a process is functioning correctly. Though instrumentation technicians may not be the primary designers of a control scheme, they are often involved in the start-up and tuning of the process.

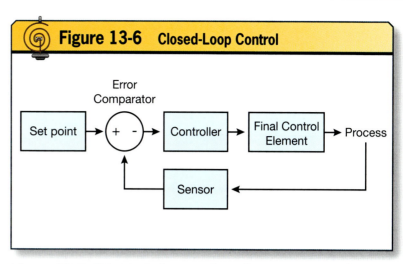

Figure 13-6. This block diagram shows the components of a closed-loop control system.

Two-Position Control

Two-position control results in the final control element being set to one of two discrete positions. An example is a valve that is fully open or fully closed. Two-position control is the easiest and least expensive method of control. It is a good choice for use in processes that can be safely and efficiently cycled on and off. A variety of devices with mechanical, pneumatic, hydraulic, and electronic components can be used to perform on/off control.

On/Off Control

In on/off control, when the set point is reached by a process variable from either direction, the final control element changes state. Processes that use on/off control typically exhibit two similar characteristics. The first characteristic is that precise control is not needed. The second characteristic is that the process has a relatively large capacity to resist change. If the system capacity is too small, it can be easy for this control method to regularly overshoot the desired output.

On/off control often results in oscillation around the set point. The addition of dead band prevents the process from cycling incessantly. However, the intentional addition of dead band to the controller results in lowered accuracy. Instead, an acceptable range of the

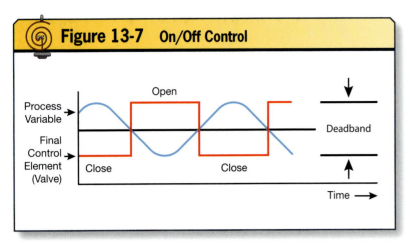

Figure 13-7. On/off control provides affordable control for processes with large capacities that do not require a high degree of accuracy.

process variable is defined. **See Figure 13-7.**

Time-Cycle Control
In time-cycle control, a time period is used to regulate the process variable. Appliances such as an automatic clothes washer use time-cycle control. Each wash cycle is based on a fixed amount of time that has been predetermined to produce the desired outcome. The washer completes its cycle in the allotted time set by the manufacturer, or in some cases a user-selected value. Once the cycle has been completed, the machine begins the next process, such as a rinse cycle, again following a predetermined amount of time. With time-cycle control, there is not necessarily a measure of the process outcome. If the results are unsatisfactory, the process can be repeated or the individual cycle times can be adjusted to correct the error.

Continuous Throttling Control

A process that requires a high degree of accuracy is often not properly controlled using on/off control. Instead, a controller algorithm that solves for the optimum position of the final control valve within the range of 0% to 100% is needed to provide adequate control. Due to the variable nature of this type of control, it may be called continuous throttling control or modulating control. Continuous throttling control is most commonly implemented in the controller by some form of a PID algorithm. As the name implies, the individual components of the PID algorithm are proportional, integral, and derivative action. **See Figure 13-8.**

Proportional Control
Proportional control is a control method that causes a control response to be made in proportion to the error amount between the set point and the measured value. The word "proportional" is appropriate to describe the control

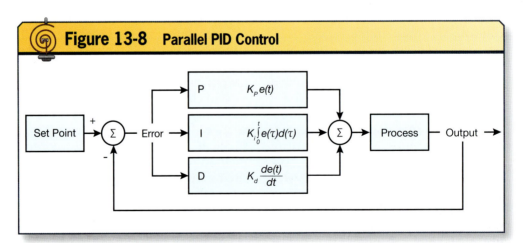

Figure 13-8. The parallel PID algorithm is a feedback control method that uses proportional, integral, and derivative functional blocks to determine the controller output.

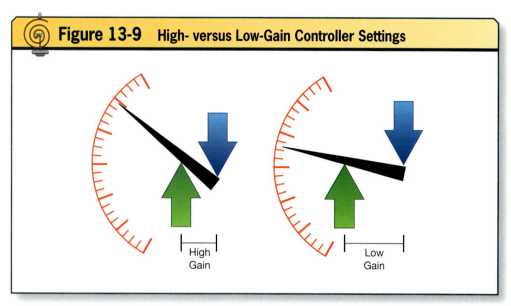

Figure 13-9. *The gain of a controller affects the degree of response, indicated on the red scale, to an error in the system, represented by the blue arrow. The gain setting works similarly to moving the fulcrum point of a balance beam, resulting in a mechanical advantage.*

action, because a large error in the system results in a large change in the controller output and a small error results in a small change in the controller output. The degree of proportional control, often referred to as the gain, is adjustable in the controller. A controller with a high gain setting responds very aggressively to an error in the system. In contrast, a controller with a low gain setting responds less aggressively to an error in the system. **See Figure 13-9.** A proportional controller with a sufficiently high gain setting functions as an on/off controller, driving the controller output signal to either extreme of the output range.

When sufficient gain is applied to a proportional-only controller, the output will stabilize; however, it will not quite reach the set point. This is the major drawback of using proportional-only control. This error is referred to as *offset*, the difference between the process set point and the steady-state response of the process variable. **See Figure 13-10.** At this point, if the gain value is increased further, the controller will begin to oscillate, because the controller output will be approaching zero.

Integral Control

Integral control is a control method that finds the sum of error in the process system over time. The word "integration" is used in calculus to describe the method used to find an area under a curve. This control method measures the amount of residual error, or offset, that is a result of proportional-only control.

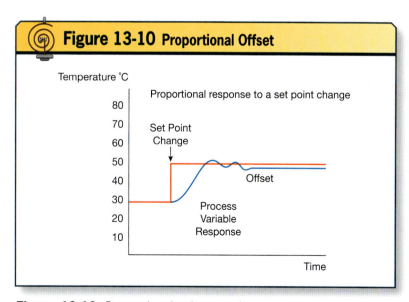

Figure 13-10. *Proportional-only control can result in an offset between the process variable and the set point.*

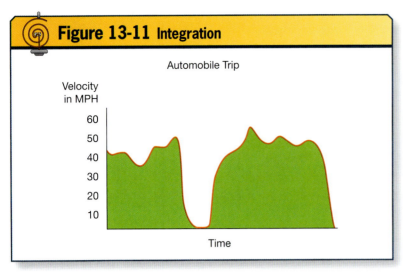

Figure 13-11. The total area (green) under the velocity line (orange) is summed, or integrated, to find the distance traveled.

Consider the example of the odometer in an automobile that is used to determine the total number of miles traveled. Whereas the speedometer gives an instantaneous value of the velocity of the car, the odometer provides data relating to the distance over time that car has traveled. As the car accelerates, decelerates, or stops to refuel the gas tank, the odometer sums, or integrates, the miles traveled. This is the concept of integration: measuring a quantity over a time period. **See Figure 13-11.**

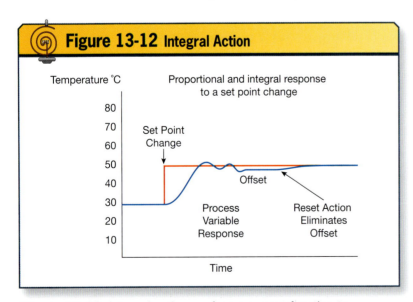

Figure 13-12. Integral action performs a reset function to eliminate persistent offset in a process variable.

The settings in an integral controller determine how often the controller resets the controller output to eliminate any offset between the process variable and the set point. Thus, a frequently used industry term for integral settings is reset. The frequency of the reset function is determined by the time setting in the controller. This period can be set to repeats per minute (RPM) or minutes per repeat (MPR). This setting tells the controller the period over which to sum the total error and reset the controller output. This integral action attempts to eliminate the offset in a control system. **See Figure 13-12.**

An unwanted consequence of using integral action is integral windup. *Integral windup* occurs when an error exists for long periods of time or when large set point changes are made, resulting in the integral action measuring a large offset. This, in turn, causes the integral action to drive the controller output to a high correction value. In some cases, a valve is moved to the 100% open position in an attempt to increase flow to the process but the integral action keeps calling for the valve to open even more, an impossible action. The result of this integral windup is that the controller may overshoot or never reach the set point. Reducing the set point back to a value that is achievable by the valve may fix the problem. Another solution to integral windup is integral windup protection logic inside the controller that limits or disables the integral action when the actuator reaches the maximum value.

Derivative Control

Derivative control is a control method that measures the instantaneous rate of change in the process system. The concept of derivatives is used in calculus to determine the slope of a point on a line. This control method is also called the rate because it measures the rate of change within a system. Derivative control is well suited to slow-responding processes that can experience overshooting of the set point. Because derivative control measures the rate of change, the controller output is adjusted to cause the process variable to gently settle at the set point instead of building up too much speed and continuing past the set point.

An example of derivative control action is observed in the operation of an automobile. Consider a human operator applying pressure to the brake pedal as the car approaches a stop sign. The rate of change in the speed of the car is used to determine how much pressure should be applied to the brake pedal. While a novice driver may stop too soon or depress the pedal abruptly due to a miscalculation of the rate of deceleration, an experienced driver is able to calculate the rate of deceleration and smoothly bring the car to a stop precisely in line with the stop sign. **See Figure 13-13.**

Derivative action is useful for processes that change relatively slowly and do not produce significant signal noise. For example, the flow through a heat exchanger can exhibit significant variability as measured by a flowmeter. This rapidly changing signal is interpreted by the derivative action as a rapid rate of change. In response, the derivative action may cause the process to become unstable. Therefore, derivative action is commonly turned off in flow applications.

PID Control Action

When an error is introduced to a PID controller, the controller's response is a combination of proportional, integral, and derivative actions. As the error increases, the proportional action of the PID controller produces a change in controller output that is proportional to the error signal. The integral action of the controller produces an output that is determined by the size of the error over a period of time. The derivative action of the controller produces an output whose magnitude is determined by the instantaneous rate of change. These three elements combine to produce a process variable response that approaches the set point with minimal overshoot and zero offset. **See Figure 13-14.**

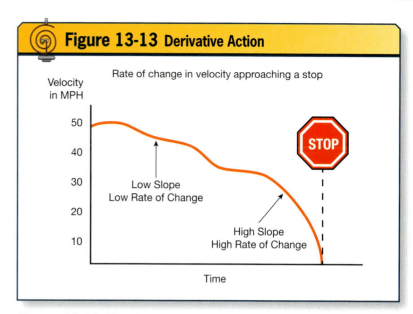

Figure 13-13. The derivative action of a controller measures the rate of change and applies the appropriate output to drive the process variable to the set point.

For additional information, visit qr.njatcdb.org
Item #2577

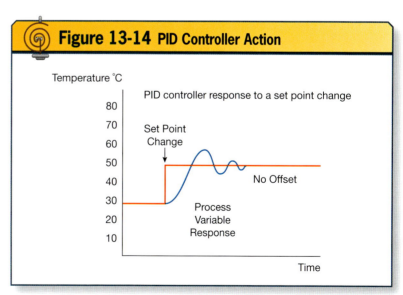

Figure 13-14. The PID controller responds to a set point change by causing the process variable to adjust to the new set point with minimal overshoot or offset.

Figure 13-15 PID Control Modes

Action	Effect
Proportional action	Causes the controller to react to the magnitude of the calculated error based on present information
Integral action	Causes the controller to react to the residual error that accumulated in the past
Derivative action	Causes the controller to react to the rate of change in the process variable to predict the future value and prevent overshooting of the set point

Figure 13-15. *PID control modes can be described in simple terms.*

ThinkSafe! Extreme care should be taken when adjusting any of the PID settings. Changes can cause a process to become unstable and potentially unsafe to those present around it.

The PID controller algorithm can be confusing for the instrument technician to fully comprehend. Some simple heuristics are helpful in remembering the purpose of each action. **See Figure 13-15.**

ADVANCED CONTROL

In addition to single feedback control loops, control strategies such as cascade loops, ratio control, and feed-forward control may be used to control a process. A particular process often benefits from using one of these advanced system designs to achieve the desired results.

Cascade Control

Cascade control is a type of control where two controllers, each on their own control loop, control a single process variable. An inner loop and an outer loop are designated, as well as an inner loop controller and an outer loop controller. The inner loop in a cascade control system must have a faster response time than the outer loop. The inner controller receives its set point from the outer loop controller. The outer loop controller receives its set point from the operator or programmed set point. **See Figure 13-16.**

Consider a continuously stirred tank reactor with a water jacket to cool the product. The primary objective is to ensure the product leaving the tank is the correct temperature. This relatively slow process is controlled by the outer loop in this cascade control scheme. An additional variable in this system is the cooling fluid supply. Depending on the demand of the cooling system, the flow rate of cooling fluid to absorb heat and transfer it from the process is unpredictable. Suppose the process plant had several operations that intermittently used the same cooling fluid supply. There must be a way to directly measure and control the fluid flow to control the product temperature. Cascade control is well suited for this application. An inner loop that measures the flow rate of the incoming cooling fluid could be used to regulate the flow of fluid to the

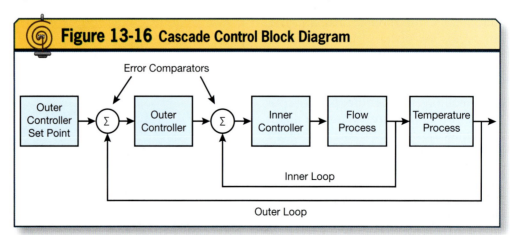

Figure 13-16 Cascade Control Block Diagram

Figure 13-16. *Cascade control is a combination of an inner and an outer control loop.*

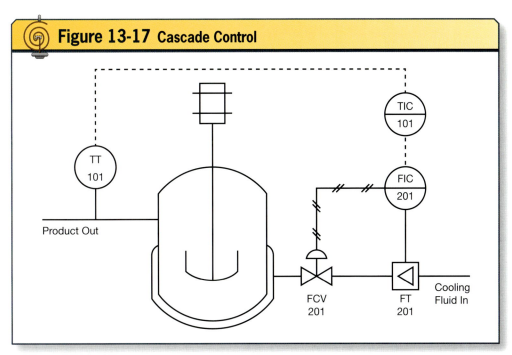

Figure 13-17. The outer loop sensor (TT 101) measures the outflowing product temperature. The inner loop sensor (FT 201) measures the cooling fluid flow rate. The outer controller (TIC 101) provides the set point for the inner controller (FIC 201).

cooling jacket. The outer loop would continually measure the temperature of the product leaving the tank and update the set point in the inner loop controller. This control scheme removes significant lag from the system by making quick adjustments based on system demand. **See Figure 13-17.**

Ratio Control

Ratio control is used to control two or more components of a process in a predetermined ratio. When a process involves the combining of two or more products, ratio control is an effective control method. Ratio controllers can be stand-alone or integrated into programmable controllers, depending on the application. Consider the example of a making a carbonated soft drink. The continuous process of blending the proper ratio of carbonated water to flavoring syrup requires precise control. Assume the ratio is nine parts carbonated water to one part flavoring syrup. The carbonated water flow through the system represents the uncontrolled flow, and the flavoring syrup is added in proportion to any flow rate. An increase in the flow of carbonated water therefore requires an increase in the flavoring syrup. **See Figure 13-18.**

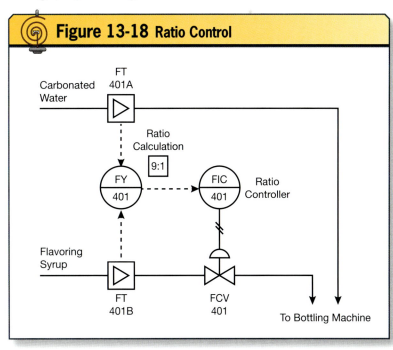

Figure 13-18. The ratio of 9:1 is measured by the ratio calculator (FY 401) and controlled by the ratio controller (FIC 401).

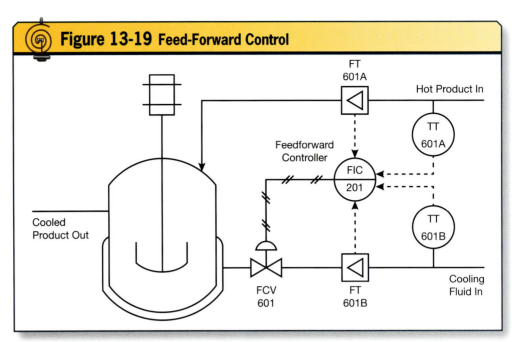

Figure 13-19. *The feed-forward controller measures the load changes, consisting of the cooling fluid temperature, hot product temperature, and hot product volume flow rate. The controller then predicts the proper amount of cooling fluid to flow into the cooling jacket.*

Feed-Forward Control

Feedback control is limited in the ability to control a process by any dead time in the system. *Feed-forward control* is a method of control that attempts to correct for a disturbance to a process before the disturbance makes any effect on the process. This method of control requires the process to be well understood and the system to account for all energy flows. A perfectly modeled process that always responds as designed would be the ideal candidate for feed-forward control. The manipulated variable would be adjusted to offset any anticipated disturbance to the system, resulting in a stable process variable.

Consider an example of a process that removes heat from a process fluid by using a cooling fluid. **See Figure 13-19.** A perfectly modeled system would account for the rate of heat transfer from the cooling jacket to the process fluid, the specific heat of the cooling fluid and the process fluid, and a constant ambient temperature and pressure. If all of these variables were known, the precise amount of cooling fluid could be metered out and placed in the cooling jacket, resulting in a product with a temperature that exactly matches the set point. In practice, feed-forward control commonly relies on a feedback loop to verify that the implemented control achieves the desired outcome.

SUMMARY

Process controllers rely on control algorithms to achieve control of the process variables. Control ranges in sophistication from simple on/off control to PID algorithms that provide control based on past, present, and future process values. The instrument technician should be familiar with various methods of control schemes to start up, loop tune, and troubleshoot control systems.

REVIEW QUESTIONS

1. What is the specific set of rules used by a process controller to solve a problem?
   a. Algorithm
   b. Dead band
   c. Final control element
   d. Hysteresis

2. A steam heat exchanger has a flow rate of 20 gallons per minute and a volume of 1 gallon per minute. The process fluid leaving the heat exchanger exhibits unstable temperature fluctuations around the set point. What is the likely cause of this instability?
   a. The controller is set for proportional control.
   b. The controller is set for reverse action.
   c. The process has a large capacity for energy storage.
   d. The process has a small capacity for energy storage.

3. A process is maintained at a level of 152 inches. A set point change is made to increase the level to 167 inches. The process time constant is found to be 31 seconds. What is the level in the tank after 1 time constant?
   a. 156.1 inches
   b. 159.7 inches
   c. 161.5 inches
   d. 164.3 inches

4. An instrument technician is installing the control system for an industrial parts washer. The parts washer contains a cleaning solvent that is supplied to spray nozzles by a pump. The operator sets the parts washer to run for 30, 60, or 90 minutes depending on how many parts are put into the washer. What type of control is used by this machine?
   a. Cascade control
   b. Closed-loop control
   c. Open-loop control
   d. Proportional control

5. An injection molding machine is used in a plant to manufacture car door panels. A conveyor assembly moves bulk plastic beads to a hopper, which is equipped with an optical level sensor. When the hopper is at the correct level, the conveyor stops feeding plastic beads. When the level drops below the level sensors, the conveyor starts feeding plastic beads into the hopper. What type of control is used by this machine?
   a. Cascade control
   b. Closed-loop control
   c. Open-loop control
   d. Ratio control

6. A water pump is installed to maintain water system pressure. A pressure switch is installed to control the pump. The switch turns the pump on when the pressure reaches 30 pounds per square inch and turns the pump off when the pressure reaches 50 pounds per square inch. Which statement best describes the operation of this system?
   a. The pump will run at a speed proportional to the pressure.
   b. The switch is used to precisely control the water pressure at a single pound per square inch value.
   c. The switch uses a dead band of 20 pounds per square inch to prevent the pump from cycling excessively.
   d. The switch uses open-loop control.

7. A PID controller has four adjustments on the user interface: gain, reset, rate, and damping. Which adjustment would be used to increase the proportional action?
   a. Damping
   b. Gain
   c. Rate
   d. Reset

8. A process operator recently changed the settings on a PID controller. The digital chart recorder now shows a consistent offset between the set point and the process variable. What is the likely cause of this offset?
   a. The operator set the controller to be direct acting.
   b. The operator set the gain too high.
   c. The operator turned the rate setting on.
   d. The operator turned the reset off.

9. A process vessel with a large capacity exhibits a large overshoot of temperature during a 12-hour process. What setting in the PID controller could help reduce such an overshoot?
   a. Feedback
   b. Gain
   c. Rate
   d. Reset

10. A control system uses two process controllers that interact to achieve a control function. What type of control action is used in this process?
    a. Cascade
    b. Feedback
    c. Feed-forward
    d. Ratio

Installation of Control Systems

The proper installation of an instrument is as vital as the proper selection and maintenance of that instrument. Improperly mounted instruments can result in inaccurately sensed data, premature device failure, and difficulty in servicing or calibration. A properly planned and executed installation leads to trouble-free service of the instrumentation.

Objectives

- » Identify proper mounting orientation.
- » Identify proper impulse tubing routing.
- » Distinguish the features of instrument manifolds.
- » Understand the pressure ratings of an instrument.
- » Properly route and connect signal wiring.

Chapter 14

Table of Contents

Systems and Applications 304
    Mounting 304
    Tubing 306
    Instrumentation Manifolds and Device Pressure Ratings 308
    Raceway or Conduit 311
    Wiring Terminations 313

Summary 319
Review Questions 319

For additional information, visit qr.njatcdb.org
Item #2690

SYSTEMS AND APPLICATIONS

Instrument installation consists of four separate functions: mounting, tubing, wiring raceway or conduit installation, and wiring termination. Each function should be performed in a manner that causes minimal process deviation and signal value variation.

Several steps are used to install a differential pressure-type transmitter. These steps can be applied generally to all devices used in a normal process environment. The technician must be sure to adhere to site-specific procedures.

Mounting

Installers sometimes have little choice about mounting position, wiring, and tubing, but at other times they may have choices for all three. Each part of an instrument's installation is important to the overall function of the device.

The accuracy of any field-mounted transmitter depends on the installation of that device. If a transmitter is mounted in a position that hinders proper operation, the overall goal of maintaining an efficient, accurate, and automatic process cannot be achieved. Likewise, if the wiring is run in a way that introduces stray signals, the highest quality and most accurate transmitters will result is an incorrect process variable being transmitted.

The installation requirements for specific sites depend on the location in which the instrument will be installed. Installations in certain process environments, such as those in the food, beverage, and pharmaceutical industries, may need to have special seals and meet other requirements to satisfy regulations. Specific requirements such as jobsite specifications and installation details along with manufacturer guidelines for a particular application should be available from the worksite supervisor. **See Figure 14-1.** As an instrumentation technician gains experience, the ability to identify good mounting practice means and methods becomes more intuitive. Still, general instrument installation practices should be followed by each installation tradesperson as their trade specific work is performed.

Transmitters should be installed to minimize vibration, shock, and temperature fluctuations. Most transmitters can be mounted in one of three ways: mounted on a wall, mounted on a pipe stand, or attached using a panel mount. Pipe mounts typically utilize steel plates or specialized brackets and U-style bolts to mount to a two-inch rigid pipe. **See Figure 14-2.**

Plate style mounts allow a transmitter to be attached to a flat surface such as a wall or control panel back plate. **See Figure 14-3.** Regardless of the mounting method, a technician must consider the ease of getting to the transmitter's electronics and its field-wiring terminations. A well-mounted instrument may be useless if the components of the instrument become inaccessible.

The technician should consider the process connections and the drain or vent valves on the transmitter so that

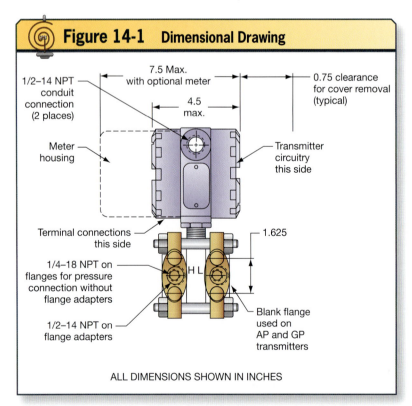

Figure 14-1. Drawings and cut sheets illustrate key information for mounting clearances and piping information.

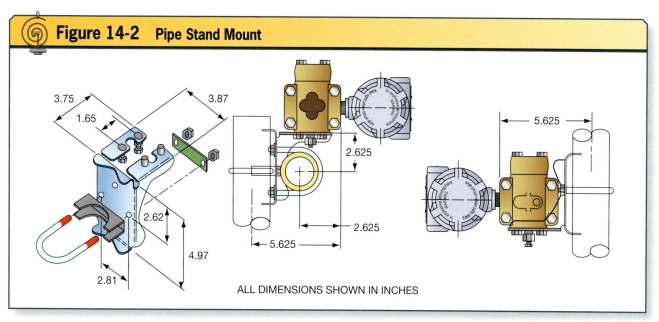

Figure 14-2. Pipe stand mounts offer a convenient secure mounting method, especially in the field.

when they are opened, they are directed away from personnel and critical process devices. In many applications, vents need to be piped to specific locations to control the vented product for safety. Before installation, or any service activity, a safety data sheet (SDS), as well as plant procedures, should be reviewed so that the technician thoroughly understands safety concerns.

Manufacturers use specific features in their products to assist in the installation. For instance, the electronics housing can often be rotated, in some instances a full 360°. It is especially important to follow the manufacturer's instructions for this feature, because some manufacturers may only allow a rotation of up to 90°. Rotating the housing more than 90° may

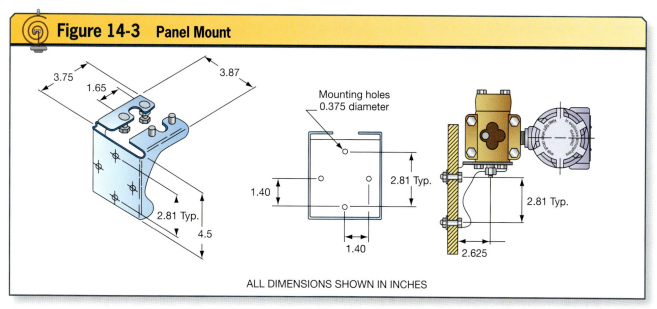

Figure 14-3. Panel mount brackets and accessories allow for a transmitter to be installed in a cabinet or anywhere there is a suitable flat surface.

> **ThinkSafe!**
> A device should never be mounted in a position that causes an unsafe working condition for the field technician.

damage the internal sensor wiring. When rotating the electronics housing, there are typically jam nuts or set screw locking mechanisms to prevent inadvertent rotation. These need to be loosened so that the housing rotates relatively freely. If the housing needs to be turned more than 90°, the transmitter must be disassembled and reassembled with the housing correctly positioned.

After the process side has been established, care must be taken so that proper access is available for the electronics side. Typically only a minimal amount of room is required to remove most electronics housings, but more space is preferable so that terminals, switches, and information may be viewed and tools may be inserted for wiring connections or adjustments. The electronics side is rarely opened after installation, but care still should be taken to ensure access. Smart devices often require access to both sides.

Wiring connections are made through the conduit openings located on the housing. Some devices have multiple conduit entry fittings, which may serve specific wiring purposes. Investigation should be completed to confirm that the correct wiring methods are entering the housing in the proper locations. It can be unclear that a conduit fitting only applies for line voltage wiring while another conduit fitting is provided for low-voltage signal wiring. **See Figure 14-4.**

Tubing

Instrumentation tubing that is properly installed is essential for a pressure transmitter to function correctly. *Impulse tubing* is tubing that carries all or a portion of the process fluid to the sensor for measurement. Proper routing, support and installation of impulse tubing is critical to make certain that the material within the tubing accurately represents the process. The field installer also should consider accessibility, safety for field personnel, ease of field calibration, and a practical working environment. The means and methods used to install the tubing will often dictate how a transmitter must be

Figure 14-4 Wiring Separation

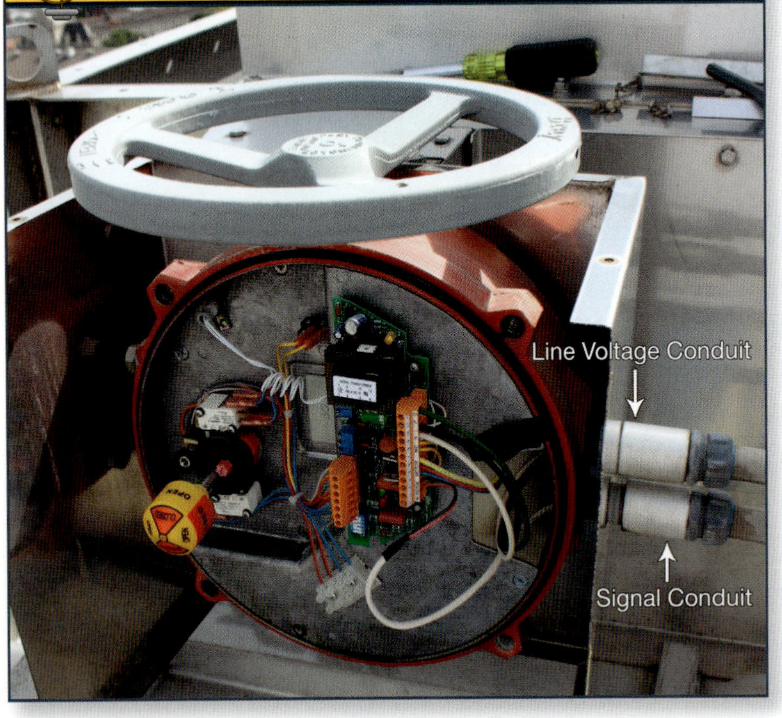

Figure 14-4. Larger electric actuators often require a higher voltage supply for operation. The high voltage wiring and low voltage signal wiring typically enter the housing through separate raceways.

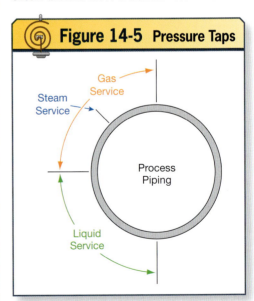

Figure 14-5. A cross section view of a process pipe is used to indicate the different locations suitable for a process tap based on the process medium.

mounted and supported. Proper layout of the entire tubing system will help to avoid issues in the future. The process line *pressure tap* location, or point where the instrumentation tubing connects to the process line, is often dictated by the process medium as well as the piping configuration. Steam, liquid, and gas are the three process mediums that are common throughout industry. **See Figure 14-5.**

For liquid measurement, the technician should mount the transmitter so that the process-measuring taps are to the side of the line, which keeps sediment from becoming trapped. Mounting the transmitter to the side or below the taps ensures that air does not become trapped in the impulse tubing.

Additionally, the tubing should be sloped up from the transmitter to the process taps using an appropriate amount of slope. This slope will assist in purging trapped gas or air. **See Figure 14-6.**

For measuring gas flows or gas levels, the process taps should be located at the top or to the side of the line to ensure that liquid can drain into the process line. As with liquid installations, the tubing needs to be installed with a specific slope for gas measurement installations. The gas measurement requires the tubing to be sloped down from the transmitter to the process line. This prevents any residual liquids or condensate from collecting in the tubing. **See Figure 14-7.**

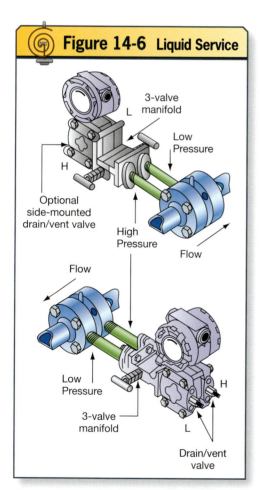

Figure 14-6. Proper mounting and routing of liquid filled impulse lines is essential to not trap air, which will result in measurement errors.

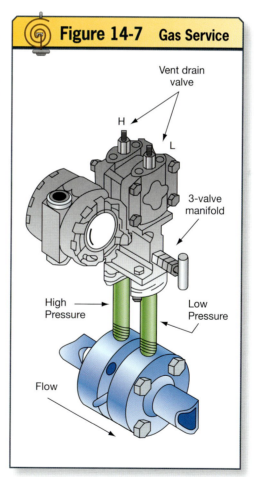

Figure 14-7. Gas flow taps must be arranged to avoid any trapped liquids which may be present, as they will hinder an accurate measurement.

Figure 14-8 Steam Service

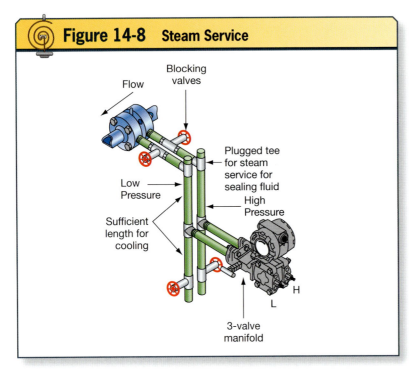

Figure 14-8. Transmitters for use in steam measurements must be installed properly to handle condensation build-up.

For additional information, visit qr.njatcdb.org Item #2627

TechTip!
When field calibration of transmitters used in steam applications occurs, the liquid head pressure on the impulse lines is often accounted for naturally. However, when bench calibrating these same devices, the liquid head must be accounted for to achieve accurate calibration once installed in the field.

Figure 14-9 Tubing Installation

Figure 14-9. Pressure taps, related impulse tubing, and instruments must be installed correctly to provide accurate measurements.

For steam measurement, the process taps should be located to the side of the line and the transmitter should be mounted below the taps to ensure the process tubing remains filled with condensate. The filled condensate helps to buffer the transmitter from the hot steam vapor. **See Figure 14-8.**

General tubing tips:
1. Process tubing length should be as short as possible.
2. High and Low pressure tubing should be run together following the same path with equal numbers of bends.
3. Proper slope should be maintained with respect to the process conditions.
4. Tubing should be routed to avoid high spots or crowns in a liquid measuring line where gasses can become trapped.
5. Tubing should be routed to avoid valleys where liquids can collect and act as a trap.
6. Instrumentation tubing may need to be insulated and/or heat traced, just as a process line does, in order to ensure proper operation.
7. Provide isolation valves at the process taps to facilitate maintenance of the instrument tubing.
8. Maintain proper support of instrumentation tubing.
9. Remember, the material present within the small tubing is just as dangerous as the material in the process line.

Following these general tips along with jobsite details and the tubing/fitting manufacturer's instructions will provide a quality installation. **See Figure 14-9.**

Instrumentation Manifolds and Device Pressure Ratings

Instrumentation blocking manifolds come in a variety of types and configurations, depending on the application. Although manufacturing standards have evolved to allow interchangeability among equipment from various manufacturers, proper installation is still key. When mating instruments to piping

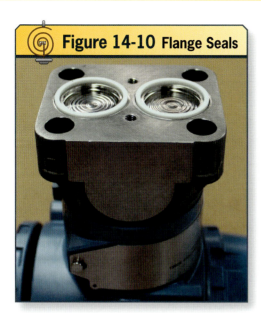

Figure 14-10. Flange bolts must be torqued properly, in order to adequately compress the seals to offer leak free service.

Figure 14-11. Burst Pressure or Maximum Working Pressure (MWP) is shown on the nameplate as well as the working pressure ranges of a transmitter.

accessories such as manifolds or flanges, the technician should ensure that the proper sealing rings, compatible with the process medium, are installed and that all bolts are tightened evenly and properly torqued. This creates a leak-free, reliable seal. **See Figure 14-10.**

Pressure transmitters and their associated fittings carry a few distinctly different pressure ratings. **See Figure 14-11.**

- Normal operating range—The range in which the device has been designed to operate.
- Burst pressure—The rating at which a device can briefly withstand pressure before the device separates, a component leaks, or both.
- Proof pressure—The rating to which a device may be overpressurized and still operate within specification once the overpressurized condition is corrected. Proof pressure is generally a multiple of normal operating pressure, such as 1.5 or 2 times the upper range.

Manifolds are designed with different features depending on the application. A three-valve version may be acceptable for a differential pressure application related to water; however, this same manifold is not acceptable for a hazardous medium such as a natural gas installation. The combined risk of a spark and the inability to properly contain vented medium, prohibits the vent fittings commonly used on a three valve manifold. **See Figure 14-12.**

Figure 14-12. Three-valve manifolds offer process isolation and equalization, while venting is provided using specialized fittings and a wrench.

Figure 14-13. *Five-valve manifolds offer tool-less isolation, equalization and venting of a transmitter's connection to process lines.*

A five-valve manifold listed for natural gas use has some obvious differences from the three-valve version, most notably the addition of two handles. These additional handles operate the vent ports. Although the three-valve manifold may also have vent ports, they must be operated by a wrench; in addition, there is no way to contain the vented medium. The five-valve manifold eliminates the need for tools, which minimizes the chance of an errant spark. In addition, the vent ports are threaded, so tubing or piping can be installed to properly divert the vented medium to the proper containment vessel. **See Figure 14-13.**

To properly block and vent an instrument, specific procedures must be followed; otherwise, catastrophic results may occur. The differential pressure cell on a 1,500-pound per square inch gas transmission line may only be designed for 500 inches of water differential. Accidentally venting one side to atmosphere while leaving the other side in service introduces a differential of 41,550 inches of water to the instrument. This will most likely destroy the cell. **See Figure 14-14.**

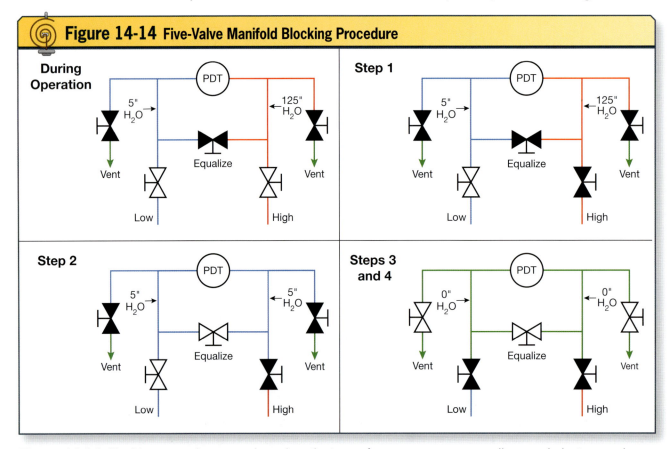

Figure 14-14. *Blocking procedures vary based on the type of process, process medium, and plant procedures.*

Figure 14-15 Piping Accessories

Item	Image	Description
Syphons or Pigtails		Looped pieces of pipe which trap water in order to insulate the steam from sensitive devices. These devices are installed between a process line and the device.
Snubber		Device with a very small bore, which slows the change in pressure experienced by a pressure device.
Isolating Diaphragm		Devices which use a small flexible barrier, often made of a thin metal alloy, which isolates an instrument from the affects of a process.

Figure 14-15. A variety of piping accessories can be used to protect instruments.

One common blocking procedure is as follows:
1. Close (block) the high-pressure process port.
2. Open the equalizing valve, slowly at first, until fully open.
3. Close (block) the low-pressure process port.
4. Open the vents on the valves to bleed off fluid (liquid or gas) trapped in the manifold.

In addition to blocking valves, a variety of piping accessories and practices are used to isolate delicate instruments from the harsh environment of the process medium. **See Figure 14-15.**

Raceway or Conduit

A variety of methods exist to connect necessary wiring to an instrument. While it is common to see vast systems of conduit run from control cabinets to instruments, the use of a cable tray is also a common method. **See Figure 14-16.**

TechTip!

Unique installations are prevalent throughout industry. Technicians should always follow facility procedures and guidelines when performing instrument blocking procedures, asking for this information if necessary.

Figure 14-16 Cable Tray

Figure 14-16. Cable tray offers a convenient means of supporting multiple cables used in industrial systems.

Regardless of the protection system implemented, it should be installed so that it does not hinder maintenance activities associated with both the process and the instruments.

Conduit runs that contain signal loops should be run so that high-frequency switching is not impressed on the signal wiring. If a high-frequency switching load is present in the area, the conventional wiring loop of 4 to 20 milliamperes direct current (DC) should not be run in the same conduit as the high-frequency switching source. In most cases, it is acceptable to run multiple analog signal loops in the same conduit if shielded, twisted-pair wiring is used.

The use of flexible conduit and wiring methods is often required to facilitate maintenance and guard against the effects of machinery vibration. Safety codes, facility standards, and design specifications typically specify the proper materials and allowable minimum and maximum flexible raceway lengths. An overly short raceway may not provide an adequate degree of vibration isolation, whereas an excessively long raceway is difficult to properly support and may pose an entanglement hazard. **See Figure 14-17.**

Instrumentation devices, control cabinets, junction boxes, and conduits are all items to which a great deal of effort is made to keep foreign objects (dust, dirt, gases, vapors, and fluids) out of the wiring system. These necessary efforts often form a well-sealed system, thereby creating an internal atmosphere. Conduits passing from one atmosphere to another, such as a cold area to a hot area, are often unavoidable; however, in such a passage, moisture is condensed inside the conduit with the changing temperatures. Provisions for sealing the internal area of the conduit around the wiring are necessary to isolate air migration through the conduit. **See Figure 14-18.**

Hazardous location seal-off fittings are commonplace throughout industry to reduce the chance of dusts, gases, and vapors migrating to sources of ignition. These devices provide an accessible point through which the proper fill material and sealing compounds can be placed and inspected at a future time. **See Figure 14-19.**

Design specifications and manufacturer instructions must be followed well in advance of the sealing process. When the initial raceways are installed, it is necessary to understand how the complete system will function. Depending on the application and type of wiring being used, oversized fittings may be required to allow adequate room for the fill material and sealing compound to work properly. Once the seal fitting is complete, any changes to the wiring that are needed may mean

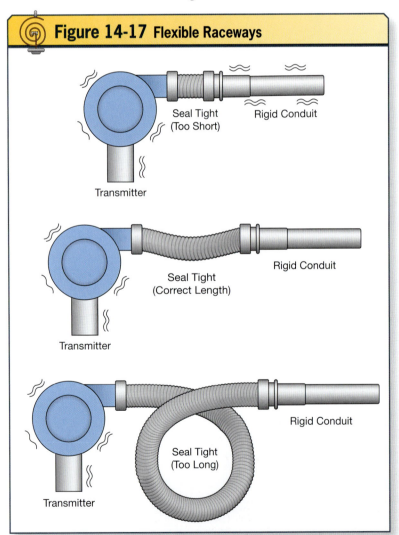

Figure 14-17. The proper length flexible raceway provides adequate vibration isolation without causing an entanglement hazard.

TechTip!

It is often necessary to insulate process lines and components for a variety of reasons. Because of this, care must be taken to ensure that adequate space is allowed around process piping and impulse tubing or piping.

that the conductors, raceway, and fitting need to be sacrificed, resulting in costly changes. **See Figure 14-20.**

Wiring Terminations

In addition to correctly running process tubing and determining the mounting location for the instrument, wiring considerations must allow efficient and accurate process measurement signal transmission. The electrical signal transmits the process variable to a controller. Even if all other considerations have been made correctly, an error can be introduced on the signal side of the transmitter rather than on the sensor side.

Most transmitters have the main electrical terminals marked as signal, loop, power, or power and communication. Typically, signal, loop, and power are used to power the device and

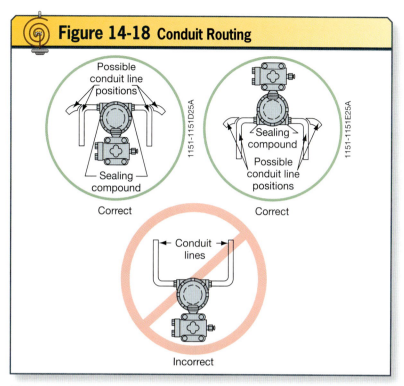

Figure 14-18. Proper conduit routing is necessary to avoid moisture or volatile gases and liquids from damaging the electronics or posing an explosion risk.

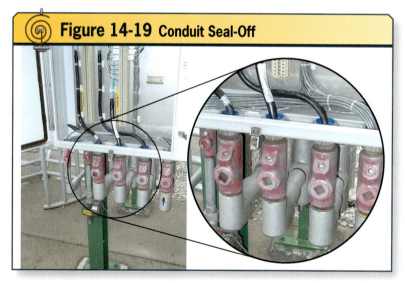

Figure 14-19. Conduit seal-offs are utilized to avoid combustibles or other foreign materials from passing through a conduit into electrical cabinets or devices.

 TechTip!

Low point drain ports on conduit alleviate moisture that may develop from condensation. The technician should always install these where required.

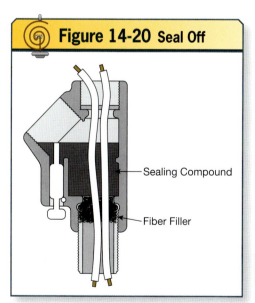

Figure 14-20. A properly listed, installed, and sealed seal-off style fitting is essential for safety.

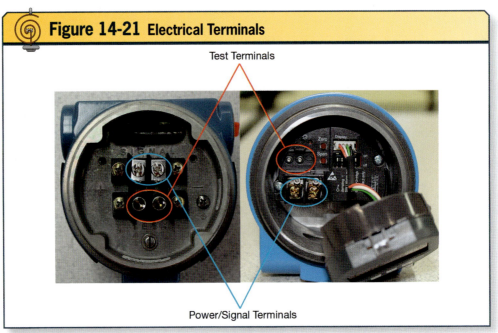

Figure 14-21. Manufacturers use a variety of electrical terminal arrangements. The technician should verify instructions before making connections and powering the device.

transmit the 4- to 20-milliampere DC signal. Power and communication terminal markings are found on smart devices for which digital communication is also taking place on the 4- to 20-milliampere DC terminals.

Some transmitters may have additional terminals marked as test terminals. *Test terminals* are electrical terminals on a transmitter used to make quick signal measurement checks or to power local indicator displays. **See Figure 14-21.**

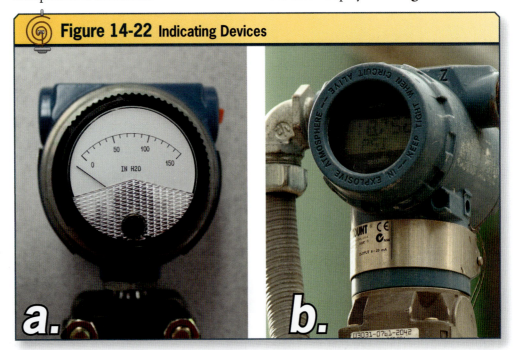

Figure 14-22. An indicating display may range from a simple needle movement (a) to a digital display offering a variety of information (b).

The test terminals have an output for a 4- to 20-milliampere DC loop that is identical to that of the signal terminals. The only wiring that should be connected to any test terminal is an integral meter for the device or a properly rated test meter used in the correct manner. **See Figure 14-22.**

For an instrument to operate continuously and allow test terminal operation, a diode is installed when the instrument is manufactured. A diode is a semiconductor device that only conducts current in one direction based on polarity. If the polarity is reversed, the diode will not conduct. When the diode is placed in this circuit and conducting properly, small resistance is created that has a resulting voltage drop of around 0.7 volts. When a current-measuring device is place in parallel around the diode, the low resistance of the measuring device allows current to flow through itself, thereby reading the circuit current. **See Figure 14-23.**

Test terminals are unique devices that offer a convenient method of checking a loop current without disturbing the wiring.

To measure the loop current, the technician should place the leads of a meter positive to positive and negative to negative with the test terminals. This should make it possible to measure a current. However, because of their unique properties, inadvertently connecting power or signal wiring to these terminals can damage the internal electronics of the instrument, as well as result in a potential short circuit. **See Figure 14-24.**

Signal wiring carries the 4- to 20-milliampere DC loop current that is used by a device or controller to indicate the process measured. Care must be taken when routing signal wiring so that it does not pick up stray voltages or noise. If signal wiring is routed in conduit or

TechTip!
Using a digital multimeter to make a connection to the test terminals could produce a small arc or ignition source.

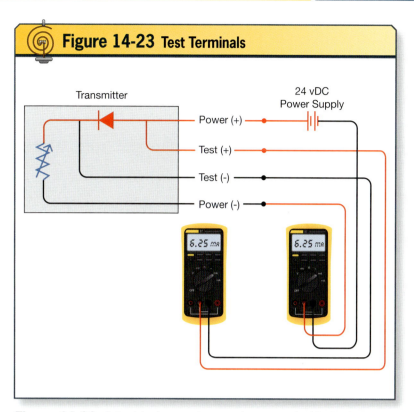

Figure 14-23. A test instrument properly connected to test terminals will measure the loop current without disturbing the loop wiring.

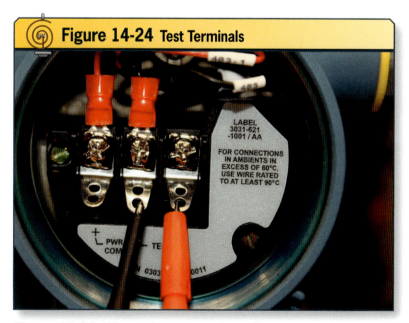

Figure 14-24. When using a properly configured test meter for loop current measurements, be sure to connect to the proper terminals on a transmitter. Connecting across the power connection will result in a short circuit.

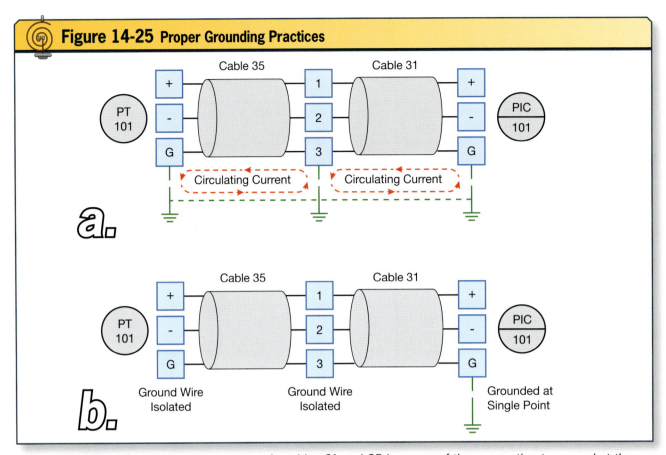

Figure 14-25. (a) Ground loops are present in cables 31 and 35 because of the connection to ground at the pressure transmitter (PT 101), junction block, and pressure-indicating controller (PIC 101). Current circulates on the ground in response to the parallel current paths. (b) When the ground is connected a single point, there is no path for circulating current.

open trays near heavy power loads or high voltages, unwanted or stray signals will be induced in signal wiring.

Most methods for signal wiring incorporate the use of a twisted pair with a shield. The signal conductors are twisted together to cancel magnetic field effects, while the shield cancels capacitive coupling effects. For the shield to work properly, it must be grounded properly. This typically occurs at the controller or power supply end of the loop, and it is grounded only once; however, project documents and installation details should be reviewed for special instructions. If the signal wiring becomes grounded at both ends, it is likely a ground loop will be formed. A ground loop is a condition in which current flows on the grounding portion of a circuit, leading to electrical noise and signal interference. **See Figure 14-25.**

Many process systems include areas of hazardous classification. The use of appropriately rated devices, as well as proper installation of raceways and required seals, limits the migration of combustibles to areas where sparks may be prevalent. A very small arc, possibly not even able to be seen, can have disastrous consequences in an explosive atmosphere. The use of an *intrinsic barrier*, a device used to control fault current and maintain an electrically safe condition within a hazardous location, mitigates this risk. By properly selecting and installing a barrier within the control circuit, the energy present at the instrument terminals within the dangerous area is significantly diminished. Very specific rules

TechTip!

When grounding a transmitter, it is vital that the technician follow the manufacturer's details and facility procedures, because proper grounding procedures can change based on a process and facility design.

for installation of these circuits must be adhered to so that the "safe" wiring cannot be compromised, such as:

1. Intrinsically Safe (IS) wiring cannot occupy conduits or raceways with non-intrinsically safe wiring.
2. IS wiring within control cabinets and enclosures must be secured separately of other wiring and devices, to prevent contact.
3. IS wiring and associated raceways all must be labelled that they are part of an intrinsically safe wiring system.

The proper design, specifications, installation and maintenance of an intrinsically safe system is paramount. **See Figure 14-26.**

When making final electrical connections on any type of terminal, care must be taken to ensure stray strands of conductor are not touching any other component. Various products, such as solderless wiring connectors, can simplify this process. Fork-style connectors, with a slot opening properly sized for the shank of a screw, offer a secure method of attaching a stranded wire under a screw head. In another application, such as a European-style terminal block, a ferrule may be used, because it contains all strands of the conductor. Regardless of the type of connector used, it must be installed properly according to the manufacturer's instructions, especially with regards to how the conductor insulation is removed or stripped. **See Figure 14-27.**

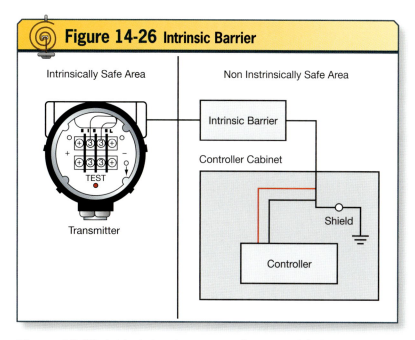

Figure 14-26. Intrinsic barriers are used to control fault current and maintain an electrically safe condition within a hazardous location.

For additional information, visit qr.njatcdb.org Item #2628

For additional information, visit qr.njatcdb.org Item #2629

Figure 14-27. Secure terminations rely on the proper conductor preparation and any necessary accessory for a quality connection.

Figure 14-28 Conductor Labeling

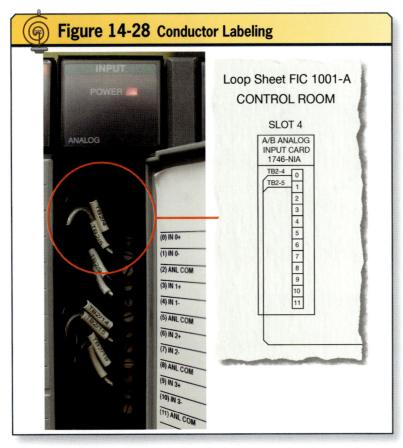

Figure 14-28. *Accurate labeling is vital. All terminations and documentation must use the same naming convention for clarity.*

Figure 14-29 Shield Isolation

Figure 14-29. *Proper isolation of unused foil shields and drain wires is required to avoid inadvertent contact with ground or energized components.*

TechTip!

Improper stripping tools and techniques can result in nicked or broken conductors, which lead to poor connections.

Cable and conductor labeling is a vital step to the installation process. Inconsistent labeling can lead to a great deal of confusion, as well as a potentially dangerous scenario. Most facilities have a labeling and tag standard in place that specifies not only the naming and identification scheme but also the label material and color. This naming convention is identical to the naming convention used on all other documents, such as loop sheets. **See Figure 14-28.**

Heat-shrinkable tubing, which is printed on, not only offers a method to adequately label the conductors but also provides an effective cover for unused shields and drain wires. Portable, handheld labeling machines are commonplace on many jobsites. These machines print directly on the heat-shrinkable tubing for legible and permanent identification. **See Figure 14-29.**

SUMMARY

Installation requirements often are dictated by the details portrayed on provided drawings. The technician must be familiar with the installation requirements related to wiring, tubing, and conduit and must understand how to address problems encountered during installation.

REVIEW QUESTIONS

1. When installing a differential pressure transmitter for liquid service, the transmitter should be placed __?__.
   a. above the process taps
   b. at the process taps
   c. below the process taps
   d. Does not matter

2. Tubing that is installed to carry process medium to the transmitter is known as __?__ tubing.
   a. conduit
   b. impulse
   c. process
   d. vent

3. The proper setting for a digital multimeter when measuring loop current at the instrument test terminals is __?__.
   a. current
   b. diode test
   c. resistance
   d. voltage

4. Shielded instrumentation cable is typically connected to earth ground __?__, unless specific requirements are identified in the project or plant standards.
   a. at every splice
   b. only at the controller
   c. only at the field instruments
   d. None of the above; shields should never be connected to ground.

5. A(n) __?__ is a piping accessory which uses a thin flexible barrier, often a metal alloy, to isolate an instrument from the effects of a process.
   a. manifold
   b. isolation Diaphragm
   c. pigtail
   d. snubber

6. Proper use of instrument manifolds requires __?__.
   a. closing both process isolation valves simultaneously
   b. following facility specific procedures as well as manufacturers' guidelines
   c. opening the equalization valve as the first step
   d. shutting down the process prior to operating any valves

7. The rating at which a device can briefly withstand pressure before the device separates, a component leaks, or both is known as __?__.
   a. burst
   b. normal
   c. proof
   d. operational

8. Intrinsic barriers are used to __?__.
   a. boost signal strength
   b. change milliamp signals to millivolt signals
   c. control fault currents
   d. provide a level of data security

Process Control Loop Checking

The integrity of the control loops within a process is critical to the overall function of the control system. When a device in the field is expected to transmit a signal to a controller, or receive a signal from a controller, it must be able to perform these functions reliably and within acceptable tolerances.

Loop-checking process control systems may seem to be a daunting task. However, with a thorough understanding of the system and the associated control loops that make up the system, the technician will be able to perform the necessary procedures to ensure the proper operation of the system and its components.

Objectives

- » Explain why loop checking is necessary.
- » Describe the three-step loop-checking procedure.
- » List all relevant documentation that must be gathered before loop checking begins.
- » Describe the visual inspection process.
- » List the two methods to verify a loop.
- » Explain the two-wire transmitter simulation setting on a loop calibrator.

Chapter 15

Table of Contents

Process Control Systems 322
Loop Checking 322
    Documentation Inspection 323
    Visual Inspection 325
    Functional Inspection 325

Example of Temperature Transmitter Loop Checking 326
Summary 328
Review Questions 329

For additional information, visit qr.njatcdb.org
Item #2690

PROCESS CONTROL SYSTEMS

Process control systems vary in complexity from a single loop to several hundred or even thousands of loops. Each loop contains multiple terminations and connections made by the installation craftsperson. Before the process may be commissioned, the integrity of each termination and connection must be verified by the instrument technician. This process is essential to verify the correct installation, wiring, and tubing connection guidelines are followed. Loop-checking procedures are followed to ensure that the operation of the various components within a control system adheres to the intent of the design.

Loop checking of process control loops is methodical, requiring the technician to apply a diverse set of skills. The technician must work hand in hand with the design team during the installation phases of the control system, as well as during the loop-checking phases, to ensure the proper operation of the components that comprise a process system. The loop-checking process requires excellent written and verbal communication skills. Neat handwriting with no stray marks is essential to ensure the loop sheets are legible. The North Atlantic Treaty Organization (NATO) alphabet (i.e., T = Tango) is commonly used for radio communication to prevent mistakes. The proper installation and operation of complex control systems require this collaborative effort to ensure a safe and efficient outcome that meets design criteria. **See Figure 15-1.**

LOOP CHECKING

Loop checking is broken down into three distinct inspection phases: documentation, visual, and functional. The American National Standards Institute and International Society of Automation (ANSI/ISA) 62382 standard titled *Electrical and Instrumentation Loop Check* provides a systematic approach to verifying the proper operation of process loops. A common misconception is that all loops and equipment must be installed before loop checking may begin. Due to time constraints for completing a project, loop checking is often started as soon as the equipment and wiring for a specific loop is completed. **See Figure 15-2.** This approach must follow the underlying principle of loop checking: confirmation that all components of the loop work together to achieve a control function. Though a technician may disconnect a loop to aid troubleshooting, the loop must be restored before loop checking begins. Separating a loop into sections or segments is not an acceptable method of loop checking.

The technician must have a firm understanding of a control loop's purpose within the overall system under control. The purpose of the control loop is to ensure the safety, efficiency, and profitability of a process system. To achieve this purpose, the control loops must operate in a manner that forces the process to operate within a desirable and predetermined fashion.

Figure 15-1. Technicians often work in teams to perform loop-checking tasks.

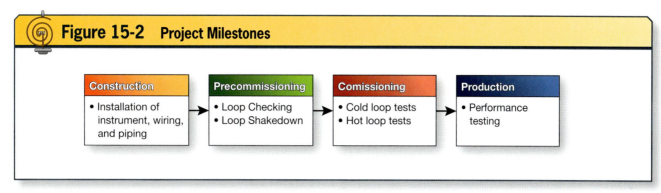

Figure 15-2. Loop checking may begin when the loop components are installed.

Documentation Inspection

The project documents serve as a guide for the technician and must be complete, accurate, and readily available. Changes made during the construction of the loop must be updated and reflected on the documents used by the technician. To ensure the proper version of installation documents are used, the technician must be familiar with the document revision system used by the project. The revision of a document is commonly found on the title block or on the front of a project guideline. **See Figure 15-3.**

An array of documents is used for loop checking. Before visual or functional checks may begin, all associated documents must be available and accurate. The process and instrumentation drawing (P&ID) serves as a map of the process and shows the interconnections between the process and the instruments. The instrument index provides information on the tag number, instrument type, and specific numbered drawing where the instrument is shown. An instrument specification sheet lists the construction of materials and other characteristics of the instrument. The instrument calibration sheet, whether factory or site generated, provides evidence of traceable and accurate calibration. The loop diagram shows all connections, both electrical and pneumatic, of the loop. Finally, the loop-checking sheet serves as a document to record the findings of the instrument technician during the loop-checking procedure. In addition, the loop-checking sheet standardizes the process to ensure consistency among technicians. **See Figure 15-4.**

Figure 15-3 Document Revisions

Revision	Activity No.	Revision Description	By
A	Review	Issued for review	
1	Approved	Issued for approval	B. Macklin
2	Revised	Revised to add column for QA	B. Macklin
3	Revised	Revised to add DCS cable tray requirements 923.16, 923.11	B. Macklin

Figure 15-3. Documents are frequently revised during a project. Technicians must make sure they are using the correct version of a document.

Figure 15-4 Loop-Checking Sheet

Loop Checking Sheet

TAG NAME(S): _____ AREA: _____ CWP: _____

LOOP: _____ SP DATA SHT. & REV: _____ SUBSYSTEM: _____

I/O TYPE: ☐ AI ☐ AO ☐ DI ☐ DO ☐ HART ☐ FFB ☐ BN ☐ F&G OTHER: _____ MFR: _____

DCS SCALING: _____ PER DATA SHEET ☐ Y ☐ N ☐ NA   SS TAG(S) PRESENT AND CORRECT: ☐ Y ☐ N

DEVICE LOCATION CORECT: ☐ Y ☐ N   KFACTOR: _____   MODEL #(S) MATCH DATA SHEET: ☐ Y ☐ N ☐ NA

DEVICE STATUS GOOD AT POWER UP: ☐ Y ☐ N ☐ NA   LOCAL DISPLAY GOOD: ☐ Y ☐ N ☐ NA   DIGITAL TAG OK: ☐ Y ☐ N ☐ NA

RANGE VALUES AND ENGINEERING UNITS: _____   PER DATA SHEET: ☐ Y ☐ N ☐ NA

FAIL DIRECTION: ☐ UPSCALE ☐ DOWNSCALE   PER DATA SHEET: ☐ Y ☐ N ☐ NA   OUTPUT: ☐ LINEAR ☐ SQ. ROOT ☐ NA

RTD/TC TYPE: _____   NUMBER OF ELEMENTS: _____   LOCAL DISPLAY AMBIENT TEMP: _____

TEMPERATURE ELEMENT(S) FAILED: ☐ Y ☐ N   HOT BACKUP CHECK: ☐ Y ☐ N ☐ NA   PRIMARY ELEMENT REINSTATED: ☐ Y ☐ N

TRANSMITTER/CONTROL VALVE CHECK:

0%	25%	50%	75%	100%	75%	50%	25%	0%

ON/OFF VALVE CHECK: ENERGIZE TO: ☐ OPEN ☐ CLOSE   POSITION SWITCH: ☐ OPEN ☐ CLOSED ☐ NA

DE-ENERGIZE TO: ☐ OPEN ☐ CLOSE   POSITION SWITCH: ☐ OPEN ☐ CLOSED ☐ NA

POSITION SWITCH INDICATOR COLOR AT OPEN: _____   AT CLOSED: _____

AIR FAIL DIRECTION: ☐ OPEN ☐ CLOSED ☐ LAST   SIG. FAIL DIRECTION: ☐ OPEN ☐ CLOSED ☐ LAST   AIR FAIL RELAY PRESENT: ☐ Y ☐ N

STROKE SPEED OPEN: _____   STROKE SPEED CLOSE: _____   PER DATE SHEET: ☐ Y ☐ N ☐ NA

DIGITAL INPUT CHECK COMPLETED: ☐ Y ☐ N   VOLTAGE: ☐ 120 VAC ☐ 24VDC   DEVICE: _____

DIGITAL OUTPUT CHECK COMPLETED: ☐ Y ☐ N   VOLTAGE: ☐ 120 VAC ☐ 24VDC   DEVICE: _____

DCS GREEN: _____   DCS RED: _____   DCS YELLOW: _____

COMMENTS, PUNCHLIST ITEMS, SAFETY ISSUES, ETC:

_____
_____
_____
_____
_____
_____

TECHNICIAN: _____   DATE: _____

Figure 15-4. *The loop-checking sheet is used to document the technician's findings on the loop installation and operation.*

Figure 15-5 Visual Inspection Checklist

Visual Inspection of Installation		
☐ Transmitter properly tagged with engraved stainless steel tag	Results Okay	
☐ Transmitter is supported securely	Date:	
☐ Indication screen is mounted between 4' and 5' above grade or platform	Name:	
☐ Sealtite drip loop is present and connected per WCG 714.35-Z	Signature:	
☐ Thermowell material matches specification sheet		

Figure 15-5. *The visual inspection checklist is used to verify that the installation matches the project guidelines.*

Visual Inspection

Visual inspection of the loop installation follows the documentation process during the loop-checking procedure. The technician is expected to verify that the electrical, pneumatic, and mechanical portions of the loop are installed in adherence with the project guidelines. This requires intimate knowledge of the project installation requirements. To assist the technician, many projects use a standard inspection and test plan that lists the specific requirements for each installation. **See Figure 15-5.** For example, if the technician is inspecting the installation of a temperature-indicating transmitter, the inspection checklist may ask the technician to verify the following items:

1. The transmitter is properly tagged with an engraved stainless steel tag.
2. The transmitter is supported securely and not subject to vibration.
3. The indication screen is mounted between 4 and 5 feet above grade or the platform.
4. The instrument electrical connection is made with liquidtight flexible conduit and the drip loop is present.
5. The thermowell material matches the specification sheet.

Functional Inspection

The functional check of the loop ensures that all devices, terminations, and computer systems are working together as designed. Before functional checking begins, a communication plan between the control room operator and the field technician should be established. The control room operator may be able to support several field teams because of the downtime required by the field team in locating the instrument and performing the visual inspection. However, if a problem is discovered, the team may need additional time to correct the problem before moving forward. Good communication helps to ensure the team is efficient and accurate.

The loop-checking sheet provides the necessary checks based on the device type. Device measurement verification can be completed using one of two main methods: forcing or simulation. Forcing the process variable verifies the sensor and transmitter simultaneously. For example, consider a differential pressure transmitter. The technician may block the high and low manifold valves and remove the bleeder from the manifold's high side. Next, a hand pump is connected to an adapter fitting to supply pressure to the sensor. **See Figure 15-6.** The technician communicates to the control room operator while applying the 0%, 25%, 50%, 75%, and 100% pressures and verifies the control system is reading the values accurately. This forced value verifies the loop components and the range of the transmitter.

Simulating the process variable does not require the transmitter to be disconnected from the process. Instead, the loop signal is either generated by a loop calibrator or simulated using a highway addressable remote transducer (HART) communicator. A loop calibrator set to operate as a two-wire transmitter simulator acts as a passive element that is powered by the normal instrument power supply. **See Figure 15-7.** In the case of a non-HART pressure transmitter, the technician removes the positive and negative leads from the transmitter and connects a two-wire loop simulator. The 0%, 25%, 50%, 75%, and 100% signal values are controlled by the loop calibrator and verified by the control room operator. This test does not verify the operation of the transmitter or sensor; however, assuming the transmitter has been properly calibrated, the system is presumed accurate. If the transmitter is HART capable, the technician can simulate an output signal using a communicator. The HART communicator is used to navigate to the loop-checking menu item where the desired milliampere direct current (DC) value is entered. This test verifies the transmitter's output signal, however, it does not verify the sensor of the transmitter.

ThinkSafe!

The technician must always anticipate process media to be present when simulating a process signal to the controller. Although every effort is made to isolate and drain sensors before inducing a test simulation, isolation is sometimes not accomplished because of valve leakage or other factors. The technician must always wear the proper personal protective equipment in case of media contact.

EXAMPLE OF TEMPERATURE TRANSMITTER LOOP CHECKING

The following scenario is an example of the loop-checking procedure for a HART temperature transmitter. The technician should always follow the site-specific loop-checking procedure, which may vary from this scenario.

The technician begins by collecting all documentation needed for the procedure. In addition, the safety plan is reviewed and any hazardous chemicals present are identified by the proper safety data sheet. Communications with the control room operator will be made on a predesignated radio channel. Next, the technician goes into the field and locates the temperature transmitter. A visual inspection of the device shows the identification tag, mounting method, and conduit connection adhere to the installation guidelines of the project. The thermowell is visually verified for installation and construction material. Because this instrument is a HART instrument, the settings of the fail mode and security switch also need to be visually inspected for proper orientation.

Figure 15-6   Pressure Transmitter Adapter

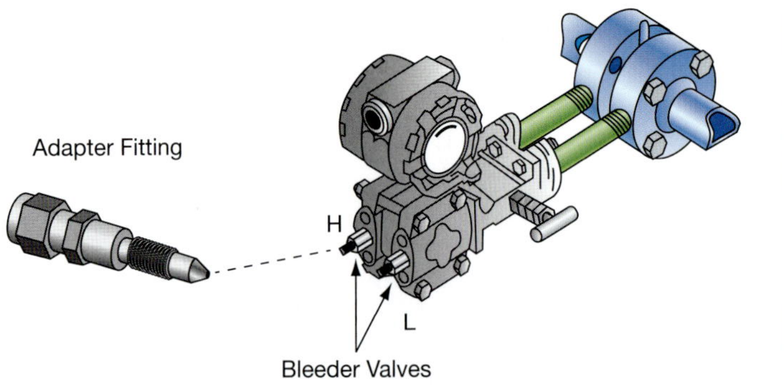

Adapter Fitting
H
L
Bleeder Valves

Figure 15-6. *The bleeder valve can be removed and temporarily replaced with an adapter fitting to force the process variable during loop checking.*

The technician then removes the positive wire from the transmitter and checks for continuity to ground. No continuity should be detected, verifying no ground fault exists on the positive. Next, the technician checks continuity between the positive and the negative wires. No continuity here means there is not a short in the circuit. The technician also checks that the shield drain wire is isolated at the instrument connection. This prevents the formation of ground loops on the signal cable.

The technician then calls the control room operator and instructs the operator to power up the loop. Once the transmitter is powered up, the control room operator should see that the instrument tag number on the distributed control system is correct and the sensor is reading ambient temperature. The technician then disconnects the temperature element. The control room operator verifies receipt of a sensor failure alarm from the control system. The technician, using a HART communicator, navigates to the loop test menu item and then simulates the transmitter operation by causing the transmitter to output 4, 8, 12, 16, and finally 20 milliamperes DC. Each point is verified by the control room operator. Next, the technician causes the output signal to exceed the low alarm point (less than 4 milliamperes DC) and the high alarm point (more than 20 milliamperes DC). The control room operator verifies the proper alarm is generated by the control system.

The final step in the loop-checking procedure is to complete all documentation and submit it to the proper authority. Loop-checking sheets should be submitted daily to prevent loss and to update project management scheduling. These sheets are combined to form a master list maintained by the instrument contractor and are ultimately submitted to the owner for approval.

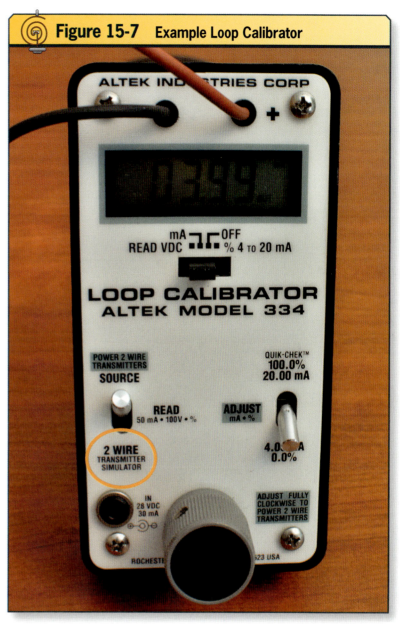

Figure 15-7. A loop calibrator may be used in the transmitter simulation (orange circle) setting to act as a transmitter during the loop-checking process.

TechTip!

Before replacing a blown fuse in a control loop, the technician should always attempt to identify the reason for the blown fuse.

SUMMARY

Loop checking is a vital part of a technician's responsibility. All loops must be verified, a process often requiring more than one technician. All team members must be thoroughly familiar with the loops to be checked and must rely on documentation, including the P&IDs.

Loop checking frequently involves the five-point tests of analog systems at 0%, 25%, 50%, 75%, and 100% signal levels, but bear in mind that other loop-checking parameters are also used. Results of these tests should be documented and analyzed before certifying the loop as operating properly.

REVIEW QUESTIONS

1. Which of the following is not part of the three-step loop-checking procedure?
   a. Documentation inspection
   b. Functional inspection
   c. Pressure inspection
   d. Visual inspection

2. At what point in a project can loop checking begin?
   a. After the cold loop testing process is complete.
   b. After the instruments are temporarily powered.
   c. Before all cables are pulled for the loop.
   d. Only after all loop components are permanently installed.

3. What radio communication standard is commonly used to prevent mistakes in alphabetical characters?
   a. ANSI/ISA process communication standard
   b. IEC global phonetic library
   c. ISO Z.210
   d. NATO alphabet

4. How can technicians ensure they are using the most current documentation?
   a. Always use initial approval drawings.
   b. Ask different trade workers about what revisions they are using.
   c. Largely ignore current documentation because it rarely affects the loop-checking process.
   d. Verify the document revision is the most current available in the project library system.

5. Which of the following ensures the loop-checking procedure is consistently performed by different teams of instrument technicians?
   a. Designating workers with "specialty roles," such as valve, temperature, or pressure loop-checking technicians
   b. Hiring certified instrument technicians
   c. Using a standardized loop-checking sheet with written procedures
   d. Issuing the same test equipment to each technician team

6. What document would the technician use to note that a valve is installed backward with regards to fluid flow?
   a. Instrument index
   b. P&ID
   c. Process flow diagram
   d. Visual inspection checklist

7. Which of the following are the two main methods for measurement signal loop checking?
   a. Continuity and pressure
   b. Forcing and simulation
   c. Shunting and forcing
   d. Simulation and grounding

8. Assume a technician uses a HART communicator to loop test a HART-capable device with a range of 32°F–212°F. What reading should the control room operator read if the technician simulates a 16-milliampere DC value?
   a. 77°F
   b. 122°F
   c. 167°F
   d. 176°F

9. What setting on a loop calibrator would be used to produce a 12-milliampere DC signal on an instrument loop that is powered from an external power supply?
   a. Milliampere DC read
   b. Percent 4 to 20 milliamperes DC
   c. Power two-wire transmitter source
   d. Two-wire transmitter simulation

10. What is the first action an instrument technician should perform if the instrument power supply immediately blows a fuse upon energization?
    a. Assume the fuse is bad and replace it.
    b. Perform a continuity check to identify a short circuit or ground fault.
    c. Replace the power supply.
    d. Reverse the red and black leads on the transmitter.

Start-Up and Tuning of Process Control Loops

A process control system is a complicated arrangement of devices and wiring methods that requires controlled responses to variables in an operating process. The technician is a valuable team member when working to ensure safe and cost-effective process control system start-up.

Objectives

» Explain the importance of performing complete loop checking before start-up.
» Discuss the importance of the technician in the safety of the start-up.
» Identify the activities of a typical start-up procedure.
» Describe loop tuning procedures.

Chapter 16

Table of Contents

Plant Start-Up Process 332
    Start-Up Roles and Responsibilities 332
    Safety During Start-up 333
    Start-Up Procedures 334

Loop Tuning 336
Summary 338
Review Questions 339

For additional information, visit qr.njatcdb.org
Item #2690

PLANT START-UP PROCESS

Start-up and loop tuning are critical procedures necessary for the proper operation of any control system. The roles and responsibilities of the technician are often clearly defined by the start-up team and may vary from plant to plant. To be an effective member of the start-up team, the technician needs a fundamental understanding of the essential components of the process system and their operating characteristics.

Start-up procedures follow the initial loop checking or loop "shakedown" process. During the loop-checking procedure, the technician verifies each control loop is complete and functioning as intended. The next step is *commissioning*, a preplanned, documented process of completing the start-up procedure for a process plant The term originates from the U.S. Navy and describes the systematic approach to verifying a new vessel performs as designed. Commissioning creates a structured path to move a job from the construction phase to turnover and production under control of the plant owner. Some process plants undergo additional *validation*, a rigorous evidence-based test to ensure the process consistently produces a product that meets quality standards.

Start-Up Roles and Responsibilities

The start-up procedure frequently involves a team of many disciplines: engineering, vendor representatives, instrument technicians, quality control and quality assurance professionals, and owner representatives. Successful start-up begins by developing a communication plan and schedule with a responsibility matrix. All parties have a vested interest in streamlining the start-up process. For the plant owner, revenue from finished product depends on a timely start-up. Similarly, the contracts that instrument contractors work under may hold back payment until commissioning is complete.

A technician's role varies based on the process design. Early start-up activities include process and instrumentation drawing (P&ID) walkdowns, leak testing, pipeline cleaning, and rotating equipment checks. The *P&ID walkdown* is the process of visually inspecting the equipment and piping to ensure it conforms to the P&ID and project specifications. This check ensures the installed equipment is free from errors in installation and construction. Any discrepancies between the design and the installation must be identified and investigated before start-up may continue. Leak testing is commonly completed using air or water. A leak is much less dangerous when discovered using air in a pipe that will carry ammonia after the process is live. A bubble test that uses soapy water to check connections assists the technician in finding air leaks.

Pipelines may contain dirt, welding slag, or other contaminants from the construction phase. Flushing the pipelines is an essential task before using process material. A pipe pig is a cleaning device used to dislodge contaminants. **See Figure 16-1.** Depending on

Figure 16-1. Pipe Cleaning Pig

Figure 16-1. A pipe pig cleans any welding slag, dirt, or foreign material from the inside of a pipe.

> **TechTip!**
> Inert gases such as nitrogen can displace oxygen. The technician should always use an oxygen concentration analyzer when working around nitrogen-purged vessels and pipelines.

the process material used in the pipeline, acid flushing or purging using an inert gas such as nitrogen may be used to prepare the pipeline for making the final product.

Rotating equipment such as agitators and pumps must be checked for proper direction. A common induction-type three-phase electric motor is reversed by switching any two stator leads. A handheld phase rotation meter can be used to determine the rotational direction before energizing the motor. **See Figure 16-2.** The technician should also verify that any gearboxes or bearing lubrication devices are filled with the proper lubricating fluid.

Safety During Start-up

Start-up can be a dangerous time for the instrument technician. Vessels are pressurized, gaskets and flanges are being stressed, and moving equipment is starting up all for the first time. A series of start-up requirements and procedures must be followed for safe, efficient control. Because start-up procedures are unique to the process to be controlled and to the devices to be used, there is no set of specific steps that a technician can follow and implement to aid in the start-up operation.

A hazard communication program, HAZCOM for short, identifies the hazard classification of a substance, as well as the safety data sheets and proper response to release or exposure. Safety showers and eyewash stations should be in good working order and clearly marked and identified. Windsocks are commonly used to indicate wind direction in the event of a chemical release. The windsock points downwind toward the danger, so the technician should relocate upwind in the event of a material release. **See Figure 16-3.**

> **ThinkSafe!**
> Start-up can be one of the most dangerous times for technicians. They must always anticipate that field devices could expose them to the media, pressures, and temperatures of the process. Prepare by using proper personal protective equipment and having emergency plans reviewed and ready to implement.

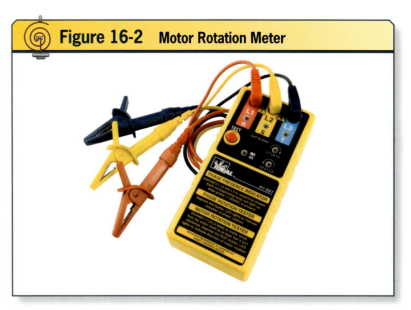

Figure 16-2. A motor rotation meter indicates the rotational direction of a motor before energization.

Courtesy of Ideal Industries

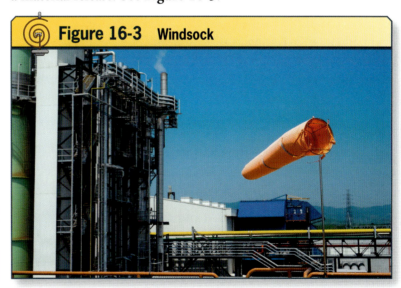

Figure 16-3. A windsock indicates the wind direction. Technicians should relocate upwind if a leak develops.

Start-Up Procedures

Input loops can be damaged during the time between loop checking and start-up. The technician in the field should make a visual inspection of the devices before start-up to ensure that the system is ready to perform. The technician should verify that all final control element devices are enabled when signaled by the controller. Alternate power sources (electrical and pneumatic) should be available for the devices and elements of the control system.

When the start-up begins, the operations team usually begins bringing the process up. The procedure the team follows often depends on the control system installed. In the work environment, the control point of operations is usually the main control room. The operator begins the sequence of bringing the process up from the control room. Often, local control stations can control the process, and in some facilities, these may be the points of control. There also are *manual loading stations*, devices or functions having a manually adjusted output that is used to actuate one or more remote devices.

Often, the technician has access to a matrix that displays the response of the system as it operates. This matrix displays the action of the system if a condition appears that has potential to produce hazardous consequences for the equipment and personnel involved. **See Figure 16-4.** The matrix displays alarm causes and the effects at the alarm points. A technician should be aware of the devices listed on such a matrix. During start-up, a technician may be asked to monitor or check such devices for errors.

Some facilities may include a visual reference that displays the sequence of operation of the process control

TechTip!
Even though technicians are typically responding to instructions from operations personnel during start-up, they must always question and discuss directions that are known to cause equipment error or damage.

Figure 16-4 Cause and Effect Matrix Example

Process Input	Alarm Tag	Descriptor	Alarm State	Process Action	Alarm Indicators
FT 1001	FAL 1001	Low flow @ 1 GPM	Light Off	Alarm only	FAL 1001
FT 1001	FALL 1001	Very low flow @ 0.5 GPM	Light Off	Pump stops, process halts, temperature element turns off	FAL 1001 UA 1001
FT 1002	FAL 1002	Low flow @ 1 GPM	Light Off	Alarm only	FAL 1002
LT 1003	LALL 1003	Low level @ 20%	Light Off	Pump stops, process halts, temperature element turns off	LALL 1003 UA 1002
LT 1003	LAH 1003	High level @ 85%	Light Off	Alarm only	LAH 1003
LT 1003	LAHH 1003	Very high level @ 95%	Light Off	Pump stops, process halts, temperature element turns off	LAHH 1003 UA 1001
TT 1004	TAH 1004	High temperature @ 85°F	Light Off	Alarm only	TAH 1004 UA 1001
TT 1004	TAHH 1004	Very high temperature @ 90°F	Light Off	Pump stops, process halts, temperature element turns off	TAHH 1004
Automatic	UA 1001	Automatic shutdown process initiated	Light Off	Pump stops, process halts, temperature element turns off	UA 1001
Manual	UA 1002	Manual shutdown process initiated	Light Off	Pump stops, process halts, temperature element turns off	UA 1002

Figure 16-4. A cause-and-effect matrix shows the actions of a process algorithm, where the volumetric flow rate is measured in gallons per minute (GPM).

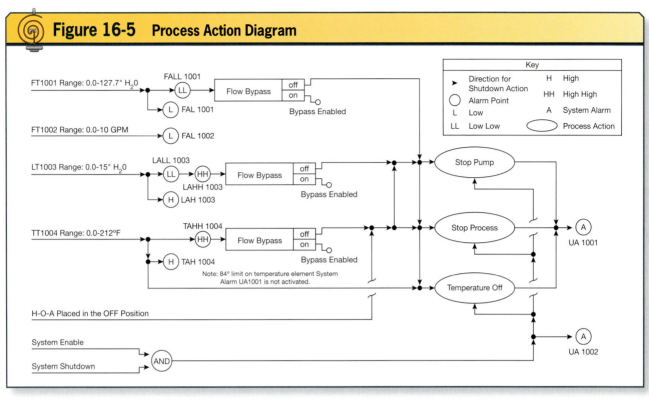

Figure 16-5. The diagram shows the logical flow of information used to control a process.

elements. This diagram provides a visual reference for operations to view how the control system controls the process. **See Figure 16-5.** Technicians can also use this diagram to aid in showing how the process is controlled by the control system. This document can provide a vital link between operations and control personnel.

Control system personnel must realize that when the start-up sequence begins, they provide support for operations. The process begins to generate dollars for the business when it is operating. An efficient start-up is the best way for a control company to present a good image to the customer.

In summary, the following steps should be undertaken to ensure that start-up is performed safely and effectively:

1. Start-up should not begin before all devices in the control system are checked out.
2. All personnel should attend a start-up organizational meeting that provides direction for all involved.
3. All personnel should be familiar with the functions and actions expected of the system, as well as the functions and actions of each person involved in the start-up.
4. All safety concerns need to be addressed before the sequence of start-up begins.
5. Potential safety hazards for all involved personnel should be discussed.
6. The start-up procedure should begin with a set of achievable goals that are understood by all personnel.
7. The facility policy for start-up procedures must always be reviewed to ensure that quality and safety procedures are well understood.

After the process is up and running, there may be a period during which the process does not stabilize. Instability occurs in proportion to the amount of work that was performed on the control system before start-up. The more work done, the more likely the process will need to be tuned. For a new system, work always remains after the process is up and running. For systems that have been retrofitted, there may be a smaller volume of work remaining.

LOOP TUNING

Processes may be oscillating, shutting down, or operating at rates other than those for which the system was designed. At such times, stable control is often established by executing *loop tuning*, a process of manipulating and adjusting the control algorithms to operate and maintain smooth process response.

The purpose of loop tuning is to adjust the control algorithm to maintain the process at its programmed set point with a minimum loop response delay. In the case of a closed loop proportional–integral–derivative (PID) control scheme, this implies that the PID control parameters need to be adjusted. Even though the control parameters affect the control valves mounted in the field, this process is not accomplished by adjusting valve parameters. Loop tuning is achieved by adjusting the parameter values contained within the controller's instructions. These instructions are part of the software configuration, and the parameters are manually entered into the controller.

The Ziegler-Nichols method of loop tuning can be used for tuning open and closed loop systems. This method requires introduction of an upset into the process, measurement of the system response, and use of this information to make loop corrections. The information gathered is plugged into a formula that modifies controller instructions sent to the output device. **See Figure 16-6.**

The Ziegler-Nichols method assumes that all necessary variables are known or can be measured. It also assumes that the person tuning the controller can obtain the variables needed. Part of the information required is the loop response. After observing the response, the loop may be tuned for optimal results. Two loop tuning techniques that use the Ziegler-Nichols method can help achieve desirable response in the process system: the open-loop technique and the closed-loop technique.

The open-loop technique introduces a change to the process with the controller in manual mode. In this tuning method, no feedback signal is sent back to the controller, hence the name "open loop." By recording the result that the introduced change has on the response of the process, calculations can be made for dead time and process gain. Dead time is a measurement of how long the process waits before reacting to the introduced change in the process. Process gain is the magnitude of change in the process variable relative to the magnitude of the

TechTip!

While performing loop tuning, the technician must be familiar with the natural response cycle of the given process. Many processes have short response cycles, whereas others have long response cycles. Patience must be exercised when performing loop tuning, because it can be a time-consuming process.

TechTip!

If the loop is not capable of being controlled with the controller in manual mode, the auto mode of the controller will not be able to achieve stable control.

Figure 16-6	Ziegler-Nichols Closed-Loop Tuning		
	K_P	T_i	T_d
P Controller	$0.5\ K_P$	∞	0
PI Controller	$0.45\ K_{P_u}$	$\dfrac{P_u}{1.2}$	0
PID Controller	$0.6\ K_{P_u}$	$\dfrac{P_u}{2}$	$\dfrac{P_u}{8}$

Figure 16-6. *The Ziegler-Nichols tuning method provides formulas to determine the settings of the gain (K_p), integral time (T_i), and derivative time (T_d). These settings are determined by finding the ultimate gain (K_{p_u}).*

introduced change. These calculations can be plugged into the PID tuning algorithm in the controller running the process system. The controller is then placed in automatic mode. The controller should exhibit a favorable response, which can be observed by noticing little oscillation of the process variable.

In contrast, the closed-loop technique is performed with the controller in automatic mode, with no integral or derivative control action present in the controller's algorithm. To perform the closed-loop technique, the process must be at a steady state, feedback must be present, and gain must be adjusted to prevent oscillation in the process variable. At this point, an increase in gain is continually introduced until sustained oscillation in the process variable is observed and the oscillations are at the same amplitude. This value is called the ultimate gain (K_{Pu}). The time the process requires to complete one oscillation is the *ultimate period* (P_u). **See Figure 16-7.** For example, assume the ultimate gain is found to be 3 and the ultimate period is found to be 4.2 minutes. Therefore, the calculated setting for a PID controller using the Ziegler-Nichols closed-loop tuning method would be as follows:

$K_p = 0.6 \times K_{Pu}$

$K_p = 0.6 \times 3$

$K_p = 1.8$

$T_i = P_u \div 2$

$T_i = 4.2 \text{ minutes} \div 2$

$T_i = 2.1 \text{ minutes}$

$T_d = P_u \div 8$

$T_d = 4.2 \text{ minutes} \div 8$

$T_d = 0.525 \text{ minutes}$

where

K_p = Gain

T_i = integral time (in minutes)

T_d = derivative time (in minutes)

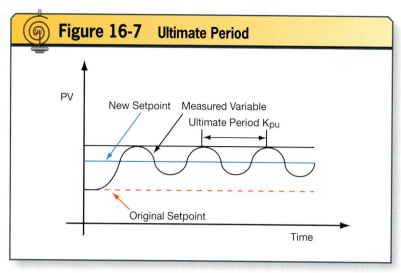

Figure 16-7 Ultimate Period

Figure 16-7. *The ultimate time period is the amount of time the process requires to complete one full oscillation of equal amplitude, where PV is the process variable.*

These calculated values can then be entered into the controller's interface to achieve the desired process response characteristics. The technician should be cognizant that different controllers and processes are tuned according to their specific design. These techniques are not inclusive, and additional loop tuning may need to occur before a desirable outcome is achieved. Although the Ziegler-Nichols method may be a good start, it does not always lead to a stable process variable. Many factors involved in the loop design can affect the tuning procedure.

With all three settings entered into the controller, the technician should be able to start the process and begin to record and evaluate the response of the system to the process actions. Additional adjustments may be needed. If the response of the control system is determined to be too slow, increasing the gain will cause a faster response action. A technician should be aware that the adjustment of one parameter of PID control often necessitates a change in one of the other variables in the controller. With practice, a technician tasked with tuning control loops often develops an intuition based on observations of the design and control objectives of the system.

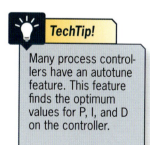

TechTip!

Many process controllers have an autotune feature. This feature finds the optimum values for P, I, and D on the controller.

SUMMARY

Technicians involved in start-up procedures ensure that a complete system loop check has been accomplished before start-up. They must be aware of all hazards to personnel and equipment and of the facility procedures to be followed. Technicians should be thoroughly familiar with all devices, wiring, and elements of the process control system.

Technicians provide support to operations during start-up and anticipate that the loop-tuning needs are usually in proportion to the number of system changes or additions made. Loop tuning usually requires changes to the controllers to achieve process stability and efficiency. Loop tuning involves the gathering of process information, resulting in modification of control settings, observation of achieved results, and perhaps further adjustment until the desired result is obtained.

REVIEW QUESTIONS

1. What is a preplanned and documented process of completing the start-up procedure?
   a. Commissioning
   b. Loop checking
   c. Loop tuning
   d. Validation

2. What is the process of visually inspecting the equipment and piping to ensure it conforms to the P&ID and project specifications?
   a. Commissioning
   b. Loop tuning
   c. P&ID walkdown
   d. Validation

3. Which of the following is an inert gas used to purge process pipes and vessels?
   a. Ammonia
   b. Chlorine
   c. Hydrogen sulfide
   d. Nitrogen

4. How can a technician reverse the rotational direction of a three-phase induction motor?
   a. Add a start capacitor.
   b. Induction motors are not directionally reversible.
   c. Reverse the polarity of the centrifugal switch.
   d. Switch any two stator leads.

5. If a technician notices a hissing sound coming from an outside vessel containing ammonia, what is the best plan of action?
   a. Attempt to locate the leak.
   b. Check the wind direction by observing a windsock and relocate upwind immediately.
   c. Continue working until an ammonia analyzer sounds an alarm.
   d. Wait for the supervisor to determine whether the condition is dangerous.

6. What procedure is used to optimize the controller settings and prevent unwanted oscillations in the process variable?
   a. Loop checking
   b. Loop tuning
   c. Startup
   d. Validation

7. What term describes the point at which the process variable exhibits sustained oscillations with equal amplitude?
   a. Baseline gain
   b. Derivative time
   c. Integration factor
   d. Ultimate gain

8. A technician observes a PID loop's ultimate gain to be 5. What is the predicted best gain setting based on the Ziegler-Nichols method?
   a. 2.5
   b. 3
   c. 3.5
   d. 4

9. A technician observes a PID loop's ultimate period to be 800 seconds. What is the predicted best integral setting based on the Ziegler-Nichols method?
   a. 100 seconds
   b. 400 seconds
   c. 800 seconds
   d. 1,600 seconds

10. A technician properly implements the Ziegler-Nichols tuning procedure. However, the process variable still exhibits excessive error. What is the likely cause?
    a. A control valve is likely sticking and should be removed from service and inspected.
    b. The controller is likely faulty and should be replaced.
    c. The technician must have made a math mistake. The Ziegler-Nichols method is designed to work on every control loop.
    d. The Ziegler-Nichols method may not be appropriate for this control loop.

Troubleshooting Process Control Loops

Performance of troubleshooting procedures on a process control system may often seem tedious, arduous, and at times difficult. However, when armed with a thorough understanding of the system and its components, data error records, plant operator reports, and instrument documentation, the technician should be capable of performing even the most complicated troubleshooting scenario.

Objectives

» Provide field assistance to troubleshoot devices.

» Explain a logical framework to troubleshoot a control system.

» Discuss and perform troubleshooting diagnostics on a smart device.

» Demonstrate knowledge of device and control-loop analysis.

Chapter 17

Table of Contents

Troubleshooting Procedures 342
    Troubleshooting Methods 342
    New Systems 344
    Active Process Control Systems 344

Troubleshooting Devices 345
    Smart Device Troubleshooting 345
    Troubleshooting Equipment 347
Summary 349
Review Questions 351

For additional information, visit qr.njatcdb.org Item #2690

TROUBLESHOOTING PROCEDURES

Before beginning any troubleshooting procedure, technicians determine which type of procedure may be necessary based on the environment. Generally, it can be helpful to consider a series of questions: Is this a new installation? Is the system undergoing renovation? Is this an active process system? Has the system undergone loop-checking procedures? Is the problem on the system localized to one particular loop, or is it affecting the entire process? The answers to these questions, among others, can provide technicians with insight into the nature of the problem and may offer a means by which to narrow down the possible causes of the error in the system.

Troubleshooting in a systematic, analytical, and logical way helps technicians achieve success. By analyzing and interpreting various forms of process documentation and alarm conditions, in addition to plant operations reports and feedback, a rational course of action can be developed.

Troubleshooting consists of finding trouble sources, studying their causes, and providing a solution. It is helpful for technicians to realize that trouble sources are rarely identified immediately. The source of the error is often isolated by a process of elimination.

The need for timely resolution of trouble conditions has a direct bearing on the safety, quality, and efficiency of a process system. Procedures often begin with verification that a problem exists within an element of the control system. Sometimes, a reported error or trouble condition is not actually an error but rather a misinterpretation of gathered data. It is the responsibility of technicians to gather as much relevant information as possible when developing a plan of attack.

Troubleshooting Methods

For technicians to become competent in troubleshooting a control system, they must first understand the operating principles of all loop components. Without working knowledge of how each instrument works and why it behaves in a certain manner, troubleshooting may seem incredibly complex. There is no replacement for a solid foundation of knowledge in the principles according to which a device operates. In addition, there is no better instruction for troubleshooting than experience. Technicians often develop intuition over time to assist in finding and correcting problems.

Before correction of a problem begins, a technician must clearly define the problem. A problem statement such as "the level in the tank is too high" or "the motor keeps blowing the fuses" is a good place to start. The technician may use a systematic approach to solve a problem. **See Figure 17-1.**

Once the problem is defined, the technician must gather information concerning the problem. In new systems, the technician may need to speak

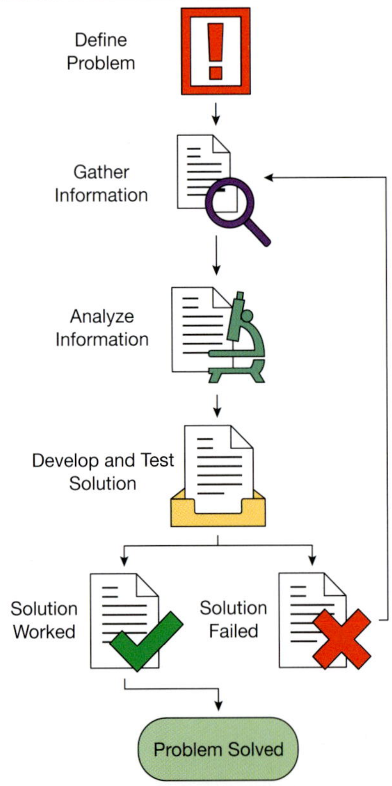

Figure 17-1. An algorithm provides a systematic approach to solving a problem.

with the installation crew or another technician who completed the loop checking. On existing systems, process operators are excellent sources of information. The technician can then analyze the information and attempt to develop a hypothesis for why the process is not acting per the design. Next, the technician should develop and implement a solution for the problem. After the solution has been implemented, the technician should seek evidence that the problem is fixed and the process is operating as designed. Oftentimes, the first solution is not correct. Even the most seasoned technicians incorrectly diagnose a problem occasionally. If the problem is not fixed, the technician should gather any additional information and restart the troubleshooting algorithm. If the problem is verified as corrected, the technician should document findings that may assist in troubleshooting similar problems in the future.

A troubleshooting *heuristic* is a method that promotes rapid problem solving by trial and error, sometimes referred to as a rule of thumb or shortcut. Troubleshooting methods such as divide and conquer or substitution methods are heuristics used to solve a problem. In the case of the divide-and-conquer method, a technician could be looking for a ground fault that has occurred and is causing a power supply fuse to blow upon energization. The technician locates the loop sheet and identifies a location to isolate the loop into segments. By breaking the loop and testing the positive and negative conductors for conductivity to ground, the technician can identify whether the ground fault is closer to the field end or controller end of the loop wiring. Through a process of dividing the remaining loop wiring in two sections while testing in both directions, the ground fault can be identified in a systematic effort. **See Figure 17-2.**

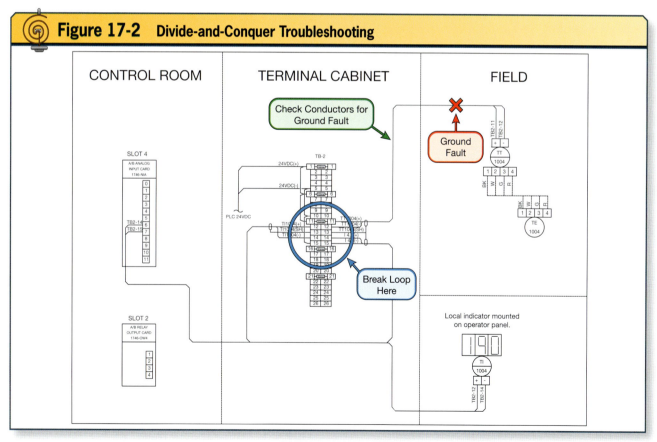

Figure 17-2. *The loop segments are isolated at terminal block 2 (TB 2) and checked for continuity between the signal wiring and the ground.*

The substitution method is used to isolate the problem to a specific device. For instance, consider a highway addressable remote transducer (HART) temperature transmitter that reports the alarm "Sensor burnout detected." After the technician tests the four-wire resistance temperature detector (RTD) and determines the sensor checks out as good, the technician may substitute an identical temperature transmitter and determine whether the new transmitter reports the same problem. If no alarm is generated, the substitution method has identified that the original temperature transmitter is behaving erratically. In this instance, the technician must always follow the proper documentation guidelines for changing out an instrument and update any plant documentation to reflect the change.

New Systems

When it comes to newly installed control systems, it is easy to jump to the conclusion that improper wiring methods may be to blame. However, most errors of this nature would have been identified during loop-checking procedures. If plant personnel are unfamiliar with a newly installed process system and its components, they may report proper functioning as an error. Reports of lights or alarms being illuminated when they shouldn't be may prove to be inaccurate. For example, local (field) indications of the process may not agree with indicators in the control room.

Troubleshooting procedures should be followed to locate operator error just as they are to locate actual errors that may exist in the control system. Elimination of all functional elements, devices, and loops as error sources may be necessary before plant personnel are convinced that a misunderstanding of data is the actual source of the perceived error.

This is not to say that newly installed systems are always without fault. Sometimes a detected error is the result of improper installation. It is often difficult to locate the source of the trouble when it immediately follows the process of loop checking. The technician should understand that mistakes can occur that allow a previously certified good loop to be faulty and should not assume that when a report is received indicating the presence of a trouble condition, it is solely operator error. The technician must use a systematic and logical approach to verify that all elements in the process loop involved are working correctly and are not the source of the error.

Active Process Control Systems

When dealing with active process control systems, the plant operations team is often the first to identify a trouble condition. Those involved in routinely operating and monitoring the process system typically have a thorough understanding of how the system should be operating, but they may not understand all components in a loop. For example, an operator may notice that fluid flow in a portion of the process is no longer under control as tightly as it had been in the past. The assumption may be that a control valve or other final control element needs repairs. The true source of the problem could be a leak in the instrument air supply upstream of the valve actuator, not the valve.

Instrument technicians should be aware that the final control element is only one of the potential points of error that may be to blame for the troublesome condition. By interpreting all associated documentation for the portion of the control system, the technician can identify all potential sources of the problem, whether it is a wiring issue or a faulty instrument. In addition, the operator reports can prove useful by providing answers to questions involving how the problem was first identified, when the problem began, any

> **TechTip!**
> Never assume an electrical system is low voltage. Some instruments use full-line voltage and present electrical shock hazards.

additional process deviations or upsets that occurred as a result of the problem, how consistent is the error on the process control system, and so on. The answers to these questions are means to identifying the source of the trouble.

Close coordination with operations personnel must take place before any repair work. The technician needs to ensure that everyone involved is fully aware of the repair procedure before beginning any work. It may be necessary to shut down the process to perform the necessary repairs, but in some cases, they may be performed while the process remains active.

TROUBLESHOOTING DEVICES

Technicians performing device troubleshooting procedures must have the ability to recognize faulty instruments. Occasionally, the device may not be the issue; instead, the error could exist in faulty wiring, terminations, configuration, or installation. Environmental conditions can also affect the operating parameters of an instrument. An instrument manufacturer is the best source for troubleshooting information for a specific instrument. While certain tests such as verifying sufficient loop power, checking impulse tubing to identify obstructions, or visually inspecting terminals to ensure low impedance connections are good first steps in troubleshooting, specific procedures for troubleshooting vary by instrument design.

For additional information, visit qr.njatcdb.org Item #2624

For additional information, visit qr.njatcdb.org Item #2625

ThinkSafe!
Do not remove transmitter covers in an explosion-proof installation without first removing the loop power.

TechTip!
Some facilities use computer maintenance management software to record all maintenance activities on a device. This is where the technician can find all device history and device-specific data, such as model number, serial number, and calibration ranges.

Smart Device Troubleshooting

The advantage of the smart device is the handheld communicator, which can save time as technicians troubleshoot devices. The functions of a loop test and transmitter test should be used to their fullest potential. Some industrial sites require the transmitter test and loop test commands to be performed on a regular schedule to verify working process loop integrity. Technicians should always perform such actions with the acknowledgment of the proper personnel. They should also take care to observe the working order of manual and automatic process loops that may be influenced.

To perform diagnostic procedures on a smart device, a communicator must be connected to the loop wiring. Sometimes, the communicator cannot recognize the device, and troubleshooting steps must be taken to establish communications. The first step in smart device troubleshooting is to verify that communication is being attempted with a smart device, not a legacy device.

Potential causes of smart device communication failures usually are associated with the loop wiring. Many related field-wiring problems can be eliminated by connecting directly at the transmitter and at a remote location and then comparing the responses of the communicator at the two locations. Even so, there are several possible explanations for why a device is not communicating. The 250 ohms of minimal loop resistance may not be present. This is a basic requirement for accurate, reliable communication to occur.

To address these problems, the technicians measure the loop voltage to see whether possible shorts or open circuits

ThinkSafe!
The technician must wear the proper PPE for the process they are working on when removing process-measuring devices for repair and should always expect and plan for pressurized process material to be present.

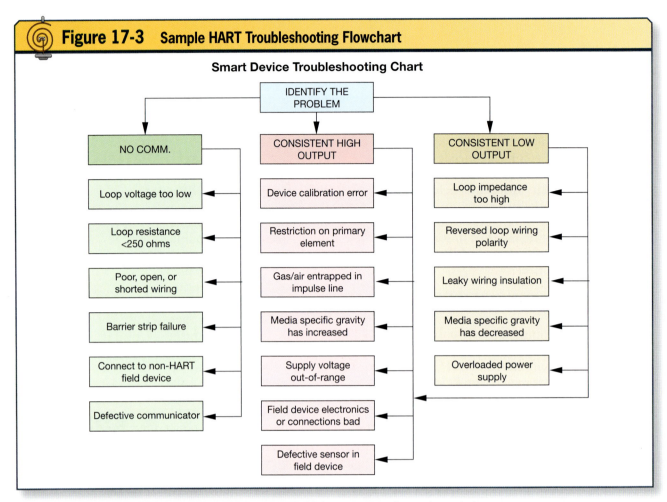

Figure 17-3. Failure to communicate, consistently high output, and consistently low output are common problems with smart instruments.

Figure 17-4. A clamp-on DC milliammeter can be used to read a 4- to 20-milliampere DC signal loop current without breaking the loop open.

Reproduced with Permission, Fluke Corporation

ThinkSafe!

The technician should always put the process loop into manual mode before connecting the communicator to a smart device. Failure to do so can cause unexpected changes in the process, which can cause injury or equipment damage.

are preventing a continuous operating voltage. Multiple ground points may be inducing loop currents that affect the signal between the communicator and the smart device.

A related troubleshooting point for loop wiring occurs at intrinsically safe wiring barriers that are in the loop wiring. Barrier troubleshooting documentation provides information to determine whether a barrier is performing correctly. The technician must follow the procedures correctly to determine whether error measurements are in the barrier. After communications with a smart device are established, the technician can isolate and diagnose the trouble spots. The smart device troubleshooting considerations can be summarized in chart form. **See Figure 17-3.**

Troubleshooting Equipment

Measuring loop signals is an important method to identify instrument problems. The traditional method to measure current is to break the loop and insert an ammeter is series with the instrument. This method may not be practical in a live process. Instead, two methods are available for the technician to use that does not require the loop to be broken. The first method is to use a clamp-on milliammeter. These meters are designed to read a milliampere direct current (DC) signal by clamping around the signal wire. **See Figure 17-4.**

A second method that measures loop current is to use the test terminals located inside the transmitter housing. A closer examination of the internal circuitry explains how the test terminals function. **See Figure 17-5.** A diode is located inside the housing and connected in a forward biased direction with the loop current. Under normal operation, the diode acts as a conductor. However, a minimum amount of voltage must be present for the diode to remain forward biased. When a DC ammeter is connected to the test terminals, the low impedance of the ammeter shorts the diode, reducing the voltage across the diode to 0 volts and causing current flow to cease through the diode.

A chart recorder provides a graphical representation of a process variable over time. Legacy versions used rolls of paper or circular graphs to document information. An electronic "paperless" version performs the same function by storing information on a local memory device or sending information across a network. **See Figure 17-6.**

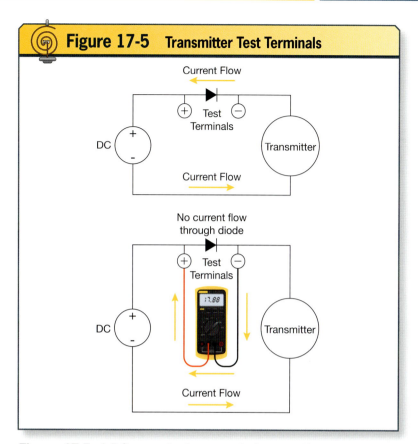

Figure 17-5. A DC ammeter is connected across the test terminals to measure the loop current.

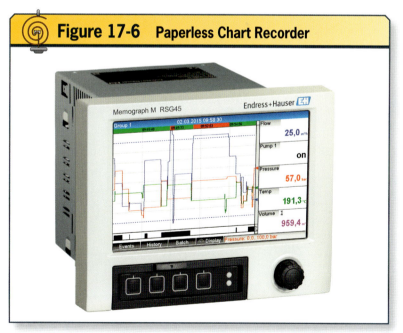

Figure 17-6. A paperless chart recorder displays process variables over time. The instrument technician uses this information to identify changes in process variables.

Courtesy of Endress+Hauser

Suppose a technician was investigating a low temperature alarm in a continuous stirred tank reactor. **See Figure 17-7.** The paperless chart recorder indicates a falling temperature signal from the temperature transmitter in loop 48 (TT 48). In addition, the recorder data shows an increased signal sent to the current to pressure transducer in loop 36 (FY 36). Flow control valve 36 (FCV 36) is coupled to an air-to-open valve actuator. After reviewing the process and instrumentation drawing (P&ID), the technician notices TT 49 measures the incoming steam temperature. A lower than normal value is measured at temperature element in loop 49 (TE 49).

Armed with this information, the technician identifies the probable cause as an issue with the steam supply, valve, or flowmeter. The technician's initial check of steam shows the system is operating properly with adequate supply pressure. The flow transmitter in loop 36 (FT 36) has an output in milliampere DC that is near 4 milliamperes DC, indicating there is a near-zero flow rate. The technician then visually inspects FCV 36. The technician notices the local stem indicator is in the closed position. A nearby pressure gauge shows 20 pounds per square inch of instrument air supplying the I/P transducer FY 36. A visual inspection of the pneumatic tubing between FY 36 and FCV 36 shows the $1/4$-inch tubing has been kinked and is severely restricted. The technician believes this is the problem. Upon repairing the tubing, the valve moves to the proper position and the temperature returns to normal.

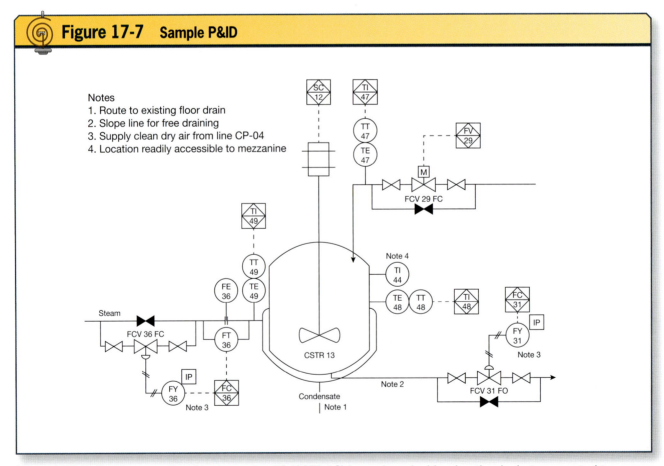

Figure 17-7. *Continuous stirred tank reactor 13 (CSTR 13) is equipped with a heating jacket connected to a steam line.*

SUMMARY

Field technicians require an understanding of the entire process of a system. A variety of troubleshooting techniques may be required, depending on the devices involved in each of the process loops. Troubleshooting instrument loops requires a thorough understanding of the process loops and associated hardware.

To be a successful troubleshooter, process control technicians must have mastered a number of skills. They must assemble documentation on related process loop elements, process error documentation, and reports from operators. They must analyze the available information to isolate the faulty process loop and faulty process element within that loop. Process control technicians prioritize repairs, coordinate work with operations personnel, perform the work, and update records.

Installation requirements often are dictated by the details portrayed on the drawings that are provided. The technician must be familiar with the installation requirements related to wiring, tubing, and conduit and must understand how to address problems encountered during installation and maintenance for smart devices and legacy devices. Troubleshooting equipment, preferred practice instructions, and documentation on calibration should be made available for the technician's use in diagnosing and repairing these devices.

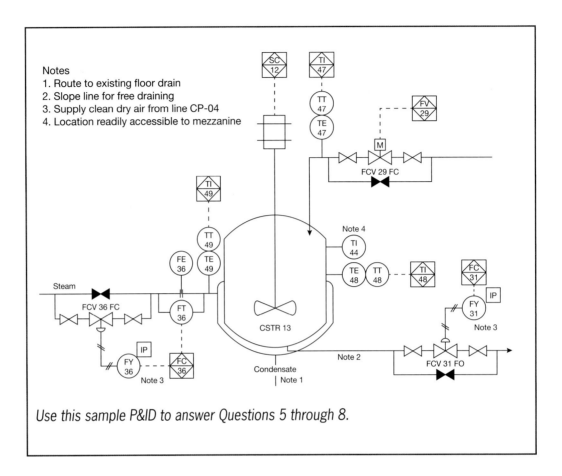

Notes
1. Route to existing floor drain
2. Slope line for free draining
3. Supply clean dry air from line CP-04
4. Location readily accessible to mezzanine

Use this sample P&ID to answer Questions 5 through 8.

REVIEW QUESTIONS

1. What step should the technician take if the initial proposed solution to fix a motor starter problem does not fix the problem?
   a. Gather more information
   b. Increase the fuse size
   c. Increase the solid-state overload setting
   d. Replace the contactor coil

2. What method is helpful when looking for ground faults on signal wiring?
   a. Analysis of variance
   b. Divide and conquer
   c. Linear regression
   d. Substitution method

3. Which factor could lead to consistently high output from a HART transmitter?
   a. Media specific gravity has increased
   b. An overloaded power supply
   c. Shorted wiring
   d. Reversed loop wiring

4. Which of the following provides a method to measure a 4- to 20-milliampere DC loop current without breaking the loop?
   a. Clamp-on AC ammeter
   b. High-potential (hipot) tester
   c. Megaohm meter
   d. Test terminal diode

Use the sample P&ID to answer Questions 5 through 8.

5. An increased air signal to FCV 31 would result in which of the following?
   a. Decreased flow out of CSTR 13
   b. Decreased temperature reading on TI 47
   c. Increased flow out of CSTR 13
   d. Increased signal from FC 36 to FY 36

6. If the equalizing valve on the manifold connected to FT 36 was left open, what would the technician expect the signal to be?
   a. 4 milliamperes DC
   b. 16 milliamperes DC
   c. 3 pounds per square inch
   d. 15 pounds per square inch

7. Suppose TT 49 is a HART instrument. What is the expected output signal if TE 49, a type J thermocouple, develops an opening?
   a. The failure signal depends on the transmitter configuration.
   b. The HART transmitter will fail high (>20 milliamperes DC).
   c. The HART transmitter will fail low (<4 milliamperes DC).
   d. There is no way to predict the signal after a failure.

8. What could be a possible cause of FCV 31 remaining in the open position irrespective of the pressure input signal?
   a. The bypass valve is currently open.
   b. The diaphragm is ruptured.
   c. FY 31 is configured to be reverse acting.
   d. The isolation valve was left closed after routine maintenance.

Distributed Control Systems

A control system can take several forms. The requirements of the process dictate which technology is applied for control of the system. Small-batch process systems successfully use programmable logic controllers or single-loop controllers to control the process. Large facilities such as chemical and petroleum plants need a more sophisticated control system to monitor and control the process. These sites are characterized by a high input/output (I/O) count and large geographical footprint. A distributed control system (DCS) is highly flexible and provides sufficient computational ability to handle the demands of these environments.

Objectives

- » Explain the difference between a conventional control system and a DCS.
- » Describe the network architecture of a typical DCS.
- » Explain the functions of the DCS I/O module, controller module, communication module, gateway, and user interface.
- » Describe the physical layer and topology used for a data highway.
- » Describe the best practices for supplying power and grounding a DCS.

Chapter 18

Table of Contents

Introduction to Distributed Control Systems 354
Definition of a DCS 354
    Components of a DCS 355
    DCS I/O Modules 356
    DCS Controller 356
    Real-Time Data Highway and Communication Modules 357
    DCS Gateways 360
    Power Supplies 361
    User Interfaces 362

Installation of a DCS 362
System Documentation 364
Summary 365
Review Questions 365

For additional information, visit qr.njatcdb.org
Item #2690

INTRODUCTION TO DISTRIBUTED CONTROL SYSTEMS

A distributed control system (DCS) is an assortment of communication, controlling, and measuring devices that the operations division uses to control the process. The architecture of a modern DCS reflects the evolution in the approach and technology used to control a process. DCS is a system of control in which the control functions are distributed over a wide area rather than being positioned at a centralized location. The control functions and instruments previously studied are a part of the DCS. The operation of the DCS requires field instrumentation, such as a level transmitter or flowmeter, to read the state of the process.

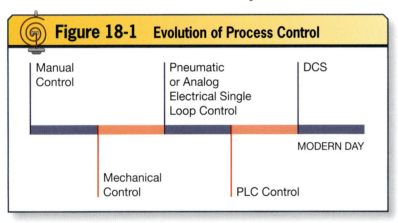

Figure 18-1. Automatic process control has evolved as technology has advanced.

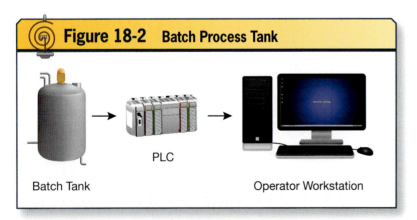

Figure 18-2. The batch process tank is controlled by a PLC connected to the operator workstation running HMI software.

DEFINITION OF A DCS

A DCS is a network of components that work together to serve a range of purposes. To better understand a modern DCS, the technician should begin by observing the history of process control. **See Figure 18-1.** Early process control was performed by operators in the field, manually reading local indicators and making adjustments to the final control elements of the process. In this early adaption, the human brain served as the controller, or decision maker, in the process. In an attempt to better regulate the process variable, mechanical devices were invented to link the process variable state to the final control element, thereby eliminating the need for constant human intervention. The advent of the pneumatic process controller continued the progression of automatic control by linking the process variable to an analog pneumatic signal. Analog electrical controllers were invented and drastically sped up the response time of the loop. Soon the advent of the digitally operated microprocessor led to the development of the programmable logic controller (PLC), which widely replaced electromechanical relay control. Today, the DCS uses digital communications across a broad network to connect hundreds or even thousands of points.

Building on the previous example, the following scenario illustrates how a DCS is implemented in a process control environment. The scenario involves a new chemical company, Triangle Chemical. The company has developed a new product that is produced in a batch process. Triangle Chemical purchases a single batch tank with all necessary instrumentation that is controlled by a PLC. The PLC is connected to an operator workstation consisting of a personal computer (PC). Software is used to create a human–machine interface (HMI) for the operator to control and monitor the process. **See Figure 18-2.**

As the company grows and sales increase, the need for two more batch tanks becomes evident. The company has a couple of options. It can purchase two more tanks and add input/output (I/O) cards to the existing PLC rack. This option is cost effective but has a limitation: all three tanks would be controlled by a single processor in the PLC. This means that if the PLC central processing unit (CPU) fails, the entire plant will shut down. The second option is to purchase two new batch tanks, each with its own PLC. **See Figure 18-3.** This option requires some variant of a plant network to connect the three PLCs to the single operator workstation.

The new tanks add capacity to the manufacturing facility and increase revenue. Sometime later, the plant business department decides it would like to access real-time and historical data to optimize the process. This change requires information from the process to be shared with both the operations department and the business department. Triangle Chemical decides to install a new DCS. The DCS will use the existing plant instruments. However, the PLCs will be replaced with individual I/O cards with redundant controllers. A new communication network will provide access to real-time data across the plant. Redundant controllers will prevent downtime in the case of a failure, eliminating the single point of failure in the PLCs' CPU.

In an example of this communication network architecture, the DCS only has two I/O cabinets. **See Figure 18-4.** The use of individual I/O cards means that individual points can be connected to the nearest cabinet. This lowers the likelihood of wire faults and reduces the installation cost for the new system.

The new DCS will allow Triangle Chemical to reduce the likelihood of control failure, share data with multiple departments within the company, and provide a scalable control system that has room for future expansion.

Components of a DCS

A DCS is a system composed of a combination of hardware and software to form a control platform. A trend toward interoperability has led to standardization of hardware and networking equipment. Each component of a DCS plays an integral role in the overall function of the system. Understanding how each component operates

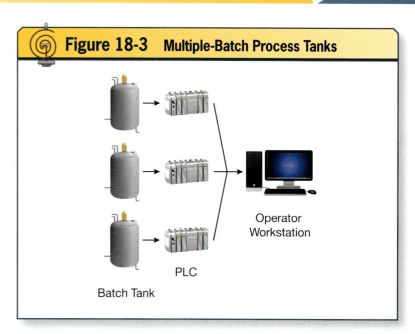

Figure 18-3. The increase to three tanks requires the addition of a plant communication network to connect the operator workstation to the batch process tanks.

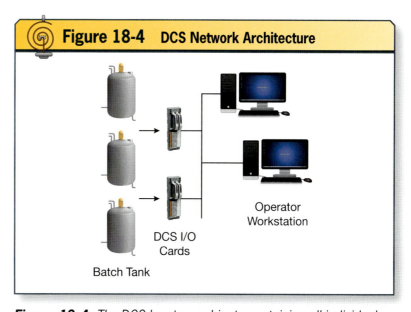

Figure 18-4. The DCS has two cabinets containing all individual I/O cards for the process instrumentation. A plant network provides access for both the operator's stations and the accounting department to monitor both real-time and historical data.

TechTip!

Although all DCSs are made up of similar components, each manufacturer has specific hardware and software designs, and components may not be interchangeable among the various systems. It is important to become familiar with the specific system design functions of a DCS before beginning maintenance activities.

within the system is essential for the technician to interact with a DCS.

DCS I/O Modules

The DCS is connected to field instrumentation through I/O modules. The DCS receives information by means of signals from field sensors, other controllers, and manual inputs from operators. After the signals are received, the DCS converts, or digitizes, them with an analog-to-digital converter (ADC), and its microprocessor performs programmed calculations and executes logic instructions to operate control loops, alarms, recorders, and other devices.

I/O modules require termination of field wiring to connect process instruments to the DCS. Termination points are indicated on loop sheets, which display the termination boxes, slots, and point addresses. DCS modules are available to connect a variety of signal types, such as analog I/Os, digital I/Os, highway addressable remote transducer (HART), Foundation Fieldbus, thermocouples, and resistance temperature detectors (RTDs). **See Figure 18-5.** An I/O module can have other features, such as electrical isolation and reverse polarity protection.

Because a DCS is a computer-based system and because all of its information is in a digital format after conversion, it can easily combine information from analog loops with discrete loops to execute logic. The I/O that the DCS uses can be as small as a few points to several thousand points. A DCS scans all primary elements or sensors, calculate loop parameters, execute logic, and send the results to its final control elements in the field. It constantly reevaluates the status of the process and performs thousands of decision-making calculations in fractions of a second.

DCS Controller

The DCS controller functions as the decision maker of the system. Much like previous technologies used a human operator or a pneumatic controller, a DCS network uses controllers to control the process. **See Figure 18-6.** The term "distributed control system" originates from the concept of moving the controlling portion of a control system from a centrally located area and distributing it across the plant. Early control systems used a single, large computer to perform all control functions. Field wiring ran from field instruments back to the centrally located computer. This design had several disadvantages. The field wiring had to be run long distances to reach the control room. This increased the likelihood of failure and the installation costs. In addition, the centrally located computer was a single point of failure for the entire system.

As technology matured and was miniaturized, controllers were produced that could be installed in field. This

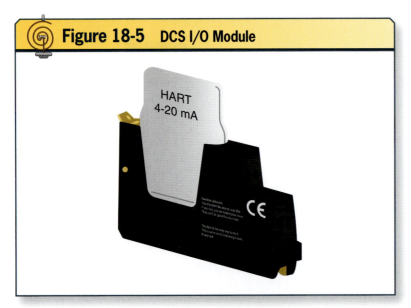

Figure 18-5. A HART module is a single-channel I/O point available to connect HART-capable field instrumentation to the DCS.

progression led to plant communication networks that linked several controllers with operator stations. As the control system's decision-making capability migrates into the field, several key advantages are realized. Field wiring can be shortened and localized to the field instrumentation, reducing costs and lessening the likelihood of faults in the wiring. If a fault does occur, it is typically isolated to a single channel. Another advantage of distributed controllers is the elimination of a single point of failure. Most DCS vendors provide the option of redundant controller installation. If a single DCS module fails, this would result in a single point, such as a thermocouple signal, that is lost in the control system.

DCS controllers can perform advanced control calculations such as proportional–integral–derivative algorithms and loop autotuning. Available memory for system configuration, scan time, I/O loading, software addressing for I/O modules, and control block memory are some of the more common considerations that must be addressed when selecting a controller.

Real-Time Data Highway and Communication Modules

Communication between two or more controllers is based on a protocol. This protocol defines how data are exchanged. No standard protocol exists for instrumentation projects. Instead, an array of proprietary and nonproprietary (open) protocols exist for implementation. A trend toward standardization and open protocols has made integrating an entire plant of different brands of controllers easier and more reliable. Many modern DCS installations use *Ethernet*, a physical layer for data transmission used in local area networks, as a standard for communication. One standard connection would be the Ethernet RJ45. **See Figure 18-7.** The use of standardized physical equipment allows use of *commercial off the shelf (COTS)*, hardware that is commercially available, typically from several manufacturers. This

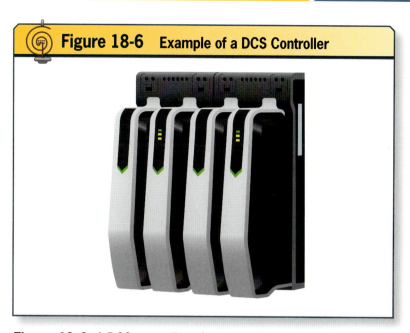

Figure 18-6. A DCS controller often consists of separate controllers installed on a common backplane. Controller modules perform control functions on I/O received using software instructions. They continuously update the I/O and perform logic functions, analog loop calculations, and other roles.

reduces custom development of hardware components and leads to systems that are easier to maintain.

A second common physical layer for communication is fiber-optic cable. Unlike all other communication methods, fiber optics use pulses of light instead of electrical impulses to transmit

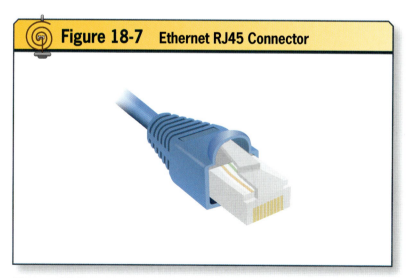

Figure 18-7. The RJ45 connector has become the industry standard for the physical layer of network communications.

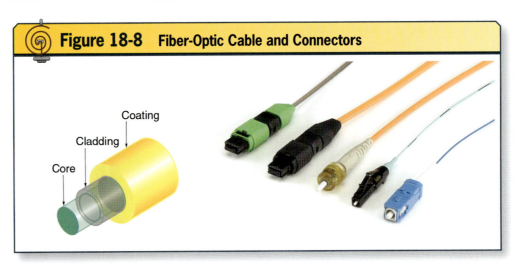

Figure 18-8. A fiber-optic cable consists of an inner core, cladding, and coating. Several connector styles are used to terminate the fiber-optic cable.

TechTip!
The installation of fiber-optic cable requires planning pipe or cable tray runs to avoid sharp bends and ensure the minimum bending radius of the cable is not exceeded.

data. Fiber-optic cables are constructed of an inner core, cladding, and buffer coating, as well as an outer jacket. **See Figure 18-8.** Light impulses are emitted from a source, commonly a light-emitting diode (LED), and travel through the core. The light waves reflect off the cladding because of its lower *refractive index*, a measurement of how easily light travels through a medium. The buffer coating and outer jacket provide protection and add strength to the cable.

Fiber-optic cable has several advantages over copper wiring. Fiber-optic cables are immune to electrical noise, such as the noise present around transformers and variable frequency drives on motors. Fiber-optic cables also have extremely high data transmission rates, generally limited only by the sending and receiving devices on either end of the cable. Finally, fiber optic cables provide a secure method to transmit information. While not impossible, tapping into a fiber optic cable is highly technical and difficult to perform without detection.

Real-time data highways can be constructed in a variety of configurations. The most common network topologies are the bus, star, and ring. **See Figure 18-9.** A benefit of a ring topology is the inherent redundancy in the event of a cable break. **See Figure 18-10.**

Similarly, two bus networks can be run in parallel to accomplish a redundant installation. **See Figure 18-11.** Redundancy is an essential consideration when installing a DCS to provide maximum uptime.

DCS communication modules are capable of transmitting information to serve both new and existing installations. These communication modules manage the flow of information between the controllers and the plant data highway. Communication modules are capable of handling large volumes of information. In addition to

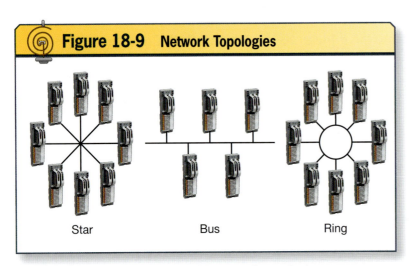

Figure 18-9. Three common DCS network topologies are star, bus, and ring.

Chapter 18 Distributed Control Systems 359

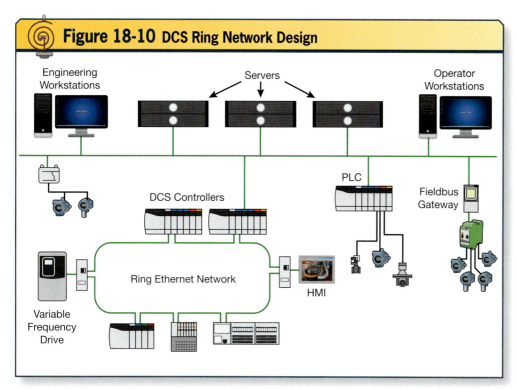

Figure 18-10. A ring network provides redundancy in the event of cable failure or damage.

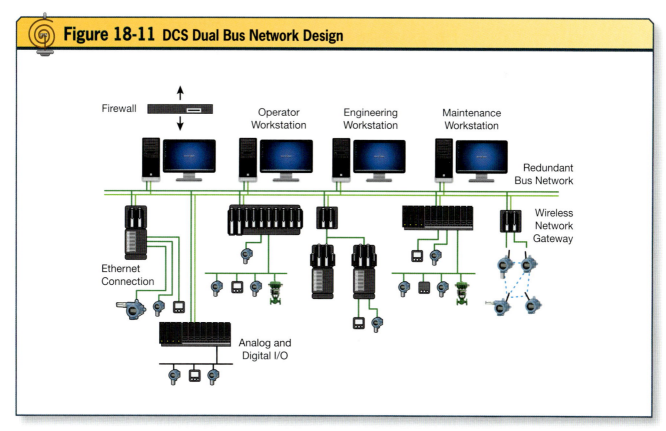

Figure 18-11. A redundant bus data highway is one method to reduce DCS downtime.

Figure 18-12 DCS Communication Module

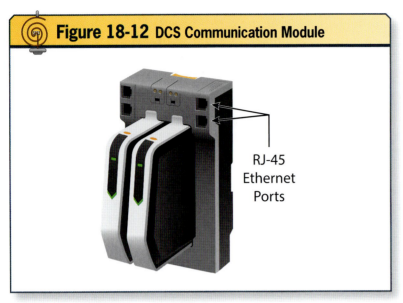

RJ-45 Ethernet Ports

Figure 18-12. A DCS communication module frequently uses redundant communication cards. RJ45 Ethernet ports on the backplane allow simple network connectivity.

TechTip!

Proper grounding is essential to allow the DCS to operate as designed. Improper grounding is one of the major causes of DCS errors. Always ensure that proper grounding techniques are used when working with a DCS.

monitoring and managing the flow of information, communication modules may establish gateways to other pieces of control equipment, such as PLCs and operator computers. The communication module provides a connection for subsystems or devices operating on the plantwide bus system to communicate. **See Figure 18-12.**

DCS Gateways

A *gateway* is network hardware that allows the interconnection of two dissimilar networks by performing a translation function. Modern control systems are often interfaced with legacy systems, requiring gateways to allow interconnection. PLCs and host computers often communicate using protocols other than that used by the control modules or the data highway communication links. Gateways are used to transfer information between two types of systems. A recent change in field instrumentation is the installation of wireless instruments. These instruments interface with the DCS via a gateway that reads the wireless signal and translates the process variables into a form suitable for transmission on the data highway. For example, an installation can have wireless instruments interconnected with Fieldbus and traditional 4- to 20-milliampere direct current instruments. **See Figure 18-13.**

Gateways usually function with relatively few problems in a system, but communication errors can occur. The transfer of information is usually seamless, but when the gateway is overloaded with information, transfer errors can occur.

Figure 18-13 Network Gateways

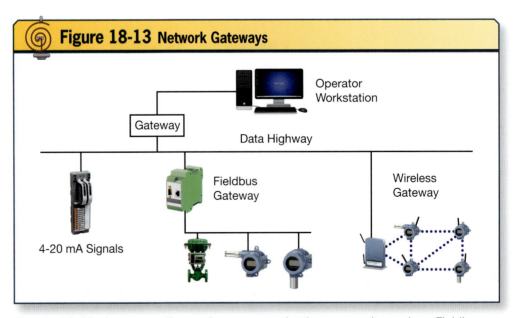

Figure 18-13. Gateways allow various communication protocols, such as Fieldbus and wireless signals, to connect to a single data highway.

Power Supplies

A *power supply* is a separate unit or part of a circuit that supplies power to the rest of the circuit or system. A DCS power supply conditions and regulates the raw electrical power into a usable and stable form. Power supplies are used for the DCS and various *peripheral components*, devices that are external to the CPU and main memory but are connected by the appropriate electrical connections. Redundant power supplies with battery backup are common and prevent DCS downtime.

Technicians may experience problems with power supplies and signal wiring. Suppose that the measurement device used had a different potential to ground than the controlling module. Errors could be introduced into the system through *ground loops*, a condition where current flows on the grounding portion of a circuit, leading to electrical noise and signal degradation. The result of a ground loop is measurement and loop-tuning corrections that do not reflect an accurate value. Ground loops are frequently introduced when the signal cable shield is grounded at multiple points in a system. **See Figure 18-14.** The best practice when grounding signal cables is to ground them at

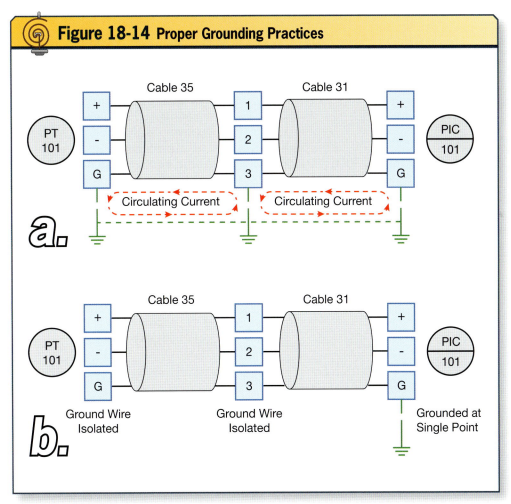

Figure 18-14. (a) Ground loops are present in cable 35 and 31 because of the connection to ground at the pressure transmitter (PT 101), junction block, and the pressure indicating controller (PIC 101). Current circulates on the ground in response to the parallel current paths. (b) When the ground is connected at a single point, there is no path for circulating current.

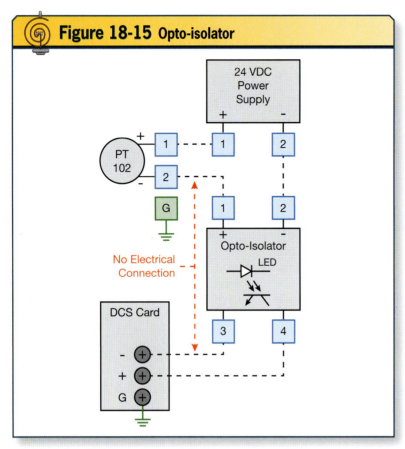

Figure 18-15. The opto-isolator is installed between the pressure transmitter (PT 102) and the DCS card. The loop signal is converted into a light signal transmitted by the LED. A phototransistor receives the light signal. The opto-isolator converts the light signal back into a loop signal that is transmitted to the DCS card.

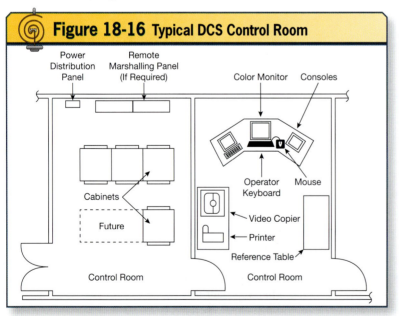

Figure 18-16. The control room provides an operator a safe area, remote from the process, to interact with the process.

one point, typically on the controller end, and then isolate the ground for the remainder of the circuit.

A device inserted into a circuit to prevent ground loops is known as an *opto-isolator*, a device that provides isolation by converting the electrical signal into a light signal and finally back into an electrical signal. The I/O terminals of an opto-isolator are not electrically connected. **See Figure 18-15.**

User Interfaces

A user interface is the part of the control system used by the operator for monitoring or controlling. It is common to hear a workstation referred to in a variety of ways: workstation, graphic user interface (GUI), man–machine interface (MMI), HMI, and others. Whatever the name, the objective is the same—to access the control system.

The use of commercially available computers as workstations has increased as proprietary DCS operator stations have become less common. A QWERTY keyboard and mouse are used, in combination with liquid crystal display (LCD) screens to serve as the interface in modern installations. A touch-screen monitor is an option allowing the operator to control the process with a simple touch. The workstation is not usually located with the DCS main controllers. Instead, the HMI is located in a control room, which provides a safer and more ergonomic workspace for operators. **See Figure 18-16.**

Graphical displays provide an animated view of the process variables and status of the process. **See Figure 18-17.** Alarming is shown on the display and is prioritized to direct the operator's attention to the most critical alarms. The graphical display may cause a change in color, cause a graphic to flash, or emit an audible notification in the event of an alarm.

INSTALLATION OF A DCS

The proper installation of a DCS requires the technician to read and understand the manufacturer-specific requirements. DCS projects are heavily documented to

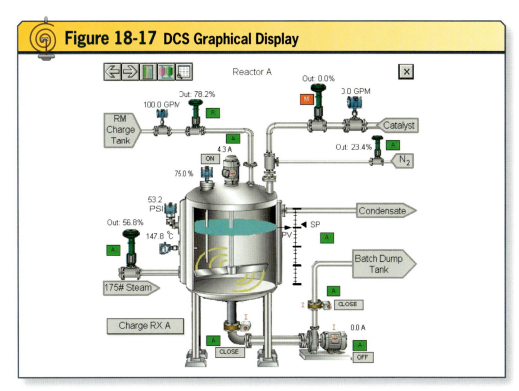

Figure 18-17. The process variable values are shown for the various instruments in a graphical display.

provide guidance during the installation phase. No single standard exists that covers all aspects of a DCS installation. Too many variables, such as manufacturer, protocol, and hardware, exist. The technician should have access to product specification sheets for equipment to determine installation data such as termination torque values, proper mounting methods, and suitable ambient temperature ranges.

Environmental conditioning is installed to prevent excessive humidity, temperature, or contaminants such as dust from infiltrating the control room. While environmental conditioning systems do not play a direct role in the control system, the failure of control modules or other electronic devices can often be attributed to inadequate control of the room's environment. The technician should pay close attention to penetrations through the control room envelope to ensure they are sealed properly.

Power distribution is primarily taken care of through the installation of the power supplies and associated wiring.

The *IEEE 1100* standard is an industry consensus for best practices when installing power and grounding electronic equipment. Installation methods such as multiwire branch circuits, as defined by *NFPA 70* Article 210, are discouraged. Instead, dedicated circuits with individual neutrals are recommended for electronic loads. Installation of feeder and branch circuits is recommended in grounded metal conduits or raceways. Signal conductors should never be installed in the same conduit as power conductors.

Always refer to the documentation of the power supply to verify that the installation meets the manufacturer guidelines. Because the power supply is a critical element in the control system, installation may include the use of an uninterruptible power supply (UPS). The UPS provides a method for seamless transfer of alternating current (AC) supply power to a battery backup supply in the event of a power failure.

To minimize the effects of electrical noise caused by outside sources, the

> **TechTip!**
>
> When signal conductors must intersect power conductors, always cross at 90 degrees perpendicular. Never run signal conductors in parallel with power conductors, because this can cause inductive coupling and create noise on the signal conductors.

primary AC source must be stable. Adequate grounding is needed to neutralize or minimize induced voltages impressed on the signal carriers and the control equipment. The entire grounding system should terminate at one common grounding electrode. **See Figure 18-18.** While grounding for safety is a familiar topic from *NFPA 70* Article 250, grounding for good electronic equipment performance is less well known to many technicians. Refer to best practices, such as *IEEE 1100*, for electronic equipment grounding details.

SYSTEM DOCUMENTATION

Complete documentation of a DCS requires all related field drawings to be included with the necessary documents for the DCS equipment and software controls. The complex arrangement of the equipment often dictates that the drawings, documents, specifications, and software documentation be highly detailed. If a company decides to expand its DCS capabilities after installation, a thorough understanding of the existing system is required.

Once the DCS is installed and commissioned, the system is turned over to the company and its technicians. Operations and maintenance departments need access to accurate documentation to use and maintain the system. In addition to the original design drawings, as-built drawings must be submitted. As-built drawings serve as a record of changes or deviations from the original design drawings. Documentation in the form of an owner's manual is compiled for use by the company's employees. In addition to an owner's manual, DCS-specific training manuals may need to be developed for the purpose of training operators to interact with the hardware and software used in the DCS design. Other common documents are parts lists with model or serial numbers, calibration records, and troubleshooting instructions.

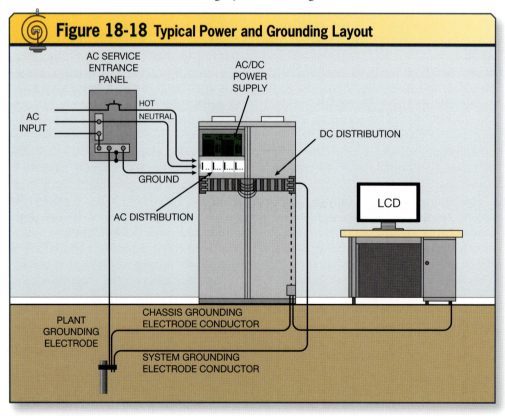

Figure 18-18. Proper grounding is essential for electronic equipment stability and performance.

SUMMARY

A DCS is a process control system that operates through the use of a computer network. Sensors, transmitters, and final control elements are connected to a network of DCS controllers. The DCS architecture leads to increased reliability, expansion capabilities, and access to data across the network. System operation may be monitored anywhere along the DCS data highway. The data bus may use one of several communication protocols dictated by the equipment manufacturer.

REVIEW QUESTIONS

1. What does a DCS I/O module use to convert an analog signal to a digital signal?
   a. ADC
   b. DAC
   c. Opto-isolator
   d. Wheatstone bridge

2. A DCS is best explained by which of the following descriptions?
   a. A network of components connected to a plant data highway
   b. PLCs connected to skid-mounted equipment
   c. A series of single-loop controllers installed in the field
   d. A single, large processing unit located in the control room

3. Which of the following is not a common topology for a DCS?
   a. Bus
   b. Ring
   c. Star
   d. Zigzag

4. What DCS component allows interconnection by performing a translation?
   a. Controller
   b. Gateway
   c. I/O module
   d. User interface

5. How can a technician prevent ground loops?
   a. Drive a ground rod near the cable termination point and connect the shield.
   b. Ground the cable shield at several points.
   c. Terminate the cable shield at one end only, typically the controller end.
   d. Use a UPS.

6. What is the name of a device that uses light to eliminate ground loops in signal cables?
   a. Ethernet communication module
   b. Intrinsically safe barrier
   c. Current-to-pressure converter
   d. Opto-isolator

7. Which of the following is not an acronym commonly used to describe a user interface?
   a. GUI
   b. HMI
   c. MMI
   d. UAT

8. What is the best method to avoid inductive coupling in signal cables that cross power conductors?
   a. Cross power conductors at 90 degrees perpendicular.
   b. Ground the signal cable shield at multiple points to ensure good grounding.
   c. Run signal cables in a nonmetallic inner duct to eliminate inductive coupling.
   d. Run signal cables parallel to power cables.

9. Which of the following statements about a DCS is false?
   a. DCS controllers are redundant.
   b. DCS is a computer-based system.
   c. A DCS program can be modified to permit expansion.
   d. A typical DCS can handle up to 50 I/O points.

10. Which DCS physical layer for communication uses light to transmit data?
    a. Ethernet
    b. Fiber optic
    c. Shielded twisted pair
    d. Wireless

Project Documentation and Management

Project management involves considering the goals and responsibilities of the project manager. It is important for all personnel involved with control systems to remember that a project is governed by standards. Project management requires a multidisciplinary approach to ensure the final product meets the customer's needs.

Objectives

- » Describe some of the project manager's responsibilities.
- » Explain the meaning of design standards.
- » List document standards common to instrument projects.
- » Describe the project life cycle.
- » Interpret project management documents such as a work breakdown structure and Gantt chart.
- » Define the scope of work.

Chapter 19

Table of Contents

Managing Projects 368
Design Standards 368
Documentation Standards 369
Quality Standards 369
Instrument and Control Project Management 371
Summary 376
Review Questions 377

For additional information, visit qr.njatcdb.org Item #2690

MANAGING PROJECTS

Project management encompasses several disciplines and crafts. Coordination among the engineering teams, project manager, operations management, and other trades is essential for a project to be successful. The fundamental concept of project management is to ensure all teams have the information necessary to bring the project to a timely and successful conclusion.

The instrument project manager uses policy and procedural standards, along with other information, to guide the project's working groups. Standards are usually assembled and grouped together by a coding system, which is supplemented by rules and regulations. Standards are often developed by an independent organization, but the rules and regulations usually are developed and implemented through the practical experience of the local facility. For example, there are Environmental Protection Agency (EPA) standards that govern the environmental impacts of an industrial site, and there are rules and regulations that provide methods to achieve the standards set by the EPA.

Before each project begins, great care must be taken to determine the applicable project standards. Care must be taken to determine whether local rules and regulations apply. A list of all relevant standards, rules, and regulations must be compiled and made available to all project personnel. The creation of a project library that contains these documents is encouraged, because it can ease the project manager's administrative burden and save time for all concerned as time progresses. Other pertinent data are entered into the library as the project progresses. The intent is to create a single point of reference for all information useful or critical to the project. A project library may be a collection of physical paper documents, though many modern projects use electronic records shared over large geographical areas using the world wide web. The electronic library provides instant access to new project related information as it becomes available. Each of several types of standards becomes a crucial reference during the project. Design, documentation, and quality are types of standards that are closely linked together.

TechTip!
Libraries are one of the most important assets available to the technician. It is important to become familiar with the location and setup of the document libraries before beginning maintenance activities.

TechTip!
Project management is not one individual's job responsibility. Every technician is responsible for a portion of the project management. They must manage their time, safety considerations, and work practices in a manner that achieves a safe and functioning control system for the project.

DESIGN STANDARDS

A *design standard* specifies the design or technical characteristics of a product in terms of how it is to be constructed. Design standards are developed by independent organizations such as the National Fire Protection Agency (NFPA), the American Society of Mechanical Engineers (ASME), or the Canadian Standards Association (CSA). These standard developing bodies may be granted accreditation by American National Standards Institute (ANSI). ANSI serves as a neutral third party to coordinate the development of standards. Design standards are voluntary by definition; however, they may be adopted by government authorities and become law. NFPA 70 is an example of a *code*, a collection of mandatory standards that have been codified by government authorities and adopted as law in many jurisdictions.

Observe the following scenario encountered by an instrument technician. The technician is tasked with installing a control valve and actuator. Many control valves are actuated using a pneumatic actuator that receives its air pressure from a current-to-pressure (I/P) transducer. The ANSI/ISA-S7.0.01 standard provides guidelines for providing instrument-quality air to the pneumatic system. The standard begins by defining the scope, purpose, and several relevant terms. The standard continues to establish four key characteristics of high-quality instrument air for pneumatic systems. These cover the pressure dew point, maximum particle size, maximum lubricant content, and prevention of contamination in the air supply system. The instrument technician should verify the air system meets these requirements to ensure the newly installed control valve meets the design standard.

The complex nature of instrumentation projects typically require adherence to several standards. The instrument technician must understand that the standards involved in the project ensure the control system is properly installed and operated. Understanding the interaction among design standards is assumed by most personnel to be a function of the engineering team. However, the instrument technician often is presented with conflicting standard requirements. An instrument technician who can identify these conflicts and relay the information to the project management team plays an important role in the project's management. For example, project equipment, components, and devices may not meet the project's relevant standards. Identifying these inconsistencies early in a project helps to reduce cost and time-consuming rework.

On certain tasks of an instrument project, there may be no design standards. Recommended practices (RPs) are documents published by organizations, such as the International Society of Automation (ISA), that provide guidance. For example, the wiring practices for the interior of control panels are covered by ISA-RP60.8. If there is a question about crucial information that will affect the assembly being installed but there are no standards, there may be an RP to provide the data needed.

DOCUMENTATION STANDARDS

Documentation provides a link between the design team and the field installers. These project documents are put together according to governing standards. A goal of the project manager is to ensure that all project team members are aware of the proper standards. If there are questions concerning the interpretation of drawings, the project manager should take steps to resolve the discrepancies before they cause delays or improper installations.

The project manager is responsible for obtaining interpretations of symbols used by other crafts that affect the work of the controls team. The use of standard symbols varies among countries and industries. In the U.S. petroleum and chemical industries, ANSI/ISA standards are commonly adopted. However, standards for documentation are published by many sources. For example, process and instrumentation drawing (P&ID) design standards can be published according to the Japanese Industrial Standard (JIS), Process Industry Practices (PIP), International Organization for Standards (ISO), or Deutsches Institut fur Normung (DIN). The project manager and all project team members who interact with the project documents should be aware which standards are in use for the project.

QUALITY STANDARDS

Quality as a measure of project success is a large part of managing instrumentation projects. Every project starts by defining the purpose or benefit of the final result. For example, the project may result in reduced emissions, increase production capability, enhance the efficiency of an existing process, or lead to another objective. Manufacturers achieve a high-quality finished product by intentionally taking steps to ensure the product is made to the design and checked for conformity. A number of systems are used by manufacturers to ensure quality is maintained in the process.

A commonly referenced quality standard in the process control environment is the International Organization for Standardization (ISO) 9000 family of quality management standards. The

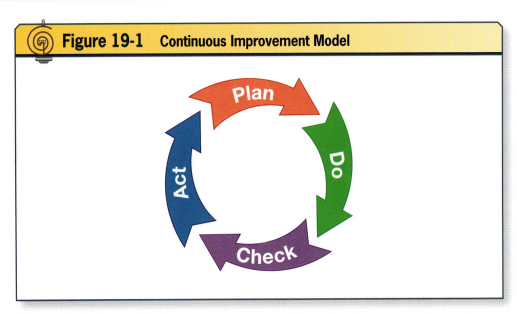

Figure 19-1. The cyclical process of planning, doing, checking, and acting on the results leads to higher-quality finished goods.

ISO is a global organization that publishes standards. Manufacturers can apply for ISO 9001 certification based on their adherence to the guidelines set out in the standard. The mnemonic model of plan, do, check, and act, or PDCA, provides an example of the continuous improvement mentality. **See Figure 19-1.**

The adage "what gets measured, gets improved" is especially relevant in this context. A properly calibrated and functioning instrument system is essential to the PDCA model.

The instrument technician often plays a role in ensuring the ISO 9001 objectives are met. The instrument technician ensures the process

Figure 19-2. Manufacturers seek ISO 9001 certification to ensure they fulfill customer needs and achieve regulatory compliance.

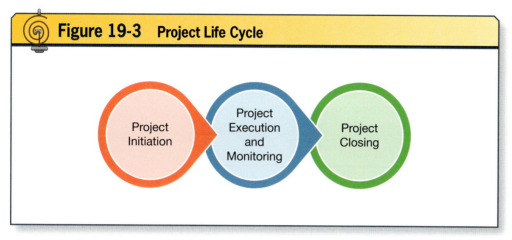

Figure 19-3. A project can be broken into distinct phases, beginning with initiation, progressing to execution and monitoring, and finishing with the closing of the project.

measurement system that performs the check function is properly installed. The ISO 9001 standard provides specific guidance on the maintenance and calibration of measurement equipment, such as a load cell or resistance temperature detector (RTD) probe. The standard also sets forth requirements for the manufacturer to maintain records that provide evidence of conformity to calibrations and other important documents. Manufacturers often display their adherence to ISO 9001 guidelines by posting signs or issuing press releases. If measurement devices are not reporting accurate and repeatable measurements, the entire quality system cannot function. **See Figure 19-2.**

INSTRUMENT AND CONTROL PROJECT MANAGEMENT

Project management is the process of ensuring the goals of the project are reached. A *project* is a temporary endeavor undertaken to create a unique product, service, or result. Each project has a project life cycle that begins with the concept of the project and ends with project turnover to the customer. **See Figure 19-3.**

In particular, instrument and control (I&C) projects frequently consist of several phases of construction during the overall project. A project manager must understand the technical requirements of the project and must be proficient in maintaining good interpersonal relationships within the overall project team. Everyone in the I&C group must be able to interact with the project manager, because the manager is the controlling authority for the project.

The I&C project manager is often confronted with different stages of work performed under different contracts. An example is the construction project that is bid separately from the contract for checking and calibration of instruments. The calibration, loop check and start up contract is usually awarded as a service contract with requirements that are independent of the construction project. The calibration contractor is usually required to have an on-site presence, and the personnel must maintain control of the instruments that are delivered to the site. Usually, this contractor is given the project under a lump-sum bid. The work performed must be checked and verified as accurate in accordance with the bid list of materials to control expenses. The verification and calibration team is often a subpart of the construction contract, but care must be taken to differentiate between the activities of the two groups during project construction.

The project management team uses a variety of documents to list and track the stages of a project. A large project can be broken into more manageable phases. Each of these phases can be further broken into smaller work packages. A *work breakdown structure (WBS)* is a project management tool used to identify the deliverables of a project and organize them in a hierarchal layout. **See Figure 19-4.**

The chronological sequence of deliverables from the WBS are shown in a network diagram chart. **See Figure 19-5.** This visual representation of the

Figure 19-4 Work Breakdown Structure

Work Breakdown Structure

Project: Heat Exchanger #3 Upgrade
PM: Jack Mc Donald
Date: 19-Nov-16
Site:

ID	Task Name	Owner	Deliverable	Logic (list IDs) Preceding	Succeeding
1	**Demo existing heat exchanger**				
1.1	LOTO	Triangle Piping	Safe worksite	none	1.2
1.2	Decontaminate	Square Environmental	Free of contaminants	1.1	1.3
1.3	Remove process piping	Triangle Piping	Piping removed	1.2	1.4
1.4	Remove existing heat exchanger	Circle Millwrights	Heat exchanger removed	1.2	
2	**Install new heat exchanger**				
2.1	Develop lift plan	Circle Millwrights	Safe work permit	1.3, 1.4	2.2
2.2	Lift heat exchanger into position	Circle Millwrights	New heat exchanger in position	2.1	2.3
2.3	Install process piping	Triangle Piping	Piping installed	2.2	
2.4	Install Instruments	Star Electrical	Instruments installed	2.2	
3	**Commission new heat exchanger**				
3.1	Pressure test	Instrument Commissioner International	Pressure test documents	2.3	
3.2	Loop test instruments	Instrument Commissioner International	Loop test documents	2.4	
3.3	Test safety relief system	Instrument Commissioner International	Safety relief test documents	3.1	
4	**Closeout**				
4.1	User Acceptance Test (UAT)	Instrument Commissioner International	UAT documents signed off and complete	3.3	4.2
4.2	Finalize as-built documents	All contractors	Project documents complete	4.1	

Figure 19-4. *The WBS lists the four phases of the project and the individual deliverables for each phase. The individual identifier (ID) numbers and logic show relationships to the project timeline.*

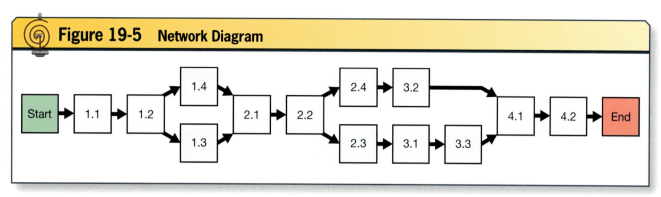

Figure 19-5. The relationship of the activities listed in the WBS and the relationship of individual deliverables can be shown in a graphical format. The items in parallel can occur concurrently, but the items in series must occur sequentially.

deliverables highlights critical path items that must be completed before subsequent work may begin.

A Gantt chart shows the start and finish dates of each task within a project. Tasks shown overlapping on a Gantt chart can be completed simultaneously. **See Figure 19-6.** Gantt charts are especially useful for tracking the schedule of a project.

Projects are initiated by determining the purpose, budget, and scope for the project. The *scope of work (SOW)* is a document that clearly defines the products or services to be delivered. This document is sometimes called the statement of work. During a project, questions of responsibility arise in determining what work will be done and who will perform the work. These questions should be addressed before the project begins and documented in the SOW. The SOW document is invaluable, because work that is not clearly defined, especially for trades, can cause unwanted confusion and ill will between the control contractor and the customer. The SOW should clearly define which tradespeople perform which work projects, and it should designate the interface requirements among trades.

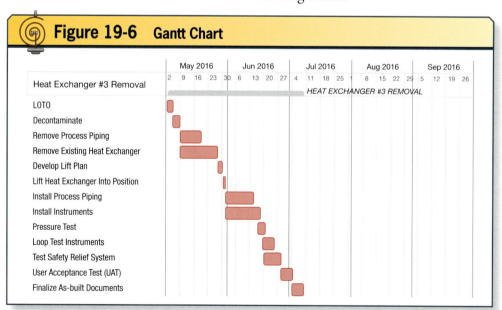

Figure 19-6. A Gantt chart provides a graphical representation of the various work tasks of the project.

While some projects are completed using a customer's employees, most projects require outside firms to be contracted to perform work to complete the project. A *request for proposal (RFP)* is a document that provides the information for a project management firm to compile a bid package to perform the work for the end user. The customer's engineering staff may work together with contracted engineering firms to develop a design and the

Figure 19-7   Request for Information

Instrument Consultants International
Request for Information No. 168-96
Date: October 25, 2016

To: Charles Vandelay   VIA EMAIL (X)   FAX ( )   MAIL ( )   HAND ( )
      Premier Construction Services
      1672 N. State St.
      Tucson, AZ 85709

From: Daniel McCarthy   (X) Jobsite   ( ) Omaha Office
Project: Batch Vessel #3
Owner Project Code: BR-95137
ICI Project No: 201597-C

Subject: Location of LT-191 on top of batch vessel #3
Specification Reference: N/A   Drawing Reference: PID-BR-37
Attachments w/ Request: (X) YES   ( ) NO   No. of Pages: 1

PLEASE RESPOND BY: Oct 28, 2016   TO THE FOLLOWING REQUEST:

The P&ID drawing indicates the radar type level transmitter located on loop 191 should be mounted on the 3" flange on top of batch vessel #3. The batch vessel's inlet fill line is situated in such a manner that the transmitter signal in unlikely to perform per the manufacturers design.

Work Delayed Pending Response (X)

RESPONSE:
Use alternate 3" flange mounted on opposite side of vessel.

By: Charles Vandelay   Date: Oct 27, 2016
Attachments w/ Response: ( ) YES   (X) NO   No. of Pages: 1

Figure 19-7. *This RFI document shows that the contractor identified a potential problem with the placement of a transmitter. The contractor is asking for more information and whether moving the transmitter is appropriate. The designer responded with an order to change the location.*

necessary documentation. The engineering approach to an instrument project usually is dictated by the performance specifications furnished by the customer. The instrument contractor is chosen based on a variety of considerations, including experience, perceived competence, engineering expertise, and quoted price. The contractor is often involved in aspects of the initial design and modifications of the control system throughout the project period. The suppliers are contacted with a *request for quotation (RFQ)*, a document soliciting a price for the equipment and material needed to complete the project.

Once the design is done and the instrument contractor begins work, the project manager's role is to monitor and control the project. As construction begins, questions concerning the project design often arise. The contractor can submit a *request for information (RFI)*, a request made by the contractor to make clear the intention of a design document or ask for guidance on a conflict in documentation. **See Figure 19-7.**

Another common controlling function is managing change control. The contractor should not deviate from the original design without submitting a change. *Management of change (MOC)* is a procedure that identifies hazards and risks associated with changes made to a process. MOC typically can be a time-consuming and paperwork-intensive process; however, the safety and functionality of the process mandate that it be managed properly. Changing one component on a process loop may result in catastrophic consequences if the change is not carefully reviewed. The cost of making a change to a project is typically low in the early stages of the project and increases as the project nears completion. The typical staffing levels also change over the course of a project. **See Figure 19-8.**

As the project nears completion, the project manager's attention shifts to the closing the project. Even though the staffing levels may decrease and the project appears to be nearly complete, the closing segment of the project life cycle requires attention to detail.

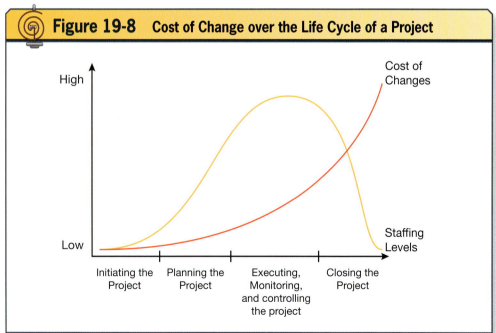

For additional information, visit qr.njatcdb.org Item #2548

Figure 19-8. Making a change in the project design is generally less expensive at the early stages of a project. As the project life cycle matures, the cost of making changes increases. Also shown are the staffing needs over the project life cycle.

During this phase, responsibility for the project is turned over to the customer. A user acceptance test (UAT) is a test that proves the equipment operates to the design specification. Once all project deliverables are satisfied and the customer approves project completion, the ownership is transferred to the customer. The owner gains control of the project and transitions into the operations and maintenance of the final product.

SUMMARY

Instrumentation projects are complex. Design standards provide guidance to ensure the system is installed to operate both safely and efficiently. Project management includes the oversight of many disciplines and crafts for every job. Because many standards exist, it is important for the technician to be aware of conflicts so that they can be addressed and resolved. Standards, rules, regulations, and project documentation should be accessible by all project team members.

An effective project manager ensures that documentation is compatible across craft boundaries and that all organizations understand their responsibilities. This concept is initiated by the SOW. The engineering aspects of a project are often shared between the customer and the contractors. Engineering approaches are generally dictated by the performance specifications.

REVIEW QUESTIONS

1. What is the role of ANSI in the development of design standards?
   a. Ensure design standards become law
   b. Hire expert engineers to write standards
   c. Maintain a project library for project managers
   d. Serve as a neutral third party to coordinate the development of standards

2. What document could be used as a reference if no design standard exists?
   a. *National Electrical Code*®
   b. RP
   c. RFI
   d. SOW

3. Which organization publishes P&ID design standards?
   a. ANSI/ISA
   b. JIS
   c. PIP
   d. All the above

4. What is the purpose of ISO 9000 standardization?
   a. To develop consensus standards for worker safety
   b. To ensure all trades have access to the project library
   c. To ensure the manufacturer makes a profit
   d. To provide a framework for maintaining quality within a manufacturer's plant

5. The acronym PDCA stands for __?__.
   a. People, design, construct, attitude
   b. Plan, describe, connect, actuate
   c. Plan, do, check, act
   d. Prevent, divide, create, assess

6. Which of the following would be considered a project?
   a. Calibrating temperature indicators during a scheduled shutdown of a heat exchanger
   b. Cleaning the filters on an air compressor as preventative maintenance
   c. Day-to-day operations in a refinery
   d. Installing a new fractional distillation column at a refinery to produce a new product

7. A document that breaks a project into a series of deliverables is called a __?__.
   a. calibration record
   b. MOC document
   c. RFI
   d. WBS

8. Before an instrument technician deviates from the project documents, a __?__ form must be submitted.
   a. calibration record
   b. MOC
   c. RFP
   d. RFQ

9. During the closing phase of a project, a UAT is completed. What is its purpose?
   a. Guarantee the equipment will make a profit for the owner
   b. Prove the equipment operates to the design specifications
   c. Prove the project followed the critical path
   d. Show that the project manager completed the project on budget

10. Where should project documentation be stored?
    a. Owner's file cabinet
    b. Project library
    c. Project management office
    d. Project manager's file cabinet

Appendix

Table of Contents

Instrumentation Formula Sheet 380
Sample P&ID 381
Instrument Letter Chart 382
Industrial Electrical Symbols 384
Standard P&ID Symbols Legend | Industry Standardized P&ID Symbols 388
Ohm's Law 395
NATO Phonetic Alphabet 395
Resistor Color Code 396
ANSI and IEC Color Codes 397

Technical Data Section 398
Metric Prefixes 400
Decimal, Hexadecimal, Octal, and Binary Conversion Chart 401
4-20 mA Transmitter Output Chart 402
3-15 PSI Transmitter Output Chart 403
Logic Gate Chart 404
NFPA Hazardous Locations 405
Geometric Formulas 406

For additional information, visit qr.njatcdb.org
Item #2690

INSTRUMENTATION FORMULA SHEET

Pressure = H * SG	Span = URV−LRV	Accuracy = $\left(\dfrac{\text{Deviation}}{\text{Span}}\right) \times 100$	Gain = $\dfrac{\text{Output span}}{\text{Input span}}$	1 PSI = 27.7"H$_2$O
Pressure = $\dfrac{\text{Force}}{\text{Area}}$	$R = K \times \dfrac{L}{A}$	PV = NRT	SG = $\dfrac{\text{Fluid density}}{\text{Water density}}$	1' H$_2$O = 0.433 PSI
Q = V * A	$Q = A_1V_1 = A_2V_2$	$Q = K\sqrt{H}$	V↑ ∝ P↓ V↓ ∝ P↑	RN = $\dfrac{\rho v d}{\eta}$
$C = \dfrac{5}{9} \times (F - 32)$	$F = \left(\dfrac{9}{5} \times C\right) + 32$	K = C + 273.15	R = F + 459.67	1 PSI = 2.036" HG
1 ft^3 = 7.4805 gallons	1" = 2.54 CM	1 ATM = 29.92" HG	1 Meter = 39.37"	1" HG = 13.6 "H$_2$O
A = π*R^2	V = π*R^2*H	$A^2 + B^2 = C^2$ $C = \sqrt{A^2 + B^2}$	ATM @ Sea level = 14.7 PSIA	1 ft^3 H$_2$O = 62.43 lbs.
E = I * R	P = I * E	$R = \dfrac{E^2}{P}$	P = I^2 * R	Efficiency = $\dfrac{\text{Output}}{\text{Input}}$

Appendix

SAMPLE P&ID

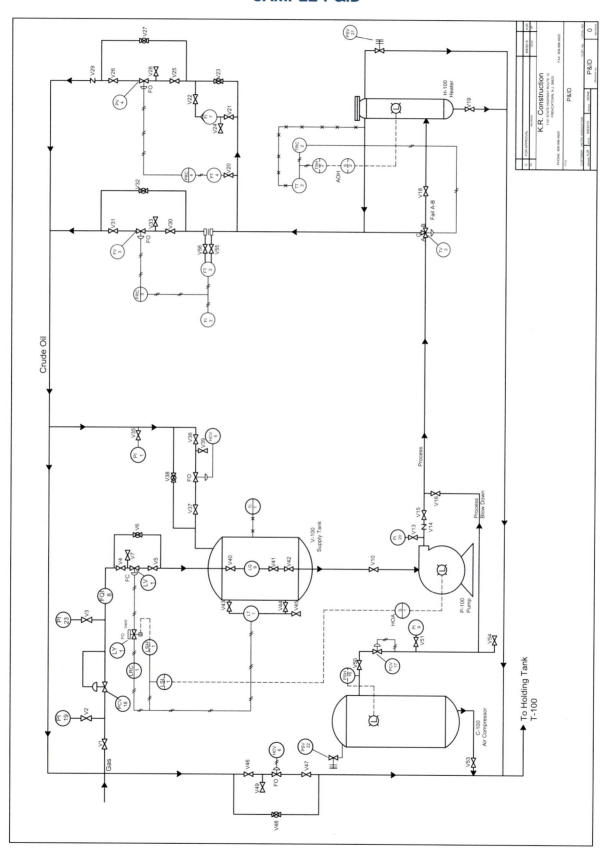

INSTRUMENT LETTER CHART

First Letters	Initiating or Measured Vaiable	Controllers				Readout Devices		Switches and Alarm Devices*			Transmitters			Solenoids, Relays, Computing Devices	Primary Element	Test Point	Well or Probe	Viewing Device, Glass	Safety Device	Final Element
		Record-ing	Indicat-ing	Blind	Self-Actuated Control Valves	Record-ing	Indicat-ing	High**	Low	Comb	Record-ing	Indicat-ing	Blind							
A	Analysis	ARC	AIC	AC		AR	AI	ASH	ASL	ASHL	ART	AIT	AT	AY	AE	AP	AW			AV
B	Burner/Combustion	BRC	BIC	BC		BR	BI	BSH	BSL	BSHL	BRT	BIT	BT	BY	BE		BW	BG		BZ
C	User's Choice																			
D	User's Choice																			
E	Voltage	ERC	EIC	EC		ER	EI	ESH	ESL	ESHL	ERT	EIT	ET	EY	EE					EZ
F	Flow Rate	FRC	FIC	FC	FCV, FICV	FR	FI	FSH	FSL	FSHL	FRT	FIT	FT	FY	FE	FP		FG		FV
FQ	Flow Quantity	FQRC	FQIC			FQR	FQI	FQSH	FQSL			FQIT	FQT	FQY	FQE					FQV
FF	Flow Ratio	FFRC	FFIC	FFC		FFR	FFI	FFSH	FFSL						FE					FFV
G	User's Choice																			
H	Hand		HIC	HC						HS										HV
I	Current	IRC	IIC	IC		IR	II	ISH	ISL	ISHL	IRT	IIT	IT	IY	IE					IZ
J	Power	JRC	JIC	JC		JR	JI	JSH	JSL	JSHL	JRT	JIT	JT	JY	JE					JV
K	Time	KRC	KIC	KC	KCV	KR	KI	KSH	KSL	KSHL	KRT	KIT	KT	KY	KE					KV
L	Level	LRC	LIC	LC	LCV	LR	LI	LSH	LSL	LSHL	LRT	LIT	LT	LY	LE		LW	LG		LV
M	User's Choice																			
N	User's Choice																			
O	User's Choice																			
P	Pressure/Vacuum	PRC	PIC	PC	PCV	PR	PI	PSH	PSL	PSHL	PRT	PIT	PT	PY	PE	PP			PSV, PSE	PV
PD	Pressure, Differential	PDRC	PDIC	PDC	PDCV	PDR	PDI	PDSH	PDSL		PDRT	PDIT	PDT	PDY	PE	PP				PDV
Q	Quantity	QRC	QIC	QC		QR	QI	QSH	QSL	QSHL	QRT	QIT	QT	QY	QE					QZ
R	Radiation	RRC	RIC	RC		RR	RI	RSH	RSL	RSHL	RRT	RIT	RT	RY	RE		RW			RZ
S	Speed/Frequency	SRC	SIC	SC	SCV	SR	SI	SSH	SSL	SSHL	SRT	SIT	ST	SY	SE					SV

Appendix

First Letters	Initiating or Measured Variable	Controllers				Readout Devices			Switches and Alarm Devices*			Transmitters				Solenoids, Relays, Computing Devices	Primary Element	Test Point	Well or Probe	Viewing Device, Glass	Safety Device	Final Element
		Record-ing	Indicat-ing	Blind	Self-Actuated Control Valves	Record-ing	Indicat-ing		High**	Low	Comb	Record-ing	Indicat-ing	Blind								
T	Temperature	TRC	TIC	TC	TCV	TR	TI		TSH	TSL	TSHL	TRT	TIT	TT	TY	TE	TP	TW		TSE	TV
TD	Temperature, Differential	TDRC	TDIC	TDC	TDCV	TDR	TDI		TDSH	TDSL		TDRT	TDIT	TDT	TDY	TE	TP	TW			TDV
U	Multivariable					UR	UI								UY						UV
V	Vibration/Machinery Analysis					VR	VI		VSH	VSL	VSHL	VRT	VIT	VT	VY	VE					VZ
W	Weight/Force	WRC	WIC	WC	WCV	WR	WI		WSH	WSL	WSHL	WRT	WIT	WT	WY	WE					WZ
WD	Weight/Force, Differential	WDRC	WDIC	WDC	WDCV	WDR	WDI		WDSH	WDSL		WDRT	WDIT	WDT	WDY	WE					WDZ
X	Unclassified																				
Y	Event/State/Presence		YIC	YC		YR	YI		YSH	YSL				YT	YY	YE					YZ
Z	Position/Dimension	ZRC	ZIC	ZC	ZCV	ZR	ZI		ZSH	ZSL	ZSHL	ZRT	ZIT	ZT	ZY	ZE					ZV
ZD	Gauging/Deviation	ZDRC	ZDIC	ZDC	ZDCV	ZDR	ZDI		ZDSH	ZDSL		ZDRT	ZDIT	ZDT	ZDY	ZDE					ZDV

Note: This table is not all-inclusive.
*A, alarm, the annuciating device, may be used in the same fashion as S, switch, the actuating device.
**The letters H and L may be omitted in the undefined case.

Other Possible Combinations:

FO	(Restriction Orifice)
FRK, HIK	(Control Stations)
FX	(Accesories)
	(Scanning Recorder)
LLH	(Pilot Light)
PFR	(Ratio)
KQI	(Running Time Indicator)
QQI	(Indicating Counter)
WKIC	(Rate-of-Weight-Loss Controller)
HMS	(Hand Momentary Switch)

INDUSTRIAL ELECTRICAL SYMBOLS

INDUSTRIAL ELECTRICAL SYMBOLS...

DISCONNECT	CIRCUIT INTERRUPTER	CIRCUIT BREAKER WITH THERMAL OL	CIRCUIT BREAKER WITH MAGNETIC OL	CIRCUIT BREAKER W/ THERMAL AND MAGNETIC OL

LIMIT SWITCHES

NORMALLY OPEN	NORMALLY CLOSED
HELD CLOSED	HELD OPEN

FOOT SWITCHES
NO / NC

PRESSURE AND VACUUM SWITCHES
NO / NC

LIQUID LEVEL SWITCH
NO / NC

TEMPERATURE-ACTUATED SWITCH
NO / NC

FLOW SWITCH (AIR, WATER, ETC.)
NO / NC

SPEED (PLUGGING)
F / R

ANTI-PLUG
F / R

SYMBOLS FOR STATIC SWITCHING CONTROL DEVICES

STATIC SWITCHING CONTROL IS A METHOD OF SWITCHING ELECTRICAL CIRCUITS WITHOUT USE OF CONTACTS, PRIMARILY BY SOLID-STATE DEVICES. USE SYMBOLS SHOWN IN TABLE AND ENCLOSE THEM IN A DIAMOND.

- INPUT COIL
- OUTPUT NO
- LIMIT SWITCH NO
- LIMIT SWITCH NC

SELECTOR

TWO-POSITION
J — K
A1
A2

	J	K
A1	X	
A2		X

X-CONTACT CLOSED

THREE-POSITION
J, K, L
A1, A2

	J	K	L
A1	X		
A2			X

X-CONTACT CLOSED

TWO-POSITION SELECTOR PUSHBUTTON

CONTACTS	SELECTOR POSITION			
	A BUTTON		B BUTTON	
	FREE	DEPRESSED	FREE	DEPRESSED
1-2	X			
3-4		X	X	X

X - CONTACT CLOSED

PUSHBUTTONS

MOMENTARY CONTACT

SINGLE CIRCUIT	DOUBLE CIRCUIT	MUSHROOM HEAD	WOBBLE STICK
NO / NC	NO AND NC		

MAINTAINED CONTACT

TWO SINGLE CIRCUIT	ONE DOUBLE CIRCUIT

ILLUMINATED
R

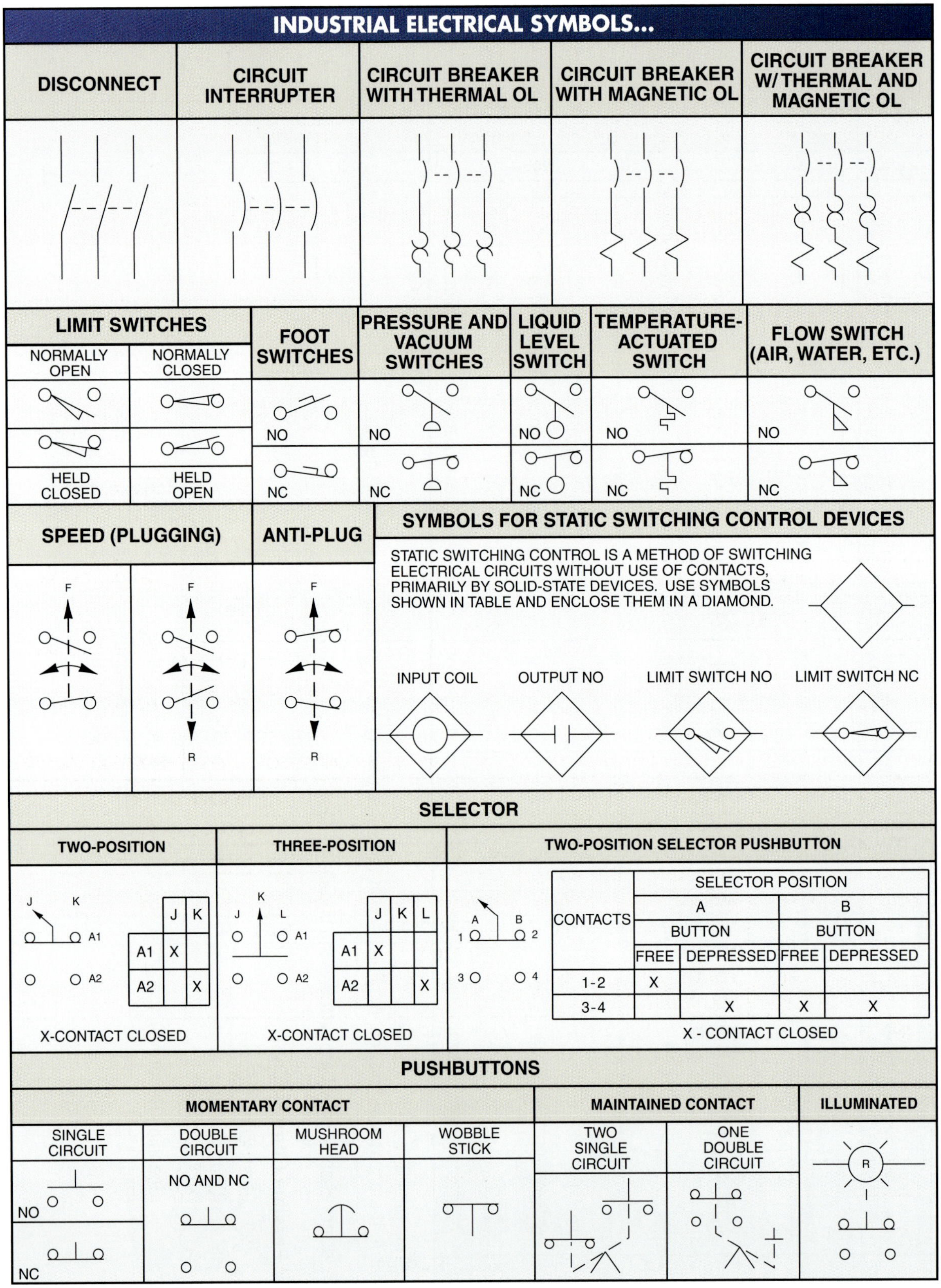

Used with permission from Electrical Motor Controls for Integrated Systems, *Fifth Edition, copyright American Technical Publishers.*

INDUSTRIAL ELECTRICAL SYMBOLS (CONT'D)

...INDUSTRIAL ELECTRICAL SYMBOLS...

CONTACTS

INSTANT OPERATING				TIMED CONTACTS - CONTACT ACTION RETARDED AFTER COIL IS:			
WITH BLOWOUT		WITHOUT BLOWOUT		ENERGIZED		DE-ENERGIZED	
NO	NC	NO	NC	NOTC	NCTO	NOTO	NCTC

OVERLOAD RELAYS

THERMAL	MAGNETIC

SUPPLEMENTARY CONTACT SYMBOLS

SPST NO		SPST NC		SPDT		TERMS
SINGLE BREAK	DOUBLE BREAK	SINGLE BREAK	DOUBLE BREAK	SINGLE BREAK	DOUBLE BREAK	SPST SINGLE-POLE, SINGLE-THROW

DPST, 2NO		DPST, 2NC		DPDT		
SINGLE BREAK	DOUBLE BREAK	SINGLE BREAK	DOUBLE BREAK	SINGLE BREAK	DOUBLE BREAK	

TERMS:
- SPST — SINGLE-POLE, SINGLE-THROW
- SPDT — SINGLE-POLE, DOUBLE-THROW
- DPST — DOUBLE-POLE, SINGLE-THROW
- DPDT — DOUBLE-POLE, DOUBLE-THROW
- NO — NORMALLY OPEN
- NC — NORMALLY CLOSED

METER (INSTRUMENT)

INDICATE TYPE BY LETTER

TO INDICATE FUNCTION OF METER OR INSTRUMENT, PLACE SPECIFIED LETTER OR LETTERS WITHIN SYMBOL.

AM or A	AMMETER	VA	VOLTMETER
AH	AMPERE HOUR	VAR	VARMETER
µA	MICROAMMETER	VARH	VARHOUR METER
mA	MILLAMMETER	W	WATTMETER
PF	POWER FACTOR	WH	WATTHOUR METER
V	VOLTMETER		

PILOT LIGHTS

INDICATE COLOR BY LETTER

NON PUSH-TO-TEST	PUSH-TO-TEST

INDUCTORS

- IRON CORE
- AIR CORE

SOLENOID

COILS

DUAL-VOLTAGE MAGNET COILS

HIGH-VOLTAGE	LOW-VOLTAGE
LINK — 1 2 3 4	LINKS — 1 2 3 4

BLOWOUT COIL

Used with permission from Electrical Motor Controls for Integrated Systems, Fifth Edition, copyright American Technical Publishers.

INDUSTRIAL ELECTRICAL SYMBOLS (CONT'D)

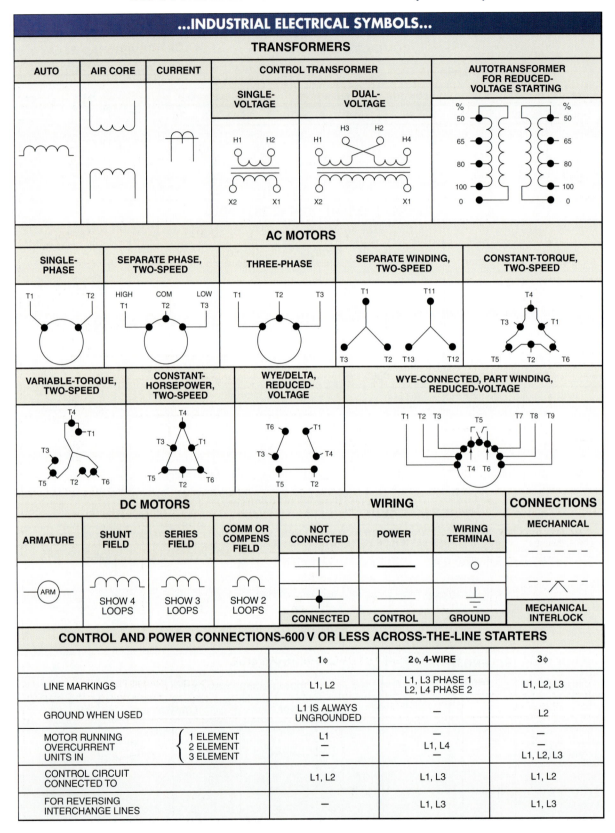

Used with permission from Electrical Motor Controls for Integrated Systems, Fifth Edition, copyright American Technical Publishers.

INDUSTRIAL ELECTRICAL SYMBOLS (CONT'D)

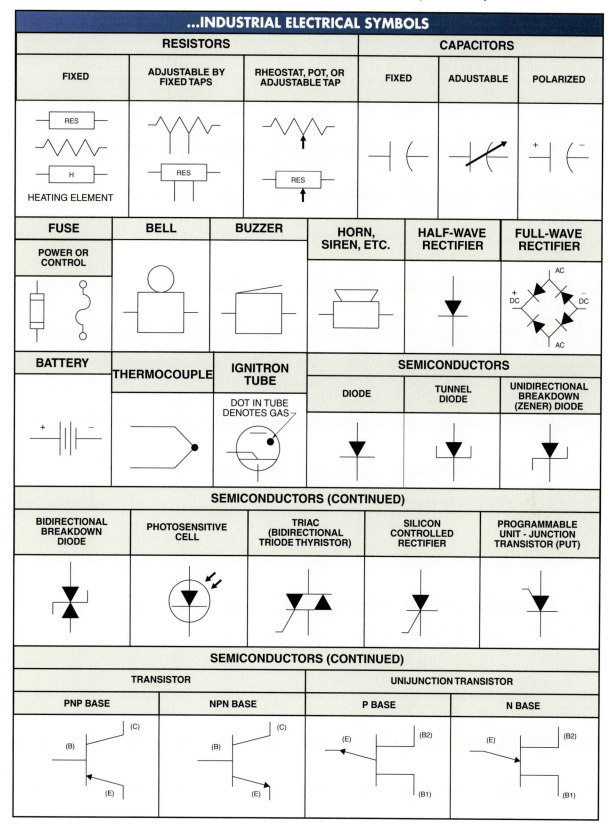

Used with permission from Electrical Motor Controls for Integrated Systems, Fifth Edition, copyright American Technical Publishers.

STANDARD P&ID SYMBOLS LEGEND | INDUSTRY STANDARDIZED P&ID SYMBOLS

Standard P&ID Symbols Legend Industry Standardized P&ID Symbols

Piping and Instrument Diagram Standard Symbols Detailed Documentation provides a standard set of shapes & symbols for documenting P&ID and PFD, including standard shapes of instrument, valves, pump, heating exchanges, mixers, crushers, vessels, compressors, filters, motors and connecting shapes.

Instrument

Symbol	Name	Symbol	Name	Symbol	Name	Symbol	Name	Symbol	Name
○	Indicator	◇	Computer Indicator	LC 65	Level Controller	M	Magnetic	▷	Or Gate
○ (dashed)	Behind Control	◇	Programmable Indicator	PT 55	Pressure Transmitter		Pitot Tube Type Flow	▷○	Not Gate
⊖	On Central Control	◇	Displayed Programmable Device	PR 55	Pressure Recorder		Pitot Tube	▽	Correcting Element
⊖	On Local Control Pane	⬡	Computer	PC 55	Pressure Controller		Wedge Meter	◇	Diamond
⊖ (dashed)	Behind a Local Control	⬢	Unit Control Panel	PIC 105	Pressure Indicating Controller		Target Meter		Pressure Gauges
○	Indicator 2	TI	Temp Indicator	PRC 40	Pressure Recording Controller		Weir Meter		Thermometers
T	Indicator 3	TT	Temp Transmitter	LA 25	Level Alarm		Ultrasonic Meter		Averaging Pitot Tube
T	Indicator 4	TR	Temp Recorder	FE	Flow Element		V-cone Meter		Level Meter
⌒	Indicator 5	TC	Temp Controller	TE	Temperature Element		Venturi Meter		Coriolis Flow Sensor
50	Odometer	FI	Flow Indicator	LG	Level Gauge		Quick Change		Flow Nozzle Meter
⦾	Pressure Gauge	FT	Flow Transmitter	AT	Analyzer Transmitter		Turbine Meter		Flume Meter
⊢⦿⊣	Flowmeter	FR	Flow Recorder	1/P	Transducer		Rotameter		Manhole
⟶	Thermometer	FC	Flow Controller	S P X	Specialty Item		Double		Socket Connection
⊕	Shared Indicator	LI	Level Indicator	S	Sampler		Steam Traced		Support Bracket
⊡	Displayed Configurable	PI	Pressure Indicator		Straightening Vanes		Venturi		Support Leg
⊕	Shared Indicator	LT 65	Level Transmitter	D	Diaphragm Meter	∞	Flowmeter		Support Ring
		LR 65	Level Recorder	R	Rotary Meter	▷	Vortex Sensor		Support Skirt
						▷	And Gate		

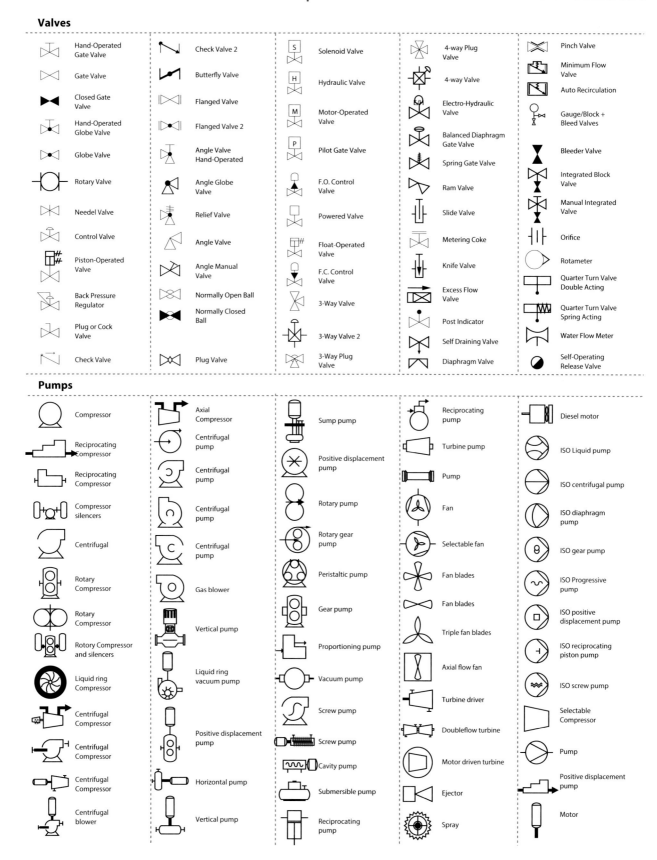

STANDARD P&ID SYMBOLS LEGEND | INDUSTRY STANDARDIZED P&ID SYMBOLS

Vessels

Vertical Vessel	Packing column	Boiler	Covered Tank	Thermal Insulation Vessel
Mixing Vessel	Drum	Dome Boiler	Floating Roof Tank	Heating-cooling Jacket Vessel
Column	Knock-out drum	Hot Liquid Boiler	Open Bulk Storage	Brackets Vessel
Tray Column	Bag	Tank	Dome Roof Tank	Dished Ends Vessel
Fluidized Bed Column	Bag (ISO)	Tank	Cone Roof Tank	Pit Vessel
Staggered Baffle Trays Column	Barrel	Tank	Internal Floating Roof Tank	Electrical Heating Vessel
Packing Column	Barrel (ISO)	Tank	Double Wall Tank	Wastewater Treatment
	Gas bottle	Open tank	Onion Tank	Bin

Filters

Mode Filter	Filter	Fixed bed air Filter	Biological Filter	High efficiency Filter
Mode Filter indication	Filter	HEPA gas Filter	Ionexchanger Filter	Primary Filter Mid-effect Filter
Filter	Gas Filter	Liquid Filter	Press Filter	Oxygen-enriched Filter
Highpass Filter	Cartridge gas Filter	Fixed bed liquid Filter	Air Filter	Filter
Lowpass Filter			Air Filter	Filter
Bandpass Filter	Roll air Filter	Rotary liquid Filter	The Filter	Rotary Filter
				Suction Filter

Compressors

Compressor	Ejector Compressor	Compressor silencers	Liquid ring Compressor	Axial Compressor
Compressor	Piston Compressor	Rotary Compressor	Centrifugal Compressor	Air Compressor
Compressor, Vacuum pump	Ring Compressor	Rotary Compressor	Centrifugal Compressor	AC air Compressor
Centrifugal Compressor	Roller vane Compressor	Rotary Compressor	Centrifugal Compressor	Screw Compressor
Diaphragm Compressor	Reciprocating Compressor	Rotary Compressor and silencers	Selectable Compressor	Turbo Compressor

Appendix 391

STANDARD P&ID SYMBOLS LEGEND | INDUSTRY STANDARDIZED P&ID SYMBOLS

Heat Exchanges

Symbol	Name	Symbol	Name	Symbol	Name	Symbol	Name	Symbol	Name
	TEMA TYPE BEM		Spray Cooler		Shell and Tube Heat		Reboiler		Oil Burner
	TEMA TYPE BEU		Forced-draft Cooling Tower		Cooler		Single Pass Heat Exchanger		Fired Heater
	TEMA TYPE AEM		Induced-draft Cooling Tower		Air-blown Cooler		Single Pass Heat Exchanger		Vertical Turbine
	TEMA TYPE AEL		Heat Exchanger		Induced Flow Air Cooler		Double Pipe Heat Exchanger		Condenser
	TEMA TYPE NEN		Heat Exchanger		Fin-fan Cooler		Hairpin Exchanger		Extractor Hood
	TEMA TYPE BKU		Heat Exchanger		Electric Heater		Spiral Heat Exchanger		Hose reel
	Plate Exchanger		Heater		Straight Tubes Heat Exchanger		Spiral Heat Exchanger		Light Water Station
	Plate and Frame Heat Exchanger		Exchanger		Coil Tubes Heat Exchanger		Air Cooled Exchanger		Combustion Chamber
	Dryer		Condenser		Finned Tubes Heat Exchanger		Briquetting Machine		Silencer
	Cooling Tower		Shell and Tube Heat		Floating Head Heat Exchanger		U-Tube Heat Exchanger		Thin-Film Evaporator
	Cooling Tower		Shell and Tube Heat		Plate Heat Exchanger		U-tube Heat Exchanger		Vent
	Cooling Tower				Kettle Heat Exchanger		Boiler		
					Reboiler Heat Exchanger				

Dryers

Symbol	Name	Symbol	Name	Symbol	Name	Symbol	Name	Symbol	Name
	Dryer		Roller Conveyor Belt Dryer		Generator		Heat Consumer		Moving Shelf Dryer
	Fluidized Bed Dryer		Motor Generator		Drying Oven		Spray Dryer		

STANDARD P&ID SYMBOLS LEGEND | INDUSTRY STANDARDIZED P&ID SYMBOLS

Appendix

STANDARD P&ID SYMBOLS LEGEND | INDUSTRY STANDARDIZED P&ID SYMBOLS

Motors

Symbol	Name	Symbol	Name	Symbol	Name	Symbol	Name	Symbol	Name
M	Motor	M	AC Motor	G	Generator	M	Motor-operated Valve		Diesel Motor
M	Step Motor	MG	Motor Generator	G	AC Generator		Motor-driven Turbine		Electric Motor
M	DC Motor	G	Gear	G	DC Generator		Motor		Turbine

Peripheral

Symbol	Name	Symbol	Name	Symbol	Name	Symbol	Name	Symbol	Name
	Impact Separator		Injector		Elevator		Elevator		Vertical Shaping Machines
	Gravity Separator		Reducer		Z-form Elevator		Skip Hoist		Pelletizing Disc
	Cyclone Separator		Rupture Disc		Lift		Conveyor		Piston Press
	Electromagnetic Separator		Viewing Glass		Hoists		Conveyor, Chain, Closed		Roller Press
	Electrostatic Precipitator Separator		Viewing Glass with Lighting		Aerator with Sparger		Conveyor, Screw, Closed		Crane
	Solidifier		Back Draft Damper		Bag Filling Machine		Roller Conveyor		Cyclone
	Permanent Magnet		Box Truck		Belt Skimmer		Conveyor		Curved Gas Vent
	Spray Nozzle		Manual Forklift		Screening		Overhead Conveyor		Firing System
	Rotary Table Feeder		Truck Forklift		Bucket Elevator		Screw Conveyor		Flame Arrestor
	Proportional Feeder		Industrial Truck				Scraper Conveyor		Palletizer
	Rotary Valve Feeder		Rolling Bin Truck		Elevator		Piston Extruder Shaping Machines		Chimney
	Metering Proportional Feeder		Ship		Gas Flare		Screw Extruder Shaping Machines		Conveyor, Vibrating, Closed
							Horizontal Shaping Machines		

393

STANDARD P&ID SYMBOLS LEGEND | INDUSTRY STANDARDIZED P&ID SYMBOLS

Piping and Connecting Shapes

Major Pipeline	Top to Top	Double Containment	In-line Mixer	Expansion Joint
Connect Pipeline	Sonic Signal	Flange	Separator	Hose
Major Straight Line Pipe	Nuclear	End Caps	Flame Arrester	Flexible Hose
Straight Line Pipe	Pneumatic Control	End Cap	Drain Silencer	Flow Indicator
Battery Limit Line	Pneumatic Binary Signal Line	Breather	Exhaust Silencer	Bell Mouth
Electronic Serial	Electric Signal Line	Drip Pan Elbow	Strainer	Removable Spool
Heat Trace	Electric Binary Signal Line	Flange	Exhaust Head	Basket Strainer
Side by Side	Electric Binary Signal Line	Union	Triangle Separator	Breather
Top-Bottom	Sleeve Joint	Socket Weld	Triangle Separator	Damper
One-to-Many	Butt Weld	Screwed Connection	Tundish	Breakthrough
Traced Line	Welded Connection	Orifice Plate	Open Vent	Orifice
Multi-Lines	Mechanical Link	Flanged Dummy Cover	Syphon Drain	Clamped Flange Coupling
Mid Arrow	Soldered/Solvent	Electrical Bounded	Hydrant	Compensate
Multi-Lines Elbow	Blind Disc	Slope Requirements Line	Swivel Joint	Coupling
Y-strainer	Spectacle Blind	Reducer	Detonation Arrestor	Electrically Insulated
Diverter Valve	Interchangeable Blind	Pulsation Dampener	Flame Arrestor	Flame Arrestor
Y-type Strainer	Open Disc	Duplex Strainer	In-line Silencer	Explosion-Proof Flame Arrestor
Rotary Valve	Orifice Plate	Vent Silencer	Steam Trap	Detonation-Proof Flame Arrestor
Expansion Joint 2		Basket Strainer	Desuperheater	Fire-Resistant Flame Arrestor
Bursting Disc		Cone Strainer	Ejector or Eductor	Fire-Resistant, Explosion-Proof Flame Arrestor
		Pilot Operated Relief Valve with Remote Sensor	Exhaust Head	Valve Manifold

Courtesy of edrawsoft

OHM'S LAW

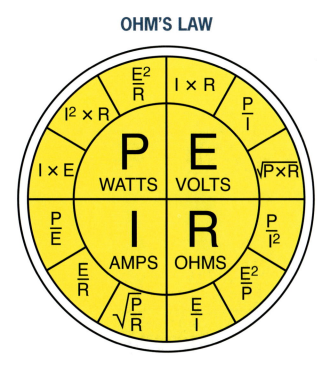

NATO PHONETIC ALPHABET

Letter	Phonetic Letter	Pronunciation
A	ALPHA	**AL** FAH
B	BRAVO	**BRAH** VOH
C	CHARLIE	**CHAR** LEE
D	DELTA	**DELL** TAH
E	ECHO	**ECK** OH
F	FOXTROT	**FOKS** TROT
G	GOLF	GOLF
H	HOTEL	HOH **TELL**
I	INDIA	**IN** DEE AH
J	JULIET	**JEW** LEE **ETT**
K	KILO	**KEY** LOH
L	LIMA	**LEE** MAH
M	MIKE	MIKE

Letter	Phonetic Letter	Pronunciation
N	NOVEMBER	NO **VEM** BER
O	OSCAR	**OSS** CAR
P	PAPA	**PAH** PAH
Q	QUEBEC	KEH **BECK**
R	ROMEO	**ROW** ME OH
S	SIERRA	SEE **AIR** AH
T	TANGO	**TANG** GO
U	UNIFORM	**YOU** NEE FORM
V	VICTOR	**VIK** TOR
W	WHISKEY	**WISS** KEY
X	X-RAY	**ECKS** RAY
Y	YANKEE	**YAN** KEY
Z	ZULU	**ZOO** LOO

RESISTOR COLOR CODE

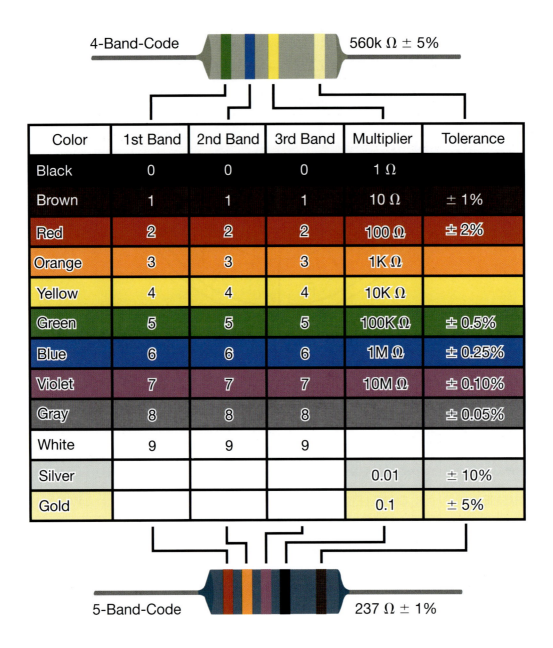

ANSI AND IEC COLOR CODES

ANSI and IEC Color Codes†
for Thermocouples, Wire and Connectors

All OMEGA™ Thermocouple Wire, Probes and Connectors are available with either ANSI or IEC Color Codes. In this Handbook, model numbers in the To Order tables reflect the ANSI Color-Coded Product.

ANSI Code	ANSI/ASTM E-230 Color Coding		Alloy Combination		Comments Environment Bare Wire	Maximum T/C Grade Temp. Range	EMF (mV) Over Max. Temp. Range	IEC 584-3 Color Coding		IEC Code
	Thermocouple Grade	Extension Grade	+ Lead	− Lead				Thermocouple Grade	Intrinsically Safe	
J			IRON Fe (magnetic)	CONSTANTAN COPPER-NICKEL Cu-Ni	Reducing, Vacuum, Inert. Limited Use in Oxidizing at High Temperatures. Not Recommended for Low Temperatures.	−210 to 1200°C −346 to 2193°F	−8.095 to 69.553			J
K			CHROMEGA™ NICKEL-CHROMIUM Ni-Cr	ALOMEGA™ NICKEL-ALUMINUM Ni-Al (magnetic)	Clean Oxidizing and Inert. Limited Use in Vacuum or Reducing. Wide Temperature Range, Most Popular Calibration	−270 to 1372°C −454 to 2501°F	−6.458 to 54.886			K
T			COPPER Cu	CONSTANTAN COPPER-NICKEL Cu-Ni	Mild Oxidizing, Reducing Vacuum or Inert. Good Where Moisture Is Present. Low Temperature & Cryogenic Applications	−270 to 400°C −454 to 752°F	−6.258 to 20.872			T
E			CHROMEGA™ NICKEL-CHROMIUM Ni-Cr	CONSTANTAN COPPER-NICKEL Cu-Ni	Oxidizing or Inert. Limited Use in Vacuum or Reducing. Highest EMF Change Per Degree	−270 to 1000°C −454 to 1832°F	−9.835 to 76.373			E
N			OMEGA-P™ NICROSIL Ni-Cr-Si	OMEGA-N™ NISIL Ni-Si-Mg	Alternative to Type K. More Stable at High Temps	−270 to 1300°C −450 to 2372°F	−4.345 to 47.513			N
R	NONE ESTABLISHED		PLATINUM-13% RHODIUM Pt-13% Rh	PLATINUM Pt	Oxidizing or Inert. Do Not Insert in Metal Tubes. Beware of Contamination. High Temperature	−50 to 1768°C −58 to 3214°F	−0.226 to 21.101			R
S	NONE ESTABLISHED		PLATINUM-10% RHODIUM Pt-10% Rh	PLATINUM Pt	Oxidizing or Inert. Do Not Insert in Metal Tubes. Beware of Contamination. High Temperature	−50 to 1768°C −58 to 3214°F	−0.236 to 18.693			S
R/SX	NONE ESTABLISHED		COPPER Cu	COPPER-LOW NICKEL Cu-Ni	Extension Grade Connecting Wire for R & S Thermocouples, Also Known as RX & SX Extension Wire.					R/SX
U*	NONE ESTABLISHED		COPPER Cu	COPPER Cu	Uncompensated for use with RTDs and Thermistors					U
B	NONE ESTABLISHED		PLATINUM-30% RHODIUM Pt-30% Rh	PLATINUM-6% RHODIUM Pt-6% Rh	Oxidizing or Inert. Do Not Insert in Metal Tubes. Beware of Contamination. High Temp. Common Use in Glass Industry	0 to 1820°C 32 to 3308°F	0 to 13.820			B
G* (W)	NONE ESTABLISHED		TUNGSTEN W	TUNGSTEN-26% RHENIUM W-26% Re	Vacuum, Inert, Hydrogen. Beware of Embrittlement. Not Practical Below 399°C (750°F). Not for Oxidizing Atmosphere	0 to 2320°C 32 to 4208°F	0 to 38.564	NO STANDARD USE ANSI COLOR CODE		G (W)
C* (W5)	NONE ESTABLISHED		TUNGSTEN-5% RHENIUM W-5% Re	TUNGSTEN-26% RHENIUM W-26% Re	Vacuum, Inert, Hydrogen. Beware of Embrittlement. Not Practical Below 399°C (750°F) Not for Oxidizing Atmosphere	0 to 2320°C 32 to 4208°F	0 to 37.066	NO STANDARD USE ANSI COLOR CODE		C (W5)
D* (W3)	NONE ESTABLISHED		TUNGSTEN-3% RHENIUM W-3% Re	TUNGSTEN-25% RHENIUM W-25% Re	Vacuum, Inert, Hydrogen. Beware of Embrittlement. Not Practical Below 399°C (750°F)−Not for Oxidizing Atmosphere	0 to 2320°C 32 to 4208°F	0 to 39.506	NO STANDARD USE ANSI COLOR CODE		D (W3)

* Not official symbol or standard designation

† JIS color code also available.

©Copyright OMEGA Engineering, Inc. All Rights Reserved. Reproduced with the Permission of OMEGA Engineering, Inc., Norwalk, CT 06854   www.Omega.com

For additional information, visit qr.njatcdb.org Item #2122

TECHNICAL DATA SECTION

TECHNICAL DATA SECTION
Reference Section

TERMINOLOGY

ACF =	Actual Cubic Feet
A/D =	Analog to Digital
ATM =	Atmospheres
BTU =	British Thermal Units
cc/min =	Cubic Centimeters per Minute
CFH =	Standard Cubic Feet per Hour (SCFH)
C_P =	Specific Heat
C.S. =	Carbon Steel
D =	Diameter
Dia. =	Diameter
Diam. =	Diameter
D/A =	Digital to Analog
EMI =	Electromagnetic Interference
EPR =	Ethylene Propylene Rubber
FDA =	Food and Drug Administration
FNPT =	Female National Pipe Thread
FPM =	Feet Per Minute
FPS =	Feet Per Second
F.S. =	Full Scale
FT =	Feet
gals =	Gallons
gpm =	Gallons Per Minute
gph =	Gallons Per Hour
H_F =	Latent Heat of Fusion
H/L =	High-Low
H_V =	Latent Heat of Vaporization
I.D. =	Inside Diameter
I/O =	Input/Output
k =	Thermal Conductivity
lbs =	Pounds
lbs/in² =	Pounds Per Square Inch
lpm =	Liters Per Minute
L/min =	Liters Per Minute
mL/min =	Milliliters Per Minute
MNPT =	Male National Pipe Thread
ms =	Milliseconds
m/s =	Meters Per Second
MSEC =	Milliseconds
NiCad =	Nickel Cadmium
NO/NC =	Normally Open/ Normally Closed
NPT =	National Pipe Thread
O.D. =	Outside Diameter
P-P =	Peak to Peak
PSIA =	Pounds Per Square Inch Absolute
PSID =	Pounds Per Square Inch Differential
PSIG =	Pounds Per Square Inch Gage
PVC =	Polyvinyl Chloride
PVDF =	Polyvinylidene Fluoride (Kynaol)
RF =	Raised Face
RFI =	Radio Frequency Interference
RMS =	Root Mean Square
SCCM =	Standard Cubic Centimeters per Minute
SCHED. NO. =	Schedule Number
SCFH =	Standard Cubic Feet per Hour
SCFM =	Standard Cubic Feet per Minute
SLM =	Standard Liters per Minute
SLPM =	Standard Liters per Minute
sq.ft. =	Square Feet
SSU =	Saybolt Seconds Universal
ΔT =	Temperature Rise
TTL =	Transistor-Transistor Logic
W =	Watts
W-hr =	Watt-Hours
W/in² =	2 Watt Density
W_T =	Weight of Material

Conversion Factors

TO OBTAIN	MULTIPLY	BY
Atmospheres	In HG@32°F	0.033421
BTU	Watt-hours	3.412
BTU	KWh	3412
Centimeters	Inches	2.540
Cm of Hg @ 0 deg C	Atmospheres	76.0
Cm of Hg @ 0 deg C	Grams/sq. cm	0.07356
Cm of Hg @ 0 deg C	Lb/sq in.	5.1715
Cm of Hg @ 0 deg C	Lb/sq ft	0.035913
Cm/(sec)(sec)	Gravity	980.665
Centipoises	Centistokes	Density
Centistokes	Centipoises	1/density
Cu cm	Cu ft	28,317
Cu cm	Cu in.	16-387
Cu cm	Gal (USA, liq.)	3785.43
Cu cm	Liters	1000.03
Cu cm	Quarts (USA, liq.)	946.358
Cu cm/sec	Cu ft/min	472.0
Cu ft	Cu meters	35.314
Cu ft	Gal (USA, liq.)	0.13368
Cu ft	Liters	0.03532
Cu ft/min	Cu meters/sec	2118.9
Cu ft/min	Gal (USA, liq.)/sec	8.0192
Cu ft/sec	Gal (USA, liq.)/min	0.0022280
Cu ft/sec	Liters/min	0.0005886
Cu in.	Cu centimeters	0.061023
Cu in.	Gal (USA, liq.)	231.0
Cu in.	Liters	61.03
Cu meters	Gal (USA, liq.)	0.0037854
Cu meters	Liters	0.001000028
Cu meters/hr	Gal/min	0.22712
Cu meters/kg	Cu ft/lb	0.062428
Cu meters/min	Cu ft/min	0.02832
Cu meters/sec	Gal/min	0.000063088
Feet	Meters	3.281
Ft/min	Cm/sec	1.9685
Ft/sec	Meters/sec	3.2808
Ft/(sec)(sec)	Gravity (sea level)	32.174
Ft/(sec)(sec)	Meters/(sec)(sec)	3.2808
Gal (Imperial, liq.)	Gal (USA, liq.)	0.83268
Gal (USA, liq.)	Barrels (Petroleum, USA)	42
Gal (USA, liq.)	Cu ft	7.4805
Gal (USA, liq.)	Cu meters	264.173
Gal (USA, liq.)	Cu yards	202.2
Gal (USA, liq.)	Gal (Imperial, liq.)	1.2010
Gal (USA, liq.)	Liters	0.2642
Gal (USA, liq.)/min	Cu ft/sec	448.83
Gal (USA, liq.)/min	Cu meters/hr	4.4029
Gal (USA, liq.)/sec	Liters/min	0.0044028
Grams	Pounds (avoir.)	453.5924

©Copyright OMEGA Engineering, Inc. All Rights Reserved. Reproduced with the Permission of OMEGA Engineering, Inc., Norwalk, CT 06854 www.Omega.com

For additional information, visit qr.njatcdb.org
Item #2574

TECHNICAL DATA SECTION (CONT'D)

Conversion Factors

TO OBTAIN	MULTIPLY	BY
Grams/(cm)(sec)	Centipoises	0.01
Grams/cu cm	Lb/cu ft	0.016018
Grams/cu cm	Lb/cu in.	27.680
Grams/cu cm	Lb/gal	0.119826
Inches	Centimeters	0.3937
Inches of Hg @ 32°F	Atmospheres	29.921
Inches of Hg @ 32°F	Lb/sq in.	2.0360
Inches of Hg @ 32°F	In. of H_2O @ 4°C	0.07355
Inches/deg F	Cm/deg C	0.21872
Kg	Pounds (avoir.)	0.45359
Kg-cal/sq meter	BTU/sq ft	2.712
Kg/cu meter	Lb/cu ft	16.018
Kg/(hr)(meter)	Centipoises	3.60
Kg/liter	Lb/gal (USA, liq.)	0.11983
Kg/meter	Lb/ft	1.488
Kg/sq cm	Lb/sq in.	0.0703
Kg/sq meter	Lb/sq ft	4.8824
KWh	BTU	.0002930
KWh	watt-hours	.001
Liters	Cu ft	28.316
Liters	Cu in.	0.01639
Liters	Cu meters	999.973
Liters	Gal (Imperial, liq.)	4.546
Liters	Gal (USA, liq.)	3.785306
Liters/kg	Cu ft/lb	62.42621
Liters/min	Cu ft/sec	1698.963
Liters/min	Gal (USA, liq.)/min	3.785
Liters/sec	Cu ft/min	0.47193
Liters/sec	Gal/min	0.063088
Meters	Feet	0.3048
Meters/sec	Ft/sec	0.3048
Meters/sec)(sec)	Ft/(sec)(sec)	0.3048
Ounces	Grams	0.035274
Pounds (avoir.)	Kg	2.2046
Pounds/cu ft	Grams/cu cm	62.428
Pounds/cu ft	Pounds/gal	7.48
Pounds/cu in.	Grams/cu cm	0.036127
Pounds/(hr)(ft)	Centipoises	2.42
Pounds/inch	Grams/cm	0.0056
Pounds/(sec)(ft)	Centipoises	0.000672
Pounds/gal. (USA, liq.)	Kg/liter	8.3452
Pounds/gal. (USA, liq.)	Pounds/cu ft	0.1337
Pounds/gal. (USA, liq.)	Pounds/cu in.	231
Sq centimeters	Sq ft	929.0
Sq centimeters	Sq in.	6.4516
Sq ft	Sq meters	10.764
Sq in.	Sq centimeters	0.155
Sq meters	Sq ft	0.0929
W-hr	BTU	.2390
W-hr	KWh	1000

©Copyright OMEGA Engineering, Inc. All Rights Reserved. Reproduced with the Permission of OMEGA Engineering, Inc., Norwalk, CT 06854   www.Omega.com

For additional information, visit qr.njatcdb.org Item #2574

METRIC PREFIXES

Prefix	Symbol	10^n	Decimal Equivalent	Name
yotta	Y	10^{24}	1 000 000 000 000 000 000 000 000	septillion
zetta	Z	10^{21}	1 000 000 000 000 000 000 000	sextillion
exa	E	10^{18}	1 000 000 000 000 000 000	quintillion
peta	P	10^{15}	1 000 000 000 000 000	quadrillion
tera	T	10^{12}	1 000 000 000 000	trillion
giga	G	10^{9}	1 000 000 000	billion
mega	M	10^{6}	1 000 000	million
kilo	k	10^{3}	1 000	thousand
hecto	h	10^{2}	100	hundred
deca	da	10^{1}	10	ten
		10^{0}	1	one
deci	d	10^{-1}	0.1	tenth
centi	c	10^{-2}	0.01	hundredth
milli	m	10^{-3}	0.001	thousandth
micro	μ	10^{-6}	0.000 001	millionth
nano	n	10^{-9}	0.000 000 001	billionth
pico	p	10^{-12}	0.000 000 000 001	trillionth
femto	f	10^{-15}	0.000 000 000 000 001	quadrillionth
atto	a	10^{-18}	0.000 000 000 000 000 001	quintillionth
zepto	z	10^{-21}	0.000 000 000 000 000 000 001	sextillionth
yocto	y	10^{-24}	0.000 000 000 000 000 000 000 001	septillionth

DECIMAL, HEXADECIMAL, OCTAL, AND BINARY CONVERSION CHART

Dec	Hex	Oct	Bin	Dec	Hex	Oct	Bin	Dec	Hex	Oct	Bin	Dec	Hex	Oct	Bin
0	0	000	00000000	16	10	020	00010000	32	20	040	00100000	48	30	060	00110000
1	1	001	00000001	17	11	021	00010001	33	21	041	00100001	49	31	061	00110001
2	2	002	00000010	18	12	022	00010010	34	22	042	00100010	50	32	062	00110010
3	3	003	00000011	19	13	023	00010011	35	23	043	00100011	51	33	063	00110011
4	4	004	00000100	20	14	024	00010100	36	24	044	00100100	52	34	064	00110100
5	5	005	00000101	21	15	025	00010101	37	25	045	00100101	53	35	065	00110101
6	6	006	00000110	22	16	026	00010110	38	26	046	00100110	54	36	066	00110110
7	7	007	00000111	23	17	027	00010111	39	27	047	00100111	55	37	067	00110111
8	8	010	00001000	24	18	030	00011000	40	28	050	00101000	56	38	070	00111000
9	9	011	00001001	25	19	031	00011001	41	29	051	00101001	57	39	071	00111001
10	A	012	00001010	26	1A	032	00011010	42	2A	052	00101010	58	3A	072	00111010
11	B	013	00001011	27	1B	033	00011011	43	2B	053	00101011	59	3B	073	00111011
12	C	014	00001100	28	1C	034	00011100	44	2C	054	00101100	60	3C	074	00111100
13	D	015	00001101	29	1D	035	00011101	45	2D	055	00101101	61	3D	075	00111101
14	E	016	00001110	30	1E	036	00011110	46	2E	056	00101110	62	3E	076	00111110
15	F	017	00001111	31	1F	037	00011111	47	2F	057	00101111	63	3F	077	00111111

Dec	Hex	Oct	Bin	Dec	Hex	Oct	Bin	Dec	Hex	Oct	Bin	Dec	Hex	Oct	Bin
64	40	100	01000000	80	50	120	01010000	96	60	140	01100000	112	70	160	01110000
65	41	101	01000001	81	51	121	01010001	97	61	141	01100001	113	71	161	01110001
66	42	102	01000010	82	52	122	01010010	98	62	142	01100010	114	72	162	01110010
67	43	103	01000011	83	53	123	01010011	99	63	143	01100011	115	73	163	01110011
68	44	104	01000100	84	54	124	01010100	100	64	144	01100100	116	74	164	01110100
69	45	105	01000101	85	55	125	01010101	101	65	145	01100101	117	75	165	01110101
70	46	106	01000110	86	56	126	01010110	102	66	146	01100110	118	76	166	01110110
71	47	107	01000111	87	57	127	01010111	103	67	147	01100111	119	77	167	01110111
72	48	110	01001000	88	58	130	01011000	104	68	150	01101000	120	78	170	01111000
73	49	111	01001001	89	59	131	01011001	105	69	151	01101001	121	79	171	01111001
74	4A	112	01001010	90	5A	132	01011010	106	6A	152	01101010	122	7A	172	01111010
75	4B	113	01001011	91	5B	133	01011011	107	6B	153	01101011	123	7B	173	01111011
76	4C	114	01001100	92	5C	134	01011100	108	6C	154	01101100	124	7C	174	01111100
77	4D	115	01001101	93	5D	135	01011101	109	6D	155	01101101	125	7D	175	01111101
78	4E	116	01001110	94	5E	136	01011110	110	6E	156	01101110	126	7E	176	01111110
79	4F	117	01001111	95	5F	137	01011111	111	6F	157	01101111	127	7F	177	01111111

4-20 mA TRANSMITTER OUTPUT CHART

Input %	Output mA	Input %	Output mA	Input %	Output mA	Input %	Output mA
0	4	25	8	50	12	75	16
1	4.16	26	8.16	51	12.16	76	16.16
2	4.32	27	8.32	52	12.32	77	16.32
3	4.48	28	8.48	53	12.48	78	16.48
4	4.64	29	8.64	54	12.64	79	16.64
5	4.8	30	8.8	55	12.8	80	16.8
6	4.96	31	8.96	56	12.96	81	16.96
7	5.12	32	9.12	57	13.12	82	17.12
8	5.28	33	9.28	58	13.28	83	17.28
9	5.44	34	9.44	59	13.44	84	17.44
10	5.6	35	9.6	60	13.6	85	17.6
11	5.76	36	9.76	61	13.76	86	17.76
12	5.92	37	9.92	62	13.92	87	17.92
13	6.08	38	10.08	63	14.08	88	18.08
14	6.24	39	10.24	64	14.24	89	18.24
15	6.4	40	10.4	65	14.4	90	18.4
16	6.56	41	10.56	66	14.56	91	18.56
17	6.72	42	10.72	67	14.72	92	18.72
18	6.88	43	10.88	68	14.88	93	18.88
19	7.04	44	11.04	69	15.04	94	19.04
20	7.2	45	11.2	70	15.2	95	19.2
21	7.36	46	11.36	71	15.36	96	19.36
22	7.52	47	11.52	72	15.52	97	19.52
23	7.68	48	11.68	73	15.68	98	19.68
24	7.84	49	11.84	74	15.84	99	19.84
						100	20

3-15 PSI TRANSMITTER OUTPUT CHART

Input %	Output PSI	Input %	Output PSI	Input %	Output PSI	Input %	Output PSI
0	3	25	6	50	9	75	12
1	3.12	26	6.12	51	9.12	76	12.12
2	3.24	27	6.24	52	9.24	77	12.24
3	3.36	28	6.36	53	9.36	78	12.36
4	3.48	29	6.48	54	9.48	79	12.48
5	3.6	30	6.6	55	9.6	80	12.6
6	3.72	31	6.72	56	9.72	81	12.72
7	3.84	32	6.84	57	9.84	82	12.84
8	3.96	33	6.96	58	9.96	83	12.96
9	4.08	34	7.08	59	10.08	84	13.08
10	4.2	35	7.2	60	10.2	85	13.2
11	4.32	36	7.32	61	10.32	86	13.32
12	4.44	37	7.44	62	10.44	87	13.44
13	4.56	38	7.56	63	10.56	88	13.56
14	4.68	39	7.68	64	10.68	89	13.68
15	4.8	40	7.8	65	10.8	90	13.8
16	4.92	41	7.92	66	10.92	91	13.92
17	5.04	42	8.04	67	11.04	92	14.04
18	5.16	43	8.16	68	11.16	93	14.16
19	5.28	44	8.28	69	11.28	94	14.28
20	5.4	45	8.4	70	11.4	95	14.4
21	5.52	46	8.52	71	11.52	96	14.52
22	5.64	47	8.64	72	11.64	97	14.64
23	5.76	48	8.76	73	11.76	98	14.76
24	5.88	49	8.88	74	11.88	99	14.88
						100	15

LOGIC GATE CHART

NAND

A	B	Output
0	0	1
0	1	1
1	0	1
1	1	0

NOR

A	B	Output
0	0	1
0	1	0
1	0	0
1	1	0

AND

A	B	Output
0	0	0
0	1	0
1	0	0
1	1	1

XNOR

A	B	Output
0	0	1
0	1	0
1	0	0
1	1	1

OR

A	B	Output
0	0	0
0	1	1
1	0	1
1	1	1

NOT

A	Output
0	1
1	0

Neg-AND

A	B	Output
0	0	1
0	1	0
1	0	0
1	1	0

Neg-OR

A	B	Output
0	0	1
0	1	1
1	0	1
1	1	0

XOR

A	B	Output
0	0	0
0	1	1
1	0	1
1	1	0

NFPA HAZARDOUS LOCATIONS

Classes
The classes define the general nature of hazardous material in the surrounding atmosphere.

Class	Hazardous Material in Surrounding Atmosphere
Class I	Hazardous because flammable gases or vapors are present in the air in quantities sufficient to produce explosive or ignitable mixtures.
Class II	Hazardous because combustible or conductive dusts are present.
Class III	Hazardous because ignitable fibers or flyings are present, but not likely to be in suspension in sufficient quantities to produce ignitable mixtures. Typical wood chips, cotton, flax, and nylon. Group classifications are not applied to this class.

Divisions
The division defines the probability of hazardous material being present in an ignitable concentration in the surrounding atmosphere.

Division	Presence of Hazardous Material
Division 1	The substance referred to by class is present during normal conditions.
Division 2	The substance referred to by class is present only in abnormal conditions, such as a container failure or system breakdown.

Groups
The group defines the hazardous material in the surrounding atmosphere.

Group	Hazardous Material in Surrounding Atmosphere
Group A	Acetylene
Group B	Hydrogen, fuel, and combustible process gases containing more than 30% hydrogen by volume or gases of equivalent hazard such as butadiene, ethylene, oxide, propylene oxide, and acrolein
Group C	Carbon monoxide, ether, hydrogen sulfide, morpholene, cyclopropane, ethyl and ethylene, or gases of equivalent hazard
Group D	Gasoline, acetone, ammonia, benzene, butane, cyclopropane, ethanol, hexane, methanol, methane, vinyl chloride, natural gas, naphtha, propane or gases of equivalent hazard
Group E	Combustible metal dusts, including aluminum, magnesium and their commercial alloys or other combustible dusts whose particle size, abrasiveness, and conductivity present similar hazards in connection with electrical equipment
Group F	Carbonaceous dusts, carbon black, coal black, charcoal, coal, or coke dusts that have more than 8% total entrapped volatiles, or dusts that have been sensitized by other material so they present an explosion hazard
Group G	Flour dust, grain dust, flour, starch, sugar, wood, plastic, and chemicals

GEOMETRIC FORMULAS

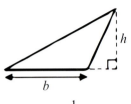

$A = \tfrac{1}{2} bh$

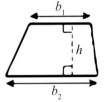

$A = \tfrac{1}{2} h(b_1 + b_2)$

$V = Bh$
$L.A. = hp$
$S.A. = L.A. + 2B$

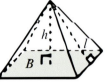

$V = \tfrac{1}{3} Bh$
$L.A. = \tfrac{1}{2} lp$
$S.A. = L.A. + B)$

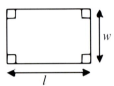

$A = lw$
$p = 2(l + w)$

$A = \pi r^2$
$C = 2\pi r$

$V = \pi r^2 h$
$L.A. = 2\pi r h$
$S.A. = 2\pi r(h + r)$

$V = \tfrac{4}{3} \pi r^3$
$S.A. = 4\pi r^2$

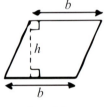

$A = bh$

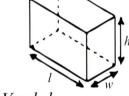

$V = lwh$
$S.A. = 2lw + 2lh + 2wh$

$V = \tfrac{1}{3} \pi r^2 h$
$L.A. = \pi r l$
$S.A. = \pi r(l + r)$

$c^2 = a^2 + b^2$

Glossary

A

Absolute Pressure: The sum of gauge pressure plus atmospheric pressure.

Absolute Zero: The temperature at which thermal energy is at a minimum and all molecular activity ceases, reported as 0°K or -273°C.

Accuracy: Conformity to an indicated, standard, or true value usually expressed as a percentage deviation (of a span reading or upper range value) from the indicated, standard, or true value.

Actuator: A part of the final control element that converts a signal into a forced action.

Aerobic: A process that requires oxygen.

Alarm: A device or function that signals the existence of an abnormal condition by means of audible or visible discrete changes, or both, intended to attract attention.

Algorithm: A detailed set of instructions executed by the central processing unit (CPU).

Alphanumeric: A character set that contains both letters and digits.

Ambient Compensation: The design of an instrument such that changes in ambient temperature do not affect readings of the instrument or compensation for ambient conditions when mounting an instrument.

Ambient Conditions: Conditions around the device that are examined (e.g., pressure, temperature).

American Standard Code for Information Interchange (ASCII): Standardized 7 digit numerical code consisting of 0s and 1s which allows a computer to recognize text character input.

Ampere: 6.25×10^{18} electrons per second flowing through a conductor.

Amplifier: A device whose output by design is an enlarged reproduction of the input signal and is energized from a source other than the input.

Analog Signal: A continuous operating signal.

Analytical Measurements: Techniques to identify the presence or concentration of specific matter within a sample.

Analyzer: An automatically-operating measuring device that monitors a process for one or more chemical compositions or physical properties.

Anchorage Point: A secure point of attachment for lifelines, lanyards, or deceleration devices.

Anemometer: An instrument for measuring or indicating the velocity of air flow.

For additional information, visit qr.njatcdb.org Item #2690

ANSI: American National Standards Institute.

Area: The measure of a surface expressed in units squared within a set of boundary lines.

As-Found Test: A calibration procedure during which the current, unaltered state of instrument operation is documented.

Automatic Control: the application of a control method for regulation of processes without direct human intervention.

Automatic Control Systems: Operable arrangements of one or more automatic controllers along with their associated equipment connected in closed loops with one or more processes.

Automatic Controller: A device or combination of devices that measures the value of the variable, quantity, or condition and operates to correct or limit deviation of this measured value from a selected reference.

B

Balloon: See *bubble*.

Baud: A unit of data transmission speed equal to the number of bits per second.

Behind the Panel: A term applied to a location that is within an area that (1) contains the instrument panel, (2) contains its associated rack-mounted hardware, or (3) is enclosed within the panel.

Bellows: A pressure sensor that converts pressure to linear displacement.

Bench set: The initial compression placed on the actuator spring with a spring adjuster with the actuator uncoupled from the valve.

Bernoulli Effect: The lowering of fluid pressure in regions where the flow velocity is increased.

Bernoulli's Principle: The sum of pressure energy and velocity energy in a line is a constant throughout the system if potential energy and friction are ignored.

Beta ratio: The ratio of the diameter of a pipeline constriction to the unconstricted pipe diameter.

Binary: A term applied to a signal or device that has only two discrete positions or states (on/off).

Bit: Short for binary digit, which is the smallest form of information, indicated by a 0 or 1.

Body: The framework that surrounds all internal components joined together inside the valve.

Bourdon Tube: A pressure sensor that converts pressure to displacement; a coiled, flattened tube that is straightened when pressure is applied.

BTU: British Thermal Unit. The quantity of thermal energy required to raise 1 pound of water 1°F, at its maximum density (1 BTU = 0.293 watt-hours = 252 calories).

Bubble: The circular symbol used to denote and identify the purpose of an instrument or function.

Buffer Solution: A solution that has a constant pH value and the ability to resist changes in pH levels.

Burst Pressure: The maximum pressure applied to a transducer-sensing element or case without causing leakage.

Bushing: A precision machined guide that provides rigidity and aligns the stem throughout the stem travel.

Byte: An 8-bit representation of a binary character.

C

Calibration: Process of adjusting an instrument or compiling a deviation chart so that the instrument output can be correlated to a known standard input value.

Capacitance: The property that may be expressed as the time integral of flow rate (heat, electric current, etc.) to or from storage divided by the associated potential change.

Capacity: The ability of a process to store energy or material.

Cascade Control: A type of control where two controllers, each on their own control loop, control a single process variable.

Catalytic Reaction: An increase in the rate of a chemical reaction due to the addition of a catalyst.

Cavitation: Boiling of a liquid caused by a decrease in pressure rather than by an increase in temperature.

Celsius: A temperature scale defined by 0°C at the freezing point and 100°C at the boiling point of water at sea level.

Centipoise: A dynamic viscosity measurement; equivalent to 0.01 poise. See also *poise*.

Central Processing Unit (CPU): The portion of a controller that decodes the instructions, performs the actual computations, and keeps order in the execution of programs.

Chatter: Rapid cycling on and off of a device in a control process.

Chromatography: The method by which individual components of a substance are separated by the rate of travel through a medium, often a column or coil.

Closed Loop: Control that uses a feedback mechanism to collect information on the effectiveness of the current controller output and makes corrections based on the process variable.

Closure Member: The movable part of a valve that forms a seal against the valve seat to control the flow of fluid.

Code: Collections of mandatory standards that have been codified by government authorities and adopted as law in many jurisdictions.

Coefficient of Resistance: Constant defining the incremental change in resistance of an RTD with respect to change in temperature: usually expressed as a percentage per degree of temperature change.

Commissioning: A preplanned, documented process of completing the start-up procedure for a process plant.

Commercial off the Shelf (COTS): Hardware that is commercially available, typically from several manufacturers.

Compiler: A program that translates a higher level language into a machine language that the CPU can execute.

Conductivity: The measurement of the relative ability to conduct electrical currents.

Configurable: A term applied to a device or system whose functional characteristics can be selected or rearranged through programming or other methods.

Continuous Emissions Monitoring Systems (CEMS): Measurement systems that determine the amount of air polluting emissions and other byproducts of industrial processes.

Control: To cause a system to act or function in a certain way.

Control Loop: A signal path in which two or more instruments or control functions are arranged so that signals pass from one to another for the purpose of measurement and/or control.

Control Mode: The type of control action used by a controller to perform control functions (e.g., on/off, PID).

Control Station: A manual loading station that provides switching between manual and automatic control modes of a control loop.

Control System: A system in which deliberate guidance or manipulation is used to achieve a prescribed value of a variable.

Control Valve: A device, other than a common, hand-actuated ON-OFF valve or self-actuated check valve, that directly manipulates the flow of one or more fluid process streams.

Controller: A device having an output that changes to regulate a controlled variable in a specific manner.

Controller Redundancy: A backup controller in place which automatically assumes control if there is a failure of the primary controller.

Controlling Means: Elements in a control system that contribute to the required corrective action.

Coriolis Meter: A recent technological development to measure mass flow rates with a high degree of accuracy.

Cubic Feet per Minute (CFM): The volumetric flow rate of a liquid or gas in cubic feet per minute.

Cycling: A periodic repeated variation in a controlled variable or process action.

D

Damping: Progressive reduction in the cycling amplitude of a system.

DCS: The distributed control system is an assortment of communication, controlling, and measuring devices that the operations division uses to control the process.

Dead Band: The change through which the input to an instrument can be varied without initiating an instrument response.

Dead Time: Time that elapses between an instrument input changes and when the instrument responds to the change.

Density: Mass per unit volume, given in units such as pounds per cubic foot.

Derivative Control: A control method that measures the instantaneous rate of change in the process system.

Design Standard: Specifies the design or technical characteristics of a product in terms of how it is to be constructed.

Deviation: Departure from a desired value.

Diaphragm: The sensing element consisting of a membrane that is deformed by the pressure differential formed across it.

Dielectric: A non-conductive medium.

Differential Pressure: The difference in static pressure between two identical pressure taps.

Digital Control: A control method based on mathematical calculations of numerical values of control signals known as bits.

Digital Signal: A term applied to a signal that uses binary digits to represent continuous values or discrete states.

Discrete Signal: A control signal that is on or off.

Distributed Control System: An assortment of communication, controlling, and measuring devices that the operations division uses to control the process.

Distributed I/O: Input or output modules that convert signal information to communication protocol that can be communicated directly to a controller.

Draft: The flow of gas through a heating system.

Drift: A change of a reading or a set point value over long periods caused by several factors, including changes in ambient temperature, time, and line voltage.

E

Eddy: Swirling motion of a liquid as it flows past an obstacle.

Electrical Current: The number of electrons that flow through a conductive material.

Electromotive Force (EMF): The electrical pressure that exists in a circuit.

Electron: A negatively-charged subatomic particle.

Endothermic: Characterized by or formed with the absorption of heat.

Energy: The capacity for doing work.

Enterprise Systems: Large scale application software packages that support business processes, information flows, reporting, and data analytics in complex organizations.

EPROM: Erasable, programmable, read-only memory that can be erased by ultraviolet light.

Equilibrium: The condition of a system when all inputs and outputs (supply and demand) are in balance.

Error: The difference between the actual and the true values, often expressed as a percentage of the span or upper-range value.

Ethernet: A physical layer for data transmission used in local area networks.

Excess air: The percentage of air supplied above the calculated requirement for complete combustion.

Excitation: The external application of electrical voltage or current.

Exothermic: Characterized by or formed with the emission of heat.

F

Fahrenheit: A temperature scale defined by 32°F at the freezing point and 212°F at the boiling point of water at sea level.

Feedback: Information about the status of the controlled variable that may be compared with the information that is desired in the interest of making them coincide.

Feedforward Control: A method of control that attempts to correct for a disturbance to a process before the disturbance makes any effect on the process.

Final Control Element: Component of a control system that directly regulates the flow of energy or material to the process. Variable of a control loop.

Flow: Travel of liquids or gases in response to a force (gravity or pressure).

Flow Straighteners: Devices that have special blades to reduce turbulence.

Fluids: Substances such as liquids, gasses, or vapors that have no fixed shape of their own and yield easily to external pressure.

Flume: A specially-shaped, fixed hydraulic structure that under free flowing conditions forces a fluid to accelerate in such a manner that the flow rate through the flume can be measured.

Flyback Diode: A diode used to eliminate a sudden voltage spike present in a control circuit when an inductive load, such as a relay or solenoid, is switched off.

Force: The push or pull along a straight line.

Formula: An equation used to explain a relationship among variables.

Forward Bias: Application of voltage or current that turns a semiconductor on.

FPM: Feet Per Minute.

FPS: Feet Per Second.

Frequency (Hz): The rate of a signal oscillation in the unit hertz (hz).

Friction: Energy lost to heat dissipation when a fluid flows through a pipe.

Functional Block Programming: A programming method utilizing a graphical diagram that contains a preprogrammed control function.

Functional Blocks: A software element that defines a specific characteristic of the process control function.

G

Gain: The ratio of output divided by input of a control device.

Gallons Per Hour (GPH): Volumetric flow rate in gallons per hour.

Gallons Per Minute (GPM): Volumetric flow rate in gallons per minute.

Gases: Fluids that have neither independent shape nor volume but tend to expand indefinitely.

Gateway: Network hardware that allows the interconnection of two dissimilar networks by performing a translation function.

Gauge Pressure: Absolute pressure minus local atmospheric pressure.

Gravity: The force that attracts a body of mass toward Earth's center or to another body of mass.

Ground loop: A condition where current flows on the grounding portion of a circuit leading to electrical noise and signal degradation.

GUI: Graphical User Interface.

H

Head Pressure: Expression of pressure in terms of the height of fluid.

Heat: Thermal energy expressed in units of calories, or British Thermal Units (BTUs).

Heat Exchanger: A vessel in which heat is transferred from one medium to another.

Heat Transfer: The process of thermal energy flowing from a body of high energy to a body of low energy.

Heuristic: A method that promotes rapid problem solving by trial and error methods, sometimes referred to as a rule of thumb or shortcut.

HMI: Human Machine Interface.

Hot Swapping: Ability to change controller components such as I/O cards without having to shut down the control system.

Hydrocarbons: Organic compounds comprised of a mixture of hydrogen and carbon.

Hysteresis: Difference between upscale and downscale results in instrument response when subjected to the same input approached from opposite directions.

I

Impedance: The total opposition to current flow (resistive plus reactive).

Impulse Tubing: Tubing that carries all or a portion of the process fluid to the sensor for measurement.

Infrared Radiation: Electromagnetic wave with a wavelength longer than the color red and therefore cannot be seen by the human eye.

Input: Incoming signal to a measuring instrument, control units, or system.

Input Test Standard: An item of test equipment used for the calibration of pressure, temperature, current, voltage, or other input measurements.

Instrument: A term used broadly to describe any device that performs a measuring or controlling function. The process variables measured by the instrument may be used for indication, recording, or control.

Instrument, Elevated: A level measurement instrument that is mounted above the measurement point.

Instrument, Suppressed: A level measurement instrument that is mounted below the measurement point.

Integral Control: A control method that finds the sum of error in the process system over time.

Integral Windup: Occurs when an error exists for long periods of time or when large setpoint changes are made, resulting in the integral action measuring a very large offset.

Interposing Relay: Relays that provide a level of isolation between two electrical circuits.

Intrinsic Barrier: A device used to control fault current and maintain an electrically safe condition within a hazardous location.

Intrinsically Safe (IS): A circuit design that does not produce any spark or thermal effects under normal or abnormal conditions that may ignite a specified gas mixture.

Input/Output (I/O): The interface between peripheral equipment and the digital systems.

I/P Transducer: A device that converts an electric current to a linear pneumatic pressure.

ISA: The International Society of Automation, formerly known as the Instrument Society of America.

J

Joule: The basic unit of thermal energy.

K

Kelvin: The unit of absolute temperature in which 0°K represents the complete absence of heat, with 273.15°K corresponding to 0°C and 373.15°K corresponding to the boiling point of water at sea level.

Kinetic Energy: The energy possessed by an object because of its motion.

L

Lag: A delay in output change following a change in input.

Laminar Flow: Straight line flow of fluid.

Linear Variable Differential Transformer (LVDT): A type of transformer used to measure linear displacement.

Linearity: Proximity of a calibration curve to a specified straight line, which is expressed as the maximum deviation of any calibration point on a specified straight line during one calibration cycle.

Local: Referring to devices or conditions where a sensor or transducer is physically located, as opposed to a central monitoring or a processing station.

Local Panel: A panel that is not a central or main panel.

Loop Sheet: A drawing used to represent the interconnection of the control loop devices, their location in the process, and a traceable signal path between the devices.

Loop Tuning: A process of manipulating and adjusting the control algorithms to operate and maintain smooth process response.

Lower Explosive Limit (LEL): The minimum concentration of a flammable substance that will support combustion and continue to burn after the ignition source is removed.

M

Machine Language: A computer programming language consisting of binary or hexadecimal instructions that a computer can respond to directly and the processor can follow.

Management of Change (MOC): A procedure that identifies hazards and risks associated with changes made to a process.

Manometer: A device that uses a liquid column to measure pressure.

Manual Loading Station: A device or function having a manually adjusted output that is used to actuate one or more remote devices.

Mass: The amount of matter in an object.

Mass Flow Rate: Volumetric flow rate times density (i.e., pounds per hour).

Matter: The substance or substances of which any physical object is composed

Measuring Junction: The point in a thermocouple where the two dissimilar metals are joined.

Meniscus: The curve in the surface of a liquid that is cereated by the surface tension of the liquid against the sides of a container.

Meter Run: A field term used to describe the upstream and downstream piping, including orifice and orifice flanging.

Monitoring: Observation of process variables for informational purposes only.

N

National Pipe Thread (NPT): Uniform standards for threads utilized on pipe connections.

Newton: SI (metric system) unit of force. Named after Sir Isaac Newton.

O

Offline Configuration: A configuration created without a connection to a device.

Offset: The difference between the process setpoint and the steady state response of the process variable.

Online Configuration: A configuration that is performed with a real time communication channel.

Opacity: The measure of how much light can pass through a substance.

Open Channel: Any fluid routing that allows the fluid to flow with a free surface.

Open Loop: Control without feedback.

Opto-isolator: A device that provides isolation by converting the electrical signal into a light signal and finally back into an electrical signal.

OSHA: The Occupational Safety and Health Administration, established by the federal Occupational Safety and Health Act of 1970 to protect the health and safety of U.S. workers.

Output: The signal provided by an instrument: for example, the signal that the controller delivers to the valve operator is the controller output.

Output Test Standard: An item of test equipment used for the calibration of pressure, temperature, current, voltage, or other output measurements.

Overshoot: The persistent effort of the control system to reach the desired level, which frequently results in going beyond the mark.

P

Parallax: An optical illusion that occurs in analog meters and causes reading errors. It occurs when the viewing eye is not in the same plane, perpendicular to the meter face, as the indicating needle.

Parts Per Million (PPM): Measurement of the mass of a chemical or contaminate per unit volume of water.

Pascal: SI (Metric System) unit of pressure. Abbreviated as Pa. (249.082 Pa = 1"H_2O.

Peripheral Components: Devices that are external to the CPU and main memory but are connected by the appropriate electrical connections.

P&ID Walk Down: The process of visually inspecting the equipment and piping to ensure it conforms to the P&ID and project specifications.

PID: Proportional, Integral, and Derivative. A three-mode control action in which the controller has time-proportioning, integral, and derivative rate actions.

Piezoelectric Effect: A change in the electrical potential of a combination of materials based on applied forces, such as pressure.

Piezoresistive Effect: A change in the electrical resistivity of a device (semiconductor,metal) when mechanical strain is applied.

Pigtails: See *syphon*.

Pilot Light: A light that indicates the status of a system.

Pitot Tube: An open-ended right-angled tube pointing into the flow of a fluid and used to measure pressure.

Pneumatics: The application of a compressed gas as fluid power to provide motion.

Poise: The unit of dynamic viscosity represented by the greek letter mu (μ) where 1 poise is equal to 0.1 Pascal-second.

Positioner: A position controller that is mechanically connected to a moving part of a final control element or its actuator and that automatically adjusts its output to the actuator to maintain a desired position in proportion to the input signal.

Positive Temperature Coefficient Thermistor: Thermistor that increases in resistance when measured temperature increases.

Post: A rigid section connected to or integral to the stem.

Potential Energy: Energy that is stored in an object because of its positional relationship to another object.

Potentiometer: A variable resistor.

Pounds Per Square Inch Absolute (PSIA): Pressure referenced to an ideal vacuum.

Pounds Per Square Inch Differential (PSID): Pressure difference between two points.

Pounds Per Square Inch Gauge (PSIG): Pressure referenced to ambient atmospheric pressure.

Power Factor (PF): The relationship of apparent power and true power, principally affected by inductive loads in a facility.

Power Supply: A separate unit or part of a circuit that supplies power to the rest of the circuit or system.

Pressure: Force exerted by fluid per unit area.

Pressure Tap: Point where the instrumentation tubing connects to the process line.

Primary Element: That part of a loop or instrument that first senses the value of a process variable and assumes a corresponding, predetermined, and intelligible state or output. The primary element is also known as a detector or sensor.

Process: The variable for which supply and demand must be balanced. It is a physical or chemical change of matter and/or conversion of energy.

Process and Instrumentation Drawing or Diagram (P&ID): Diagram which identifies the process and the related instrumentation for that process.

Process Control: The method by which we regulate a particular process.

Process Flow Diagram (PFD): A diagram used to indicate the general flow of plant processes and equipment.

Process Isolation Diaphram: A device which use a small flexible barrier, often made of a thin metal alloy, which isolates an instrument from the affects of a process.

Process Lag: The time it takes for a process to recover from a change in loading or setpoint.

Process Variable: Any variable property of a process; the part of the process that changes and therefore needs to be controlled.

Program: A series of instructions that logically solve given problems and manipulate data.

Programmable Logic Controller (PLC): A controller, usually with multiple inputs and outputs, that contains an alterable program.

Project: A temporary endeavor undertaken to create a unique product, service, or result.

Proof Pressure: The specified pressure that may be applied to the sensing element of a transducer without causing a permanent change in the output characteristics.

Proprietary Protocol: A nonstandard communication format and language owned by a single organization or individual.

Proportional Control: A control method that causes a control response to be made in proportion to the error amount between the setpoint and the measured value.

Protocol: A formal definition that describes how data are to be exchanged.

Pull-Down Resistor: A resistor used to hold the input of a controller (microprocessor based) at a low value in the absence of an input signal.

Pull-Up Resistor: A resistor used to hold the input of a controller (microprocessor based) at a high value in the absence of an input signal.

R

Random Access Memory (RAM): Memory that can be read and changed during computer operation. Unlike other semi-conductor memory, RAM is volatile. If power to the RAM is disrupted or lost, all stored data are lost.

Range: Set of values over which measurements can be made by a device without changing the device's sensitivity.

Rankine: An absolute temperature scale based on the Fahrenheit scale, with 180° between the freezing point and the boiling point of water (459.67°R = 0°F).

Ratio Control: Used to control two or more components of a process in a predetermined ratio.

Read-Only Memory (ROM): Memory that contains fixed data. The computer can read the data but cannot change the data in any way.

Reagent: A substance that is added to cause a chemical change to occur.

Reference Junction: The cold junction in a thermocouple circuit that is held at a stable known temperature.

Refractive index: A measurement of how easily light travels through a medium.

Relay, Electrical: A device that activates a set of electrical contacts in response to an electrical signal.

Repeatability: The ability of a transmitter to reproduce output readings when exactly the same measured value is applied to it consecutively, under the same conditions, and approaching from the same direction.

Request for Information (RFI): A request made by the contractor to make clear the intention of a design document or ask for guidance on a conflict in documentation.

Request for Proposal (RFP): A document that provides the information for a project management firm to compile a bid package to perform the work for the end user.

Request for Quotation (RFQ): A document soliciting a price for the equipment and material needed to complete the project.

Reset Action: See *integral control action*.

Resistance, Electrical: The opposition of current flow in an electrical circuit.

Resistance Temperature Detector (RTD): Device which changes resistance based on a change of temperature.

Resolution: The smallest detectable increment of measurement.

Reverse Bias: The application of DC voltage that prevents or greatly reduces current flow in a diode or transistor.

Reynold's Number: A unitless number used to gauge the amount of fluid flow resistance in a line.

S

Scan: Periodically sampling in a predetermined manner each of a number of variables.

Scan time: The time it takes for a controller to read the inputs, update the program and apply the new output.

Scope of Work: A document that clearly defines the products or services to be delivered. This document is sometimes called the statement of work.

Schematic Diagrams: Use graphical symbols to show connections between components and the function of the circuit.

Seat: The sealing surface for shutting off or controlling the flow of a fluid within the valve.

Self-heating: Internal heating of a transducer as a result of power dissipation.

Shared Controller: A controller containing preprogrammed algorithms that are usually accessible, configurable, and assignable. Permits a number of process variables to be controlled by a single device.

SI System: The most widely-used measuring system worldwide. It is based off the metric measures of meter, liter and gram.

Signal: Information in the form of a pneumatic pressure, an electric current, or mechanical position that carries information from one control loop component to another.

Signal Conditioners: Devices that convert one type of electronic signal into another type of signal.

Single Loop Controller: A device that performs a control function of one process variable only.

Slots: Provisions for installing control cards such as communication, CPU, and I/O.

Slurry: A fluid containing insoluble matter (e.g., mud).

Snubber Diode: See *flyback diode*.

Snubber (electrical): An electric circuit intended to suppress voltage spikes.

Snubber (piping accessory): A device with a very small bore or restriction, which minimizes the shock to a pressure instrument by slowing the pressure change.

Software: A collection of programs and routines associated with a computer.

Span: The difference between the upper and lower limits of a range which is expressed in the same units as the range.

Span Shift: An error identified as an output signal that does not reflect the desired span.

Specific Gravity: The ratio of mass of any material to the mass of the same volume of water at 4°C.

Specification (SPEC) Sheet: The primary source of instrument information when ordering, designing, installing, maintaining, and calibrating an instrument.

Static Pressure: The pressure of a fluid at rest.

Steady State: A situation in which process conditions have stabilized, or an unvarying condition in a physical process.

Stem: The device that transmits the motion of the valve actuator to the disc or plug, thereby raising, lowering, or turning it.

Stem Sealing Device: Commonly called the packing. Provides leakproof closure around the stem of the valve while still allowing the stem to rotate or move linearly.

Strain Gage: A device that changes resistance in response to an applied force.

Structured Text: A text-based control programming language, which uses statements such as "IF...ELSE...AND...THEN..." to accomplish a task.

Syntax: The rules governing the structure of a language.

Syphons: Looped pieces of pipe which trap water in order to insulate the steam from sensitive devices. These devices are installed between a process line and the measuring device.

System: Generally refers to all control components, including process, measurements, controller, operator, and valves, along with any other additional equipment that may contribute to its operation.

T

Tag Number: Alphanumeric sequence that identifies a device by a unique code.

Terminal: The point at which a conductor from an electrical component, device or network comes to an end and provides a point of connection to external circuits.

Test Terminals: Electrical terminals on a transmitter used to make quick signal measurement checks, or to power local indicator displays.

Thermocouple: A device constructed of two dissimilar metals that generates a millivoltage as a function of temperature difference between measuring and reference junctions.

Thermowell: A closed-end tube designed to protect temperature sensors from harsh environments.

Time Constant: The time required for a system to reach 63.2% of the final value.

Topology: The physical layout of a data network.

Traceability: An unbroken chain of measurements and associated uncertainties.

Transducer: A device that converts information of one physical form to another physical type in its output.

Transient: The response time required for a deviated process variable to return to within steady-state error limits.

Transistor: A semiconductor device made up of three distinct regions: the collector, the base, and the emitter. Capable of amplification, switching, and rectification.

Transmitter: A device that senses a process variable through the medium of a sensor and converts the input signal to an output signal of another form. The output of the transmitter is a steady-state value that varies only as a predetermined function of the input (process variable). The sensor may or may not be integral with the transmitter.

Trim: Internal valve components that come into direct contact with process fluid.

Turbulence: Violent or unsteady movement of air or water.

U

Ultimate Period: The time the process requires to complete one oscillation.

Upper Explosive Limit (UEL): The maximum concentration of a flammable substance that will support combustion and continue to burn after the ignition source is removed.

V

Vacuum: A space that is devoid of matter.

Vacuum Pressure: Any pressure less than atmospheric pressure.

Validation: Rigorous evidence based test to ensure the process will consistently produce a product that meets quality standards.

Valve Bonnet: Provides the pressure boundary for the upper portion of the valve and allows leakproof closure of the valve body.

Vapor: The gaseous product of evaporation.

Velocity: The rate at which a substance moves from its original position in a given amount of time.

Vena Contracta: The point downstream of an orifice plate where the fluid velocity is greatest and pressure is lowest due to the inertia of the moving fluid.

Viscosity: The extent of friction between two adjacent layers of a fluid.

Volume: A measurement of occupied space

Volumetric Flow Rate: A quantity moving past a given point in a specified time period.

Vortex: A mass of whirling air.

W

Weir: A dam-like obstruction placed within an open channel such that the liquid flows over it, often through a specially shaped opening.

Wiring Diagrams: Illustrate the physical connection between electrical components in a machine or production system.

Word: The number of bits treated as a single unit by the CPU.

Work Breakdown Structure (WBS): A project management tool used to identify the deliverables of a project and organize them in a hierarchal layout.

Z

Zero Shift: Change resulting from an error that is the same throughout the scale.

Ziegler-Nichols: A mathematical method used to determine PID control loop settings.

Index

3-wire RTD configuration, 183f
4-wire RTD configuration, 183f

A

A, 48f, 49f
Absolute pressure, 94–95, 94f
Absolute zero, 88, 172
Absorbing vessel detailed, 53f
AC. *See* Alternating current (AC)
AC input card, 268–269, 269f
Acceptable level of inaccuracy (tolerance), 67–68
Accuracy, 70, 73f
Accuracy tolerance, 67–68
Accuracy value, 71
Acetylene, 258
Acid flushing, 333
Acid Rain Program, 247
Acidic solution, 244
Actuator, 36. *See also* Valve actuator
ADC. *See* Analog-to-digital converter (ADC)
Adjustment screws, 76, 76f
Advanced control, 298–300
Aerobic, 254
AI block. *See* Analog input (AI) block
Air pressure safety valve, 268
Air-to-fuel ratio, 249
Alarm, 170
Algebra, 10
Algorithm, 22, 290
Alkaline solution, 244
Alternating current (AC), 12
Ambient compensation, 235
Ambient conditions, 235
Ambient temperature variations, 138, 165
American National Standards Institute (ANSI), 42, 368
American Petroleum Institute (API), 62
American Society of Mechanical Engineers (ASME), 368
Ammeter, 347
Ampere, 7
Amperometric sensor, 254
Amplifier, 230–231
Analog I/O, 274–275
Analog I/O power, 280f
Analog input (AI) block, 193, 193f
Analog output cards, 280f
Analog signal, 27
Analog-to-digital converter (ADC), 77, 199, 356
Analytical measurement, 23, 240–263
    analyzer, 242
    buoyancy density, 256–257
    catalytic bead sensor, 258, 258f
    combustion analysis, 248–251
    conductivity, 242–243
    continuous emissions monitoring system (CEMS), 247–248
    density measurements, 254–257
    flame ionization detector, 258–259
    flammable gas detector, 257–260
    gas chromatography, 251–253
    hydrogen sulfide monitor, 260–261, 260f
    hydrostatic density, 255–256, 256f
    infrared gas detector, 259–260
    opacity, 249–250
    oscillating tube Coriolis density, 255
    oxidation reduction potential (ORP), 246–247
    oxygen concentration, 253–254
    pH measurement, 243–246
    sampling, 242
    silica analyzer, 247
    vibronic density, 254, 255f
Analyzer, 242
Anchorage point, 5
Anemometer, 161
Angle-body valve, 213f
ANSI. *See* American National Standards Institute (ANSI)
ANSI/ISA 62382 standard, 322
AO block, 193f
AO trim, 204
API. *See* American Petroleum Institute (API)
Archimedes's law, 122, 123f
Area, 142–143
As-found test, 72–75
As-left test, 77–80
ASME. *See* American Society of Mechanical Engineers (ASME)
ASME 165. standard, 80
Assembler, 282f
Assembly language, 282f
Atmospheric pressure, 88, 89f, 90f, 91
Atmospheric pressure chart, 90f
Auto-tune feature, 290
Automatic control, 22, 23f, 266–267
Automatic control system, 29, 264, 267
Automatic controller, 22, 264, 288. *See also* Process controller
Avogadro, Amedeo, 107
Avogadro's constant, 107

B

B, 48f, 49f
Ball float, 120–121, 121f
Ball valve, 217–218, 218f
Balloon, 45
Balloon, inflated, 91, 91f
Barrier troubleshooting documentation, 346
Basic solution, 244
Batch process tank, 354f
Beer brewing, 256
Behind-panel instrument, 45f
Behind the panel, 45
Bell 202 frequency shift keying (FSK) standard, 192
Bellows, 98
Bellows pressure-sensing element, 98, 98f
Bellows-style pressure-sensing devices, 98–99
Bench calibration setup, 66f
Bench set, 220
Bernoulli, Daniel, 146
Bernoulli effect, 146
Bernoulli's principle, 146–147, 146f
Beta ratio, 150
Bimetallic element, 173, 173f
Bimetallic thermometer, 173
Binary input, 268–269
Binary output, 270–271
Binary signal, 27–28
Bit, 26
Blind temperature transmitter temperature element, 47f
Blind transmitter, 50f
Blocking procedure, 310–311, 310f
Blown fuse, 327
BMS. *See* Burner management system (BMS)
Boiler combustion control, 251, 251f
Boiler steam drum level measurement, 136, 136f
Bourdon tube, 98, 98f
Bourdon tube thermometer, 174f
Bourdon tube-type detectors, 98
Boyle, Robert, 107
Breathing, human, 90, 90f
British thermal unit (BTU), 170
BTU. *See* British thermal unit (BTU)
Bubble, 44f, 45, 48f
Bubble test, 332
Bubbler level measurement system, 129–130, 129f
Bubbler tube, 130
Buffer solution, 246
Bulk powders, 125
Bump test, 261
Bump test station, 261f
Buoyancy density, 256–257
Burner management system (BMS), 251
Burst pressure, 309, 309f
Bus topology, 358f
Bushing, 212
Butterfly valve, 217, 217f
Byte, 27

C

C, 48f, 49f
Cable and conductor labeling, 318, 318f
Cable tray, 311f
Cage designs, 216, 216f
Cage-guided sliding-stem valve, 215, 215f
Calculus, 10
Calibration, 25, 26f. *See also* Calibration procedure and documentation
Calibration accuracy, 70–72
Calibration bench, 65, 66f
Calibration data record, 67, 67f, 77–78, 79f
Calibration errors, 73–75
    dead band, 74
    error correction procedures, 75–77
    hysteresis, 74, 74f
    nonlinearity, 74f, 75
    span shift, 73, 73f
    zero shift, 71f, 73

Calibration flowchart, 75f
Calibration procedure and documentation, 25, 26f, 60–85
  accuracy of test equipment, 67–68
  accuracy value, 71
  as-found test, 72–75
  as-left test, 77–80
  bench calibration setup, 66f
  calibration accuracy, 70–72
  calibration bench, 65, 66f
  calibration errors, 73–75
  calibration flowchart, 75f
  calibration sheet (calibration data record), 67, 67f, 77–78, 79f
  calibration standards, 65–69
  calibration sticker, 68f
  certification date, 64
  certified test equipment, 68
  concepts of calibration, 63–65
  dead band, 74
  deviation, 71
  digital multimeter, 66, 67f
  error correction procedures, 75–77
  five-point check, 69–70
  gain, 72
  HART block diagram, 76f
  HART transmitter calibration flowchart, 77f
  hysteresis, 74, 74f
  input test standard, 66
  lower-range value (LRV)/upper-range value (URV), 64
  metrology laboratory, 65
  necessity of calibration, 62–63
  nonlinearity, 74f, 75
  output test standard, 66–67
  pneumatic calibrator, 66f
  range, 63
  regulatory bodies, 62
  repeatability, 72
  repeatability vs. accuracy, 73f
  resolution, 67
  smart instrument. See Smart instrument calibration
  span, 63, 64–65
  span shift, 73, 73f, 76
  specification sheets, 80–83
  test equipment, 68–69, 69f
  tolerance, 67–68
  traceability, 68
  typical calibration problems, 62–63
  upscale and downscale check, 70, 74
  valve positioner, 237–238
  zero and span shift, 76
  zero shift, 71f, 73, 75–76
Calibration sheet, 67, 67f, 77–78, 79f
Calibration standards, 65–69
Calibration sticker, 68f
Canadian Standards Association (CSA), 368
Capacitance, 102, 131
Capacitance-based transducer, 102
Capacitance cell, 102f
Capacitance level sensor measurement, 130–131
Capacitor construction, 102f
Capacity, 290, 290f
Carbon counter, 258
Cardinal points, 69
Cascade control, 298–299, 299f
Cascade control block diagram, 298f
Catalytic bead sensor, 258, 258f
Catalytic reaction, 258
Cause-and-effect matrix, 334, 334f
CCS. See Combustion control system (CCS)

Celsius, Anders, 172
Celsius (C) temperature scale, 171, 172, 172f
CEMS. See Continuous emissions monitoring system (CEMS)
CEMS analyzer, 248f
Centipoise, 148
Central processing unit (CPU), 278
Centrifugal pump, 53f
Certification date, 64
Certified test equipment, 68
Cesium-137, 133
Chain float, 121, 121f
Charles, Jacques, 107
Chart recorder, 347, 347f
Checkpoints, 69
Chlorine, 246
Chromatogram, 252f
Chromatography, 251
Cipolletti weir, 164
Circle, 143f
Clamp-on milliammeter, 346f, 347
Clapeyron, Emilie, 107
Class 1 filled bulb thermometer, 174f
Clean Air Act, 247
Closed-loop control, 31f, 292–293, 293f
Closed-loop tuning technique, 337
Closed-loop valve positioning, 233–234, 234f
Closed-tank, dry reference leg measurement, 126–127, 126f
Closed-tank, filled (wet) reference leg measurement, 127–128, 127f, 128f
Closing the project, 375–376
Closure member, 212, 212f
Cobalt-60, 133
Code, 368
Coefficient of resistance, 180
Column height, 97
Combustion, 248
Combustion analysis, 248–251
Combustion control system (CCS), 251
Combustion efficiency, 249
Commercial off the shelf (COTS), 357
Commissioning, 332
Communication module, 358, 360f
Communication ports, 280f
Compact PLC, 281, 281f
Compiler, 282
Complete immersion thermometer, 174
Computer maintenance management software, 345
Concentric orifice plate, 154, 154f
Conductivity, 242–243
Conductivity probe level detection system, 124, 124f
Conductor labeling, 318, 318f
Conductor terminations, 317f
Conduit routing, 313f
Conduit seal-off, 313f
Confined spaces, 5–6
Construction project, 371–376
Contact level instruments, 130–132
Continuity equation, 146f
Continuous atmospheric monitoring, 5
Continuous emissions monitoring system (CEMS), 247–248
Continuous improvement model, 370f
Continuous level measurement, 112
Continuous stirred tank reactor, 4, 4f, 348f
Continuous temperature measurement, 176
Continuous throttling control, 294–298
Control. See Fundamentals of control
Control cabinet verification, 322f
Control devices, 29

Control loop, 11, 290
Control loop block diagram, 11f
Control modes, 293–298
Control program, 282–284
Control room, 266f
Control signals, 24–27
Control station, 28–29, 29f
Control system, 22–23, 352
  automatic control, 22, 23f
  control valve, 36–37, 37f
  controller, 34–35
  defined, 2
  diagram showing relationship of individual instrumentation components, 33f
  efficient control, 38f
  elements, 32, 38
  final control element, 34f, 36–37
  goal, 21
  manual control, 21–22
  primary element, 32–33, 32f
  process and instrumentation drawing (P&ID), 34f, 35
  process regulation, 36, 36f
  steady state, 35
  transducer, 35–36
  transient response, 35
  transmitter, 33–34, 33f
  uses, 21
Control valve assembly, 36–37, 37f, 81f, 208–239
  actuator, 218–225. See also Valve actuator
  ball valve, 217–218, 218f
  butterfly valve, 217, 217f
  common terminology and components, 212, 212f
  electropneumatic positioner, 236, 237f
  gate valve, 216, 216f
  globe valve, 211f, 213f, 214–216, 237f
  I/P transducer, 228–230
  limit switch, 231, 231f
  linear variable differential transformer (LVDT), 232–233, 233f
  nozzle-flapper mechanism, 227–228, 228f
  open- versus closed-loop valve positioning, 233–234, 234f
  pneumatic boosters and amplifiers, 230–231, 230f
  pneumatic positioner, 235–236, 236f
  pneumatic regulator, 226–227, 227f
  pneumatics, 210–211
  position-indicating devices, 231–233
  potentiometer position indicator, 232, 233f
  reed switch, 231–232, 232f
  rotary valve, 216–218
  sliding-stem valve, 212f, 213–216
  smart positioner, 236–237, 237f
  solenoid, 226, 226f
  types of control valves, 211–218
  valve positioner, 210, 210f, 233–238
Controller, 11, 34–35, 38, 266, 290. See also DCS controller; Process controller
Controller faults, 279
Controller I/O rack, 193f
Controller redundancy, 279
Controlling means, 23
Controls engineer, 20
Conventional devices, 190
Conventional pressure gauge, 47f
Conventional slot PLC, 279
Cooling tower, 53f, 242, 242f
Coriolis meter, 161, 255
Corrosion (HCS pictogram), 9f
CPU. See Central processing unit (CPU)

Critical path items, 373
CSA. *See* Canadian Standards Association (CSA)
Cubic centimeters, 113
Cubic foot (feet), 113
Current, 12
Current-based signals, 105
Current to voltage conversion, 275f
Cut sheets, 304f
Cutthroat flume, 164, 165f

D

D, 48f, 49f
DAHS. *See* Data acquisition and handling system (DAHS)
Dall flow tube, 157–158
Damping, 203
Data acquisition and handling system (DAHS), 248
DC ammeter, 347
DC parallel circuit, 15, 15f
DC series circuit, 14, 14f
DC theory
    current, 12
    Ohm's law, 13–14, 16
    parallel circuit, 15, 15f
    resistance, 12–13
    R_T, 15, 16
    series circuit, 14, 14f
    voltage, 12, 13
DCS. *See* Distributed control system (DCS)
DCS controller, 356–357, 357f
DD. *See* Device description (DD)
Dead band, 74
Dead Sea, 256
Dead time, 291, 291f
Dedicated loop controller, 277–278
Density
    buoyancy, 256–257
    defined, 134
    fluid flow, and, 148
    hydrostatic, 255–256, 256f
    level, and, 134–138
    mass and volume, 113, 134
    oscillating tube Coriolis, 255
    temperature, and, 135f, 136f
    vibronic, 254, 255f
    water vs. mercury, 115f
Derivative action, 297, 297f, 298f
Derivative control, 296–297
Design engineer, 20
Design standards, 368–369
Design technologist, 20
Deutsches Institut fur Normung (DIN), 369
Deviation, 31, 71
Device description (DD), 195
Device history/device-specific data, 345
Device measurement verification, 325–326
Diagrams, charts, identifiers. *See* Instrumentation symbols and diagrams
Diaphragm actuator, 210, 219–221
Diaphragm seal, 138f
Dielectric, 130
Diesel vehicle fuel, 260
Differential pressure, 94, 94f
    flow, 143, 147f, 153–159
    flow measurement, 153–159
Differential pressure detector method of liquid level measurement, 125
Differential pressure flow meter, 153–159
    flow nozzle, 159
    flow tube, 157–158
    orifice plate, 153–155
    pipe elbow, 156–157, 157f
    pitot tube, 158, 158f
    pressure tap, 155–156
Differential pressure switch, 100–101, 101f
Differential pressure transmitter, 112, 127
Digital communication mapping, 282
Digital control, 278
Digital multimeter, 66, 67f
Digital signal, 28, 105
Digital-to-analog (DAC) converter, 77, 203
Dimensional drawing, 304f
DIN. *See* Deutsches Institut fur Normung (DIN)
Dip tube, 130
Direct-acting actuator, 221
Direct-acting valve bodies, 221
Direct current. *See* DC theory
Direct level measuring device, 112
Direct-mount transmitter, 186
Discrete point measurement, 112
Discrete signal, 27
Discrete temperature control, 175
Displacement meter, 152–153
Displacer method of level measurement, 122–124
Dissolved oxygen, 253–254
Distributed control system (DCS), 276, 284, 352–365
    communication module, 358, 360f
    controller, 356–357, 357f
    defined, 365
    dual bus network design, 358, 359f
    environmental conditioning system, 363
    Ethernet, 357
    example (Triangle Chemical), 354–355
    fiber-optic cable, 357–358, 358f
    gateway, 360, 360f
    graphical display, 362, 363f
    ground loops, 361, 361f
    grounding, 360, 361f, 364, 364f
    I/O modules, 356, 356f
    IEEE 1100 standard, 363
    installation, 362–364
    network architecture, 355, 355f
    network topologies, 358, 358f
    opto-isolator, 362, 362f
    overview, 354
    owner's manual, 364
    power supplies, 361–362
    redundancy, 358
    ring network design, 358, 359f
    system documentation, 364
    training manuals, 364
    typical DCS control room, 362f
    typical power and grounding layout, 364f
    uninterruptible power supply (UPS), 363
    user interface, 362
    workstation, 362
Distributed I/O, 279, 279f
Divide-and-conquer troubleshooting, 343, 343f
Document library, 368
Document revisions, 323, 323f
Document standards, 369
Documentation inspection, 323
Double-acting hydraulic actuator, 225, 225f
Double-acting piston actuator, 222, 222f
Double-seated globe valve, 214, 214f
Draft, 249
Drawings, 304f
Drift, 186
Dry solids, 125
Dual bus network design, 358, 359f
Dual-sensor RTD, 184
Dumb devices, 190

E

E, 48f, 49f
E/P transducer, 228
Ears "popping," 91
Eccentric orifice plate, 154, 154f
Eddy, 149
EEPROM. *See* Electrically erasable programmable read-only memory (EEPROM)
Efficient control, 38f
Electric (motor-operated value) actuator, 223–224, 224f
Electric solenoid, 226f
Electrical and Instrumentation Loop Check, 322
Electrical current, 12
Electrical current path, 8f
Electrical currents, 7
Electrical safety, 6–8
Electrical safety program procedures, 7
Electrical signal, 28
Electrical terminals, 314f
Electrically erasable programmable read-only memory (EEPROM), 199
Electrochemical cell principle, 260–261
Electromagnetic flow meter, 162
Electromotive force (EMF), 12
Electron, 12
Electronic signaling, 268f
Electropneumatic positioner, 236, 237f
Electrostatic discharge protective measures, 278
Elevated transmitter installation, 129f
Elevated zero and suppressed output, 129
Emergency shutdown (ESD) system, 284–285
EMF. *See* Electromotive force (EMF)
Energy
    defined, 145
    kinetic, 144
    potential, 143–144
Enterprise system, 267
Environment (HCS pictogram), 9f
Environmental conditioning system, 363
Environmental conditions, 345
EPA Method 9, 249–250
EPROM. *See* Erasable programmable read-only memory (EPROM)
Equal percentage cage, 216, 216f
Equipment symbols, 52, 53f
Equivalent pressures (different units of measure), 105f, 106
Erasable programmable read-only memory (EPROM), 199
Error comparator, 293, 293f
Escherichia coli, 246
ESD system. *See* Emergency shutdown (ESD) system
Ethernet, 357
Ethernet RJ45 connector, 357f
Evaporative cooling water system, 242, 242f
Evolution of process control, 354, 354f
Excess air, 249
Exclamation Mark (HCS pictogram), 9f
Exploding bomb (HCS pictogram), 9f
Eyewash stations, 333

F

F, 48f, 49f
F_A, 219
F_S. *See* Spring force (F_S)
Fahrenheit, Daniel, 171
Fahrenheit (F) temperature scale, 171, 172f
Fail-closed valve, 219
Fail high, 200, 200f, 201f
Fail-indeterminate valve, 219

Fail-locked valve, 219
Fail low, 200, 200f, 201f
Fail-mode switch, 200, 200f
Fail-open valve, 219
Failure mode indications, 53, 53f
Fall protection, 5
Farad, 131
Fast key sequences, 197, 197f
FC, 53
FDA. *See* U.S. Food and Drug Administration (FDA)
Feed-forward control, 300, 300f
Feedback, 224
Feedback control, 300
Ferrule conductor, 317f
FF, 48f, 49f
FI, 53
Fiber-optic cable, 357–358, 358f
Field I/O signals, 267–276
Field mounted (symbols), 46f
File library, 196
Fillage, 112
Filled bulb thermometer, 174, 174f
Final control element, 11, 34f, 36–37, 38
Fire tetrahedron, 257f
First-letter identifiers, 48, 48f, 49, 49f
Fisher molded diaphragm, 219f
Five-fold manifold, 310, 310f
Five-point check, 69–70
Five-point test of analog systems, 326, 328
FL, 53
Flame (HCS pictogram), 9f
Flame ionization detector, 258–259, 259f
Flame Over Circle (HCS pictogram), 9f
Flammable gas detector, 257–260
Flange seals, 309f
Flange tap, 156, 156f
Flexible raceways, 312, 312f
Float manometer, 96f
Flow, 35. *See also* Principles of flow
Flow control, 37
Flow control loop example, 25f
Flow energy, 144f
Flow nozzle, 159
Flow rate, 23
Flow straightener, 149
Flow switch, 152, 152f
Flow tube, 157–158
Flue gas samples, 250
Fluid flow measurement. *See* Principles of flow
Fluid viscosity, 148
Fluids, 142
Fluke 754 documenting process calibrator, 78
Flume, 163–164
Flushing the pipelines, 332–333
Flyback diode, 272–273, 272f
FO, 53
Force, 210
Forcing the process variable, 325–326
Fork-style connectors, 317, 317f
Formula, 10
Forward-biased current, 272f
Fossil fuel-powered boiler, 251
Fossil fuel-powered equipment, 248
Fossil fuels, 248
Foundation Fieldbus, 105, 193–194, 194f, 236
FQ, 48f, 49f
Friction, 148
FT, 48
Full-flow ball valve, 218, 218f
Full immersion thermometer, 174
Full trim, 202

Fume extraction hood, 53f
Functional block, 193, 193f, 283f
Functional block programming, 283, 283f
Functional inspection, 325–326
Fundamentals of control, 288–301. *See also* Process controller
 advanced control, 298–300
 algorithm, 290
 cascade control, 298–299, 299f
 closed-loop control, 292–293, 293f
 continuous throttling control, 294–298
 control modes, 293–298
 dead time, 291, 291f
 derivative control, 296–297
 feed-forward control, 300, 300f
 integral control, 295–296
 on/off control, 293–294, 294f
 open-loop control, 291–292, 292f
 PID algorithm, 294, 298
 PID control action, 297, 297f
 process capacity, 290, 290f
 process lag, 291, 291f
 proportional control, 294–295
 ratio control, 299, 299f
 time-cycle control, 294
 two-position control, 293–294

G
G, 48f, 49f
Gain, 72, 295
Galilei, Galileo, 171
Gamma rays, 133
Gantt chart, 373, 373f
Gas chromatograph, 252, 252f
Gas chromatography, 251–253
Gas cylinder (HCS pictogram), 9f
Gas flow taps, 307f
Gases, 142
Gate valve, 216, 216f
Gateway, 360, 360f
Gauge glass method, 117–120
Gauge pressure, 95–96, 95f
Gay-Lussac, Joseph Louis, 107
Gear meter, 152, 153f
Geometry, 10
Globe valve, 211f, 213f, 214–216, 237f
Glycerin, 148
Grains, 125
Graphical user interface (GUI), 28–29, 29f, 281, 362
Gravity, 88
Gray-scale imaging, 285
Ground loops, 361, 361f
Grounding, 316, 316f, 360, 361f, 364, 364f
GUI. *See* Graphical user interface (GUI)
Guided wave radar, 131–132, 131f

H
H, 48f, 49f
H flume, 164, 164f
Handheld communicator, 195f
Handheld phase rotation meter, 333, 333f
Handwriting, 322
HART. *See* Highway addressable remote transducer (HART)
HART block diagram, 76f, 199f
HART Communication Foundation, 192, 196
HART communicator, 194–198, 326
HART device. *See* Smart instrument
HART device calibration. *See* Smart instrument calibration

HART digital communication, 192f
HART input cards. *See* Highway addressable remote transducer (HART) input cards
HART module, 356f
HART transmitter calibration flowchart, 77f, 205f
HART troubleshooting flowchart, 346f
Hazard communication program (HAZCOM), 333
Hazard communication standard pictogram, 9f
Hazardous location seal-off fittings, 312, 313f
HCS pictograms and hazards, 9f
Head, 115
Head pressure, 115
Head pressure measurement, 117f
Health hazard (HCS pictogram), 9f
Heat, 170
Heat exchanger, 171, 171f
Heat-shrinkable tubing, 318
Hidden from operator (symbols), 46f
High damping values, 203
High-gain controller settings, 295f
High-level languages, 282, 282f
High pressure switch, 100
High-viscosity fluid, 148
Highway addressable remote transducer (HART), 105, 192–193, 236
Highway addressable remote transducer (HART) cards, 275, 275f
History of process control, 354, 354f
HMI. *See* Human-machine interface (HMI)
HMI graphics, 285f
Home heating system thermostat, 266–267
Homeostatic phenomenon, 2
Hot swapping, 279
Hot-wire anemometer, 161
Human breathing, 90, 90f
Human-machine interface (HMI), 29, 281, 281f, 285, 362
Humidity, 138, 165
Hydraulic actuator, 224–225, 225f
Hydrocarbon combustion, 248
Hydrocarbons, 258
Hydrodesulfurization process, 260
Hydrogen sulfide (H_2S), 260
Hydrogen sulfide monitor, 260–261, 260f
Hydrometer, 256, 257f
Hydrostatic density, 255–256, 256f
Hydrostatic pressure, 115–117
Hydrostatic pressure instruments, 126–130
Hysteresis, 74, 74f

I
I, 48f, 49f
I/O. *See* Input and output (I/O)
I/O card, 26
I/O devices, 23
I/O modules, 356, 356f
I/O rack, 193f
I/O signals, field, 267–276
I/P transducer, 31, 35, 35f, 228–230
I&C projects. *See* Instrument and control (I&C) projects
Ideal gas law, 107
IEC 61131-3, 282
IEEE 1100 standard, 363
IF...ELSE...AND...THEN..., 284
Impulse tubing, 306
Inch-of-water measurement, 92–93, 93f
Inches of mercury ("Hg), 105, 105f
Inches of water ("H_2O), 105, 105f
Inclined manometer, 96f

Inconel, 181
Index, 56
Indicating devices, 314f
Indicating differential pressure transmitter, 47f
Indicating gauge pressure transmitter, 47f
Indicating temperature transmitter, 47f
Indicating transmitter, 50f
Inductive coupling, 363
Industrial pressure transmitter, 101
Inferred level measuring device, 112
Inflated balloon, 91, 91f
Infrared gas detector, 259–260
Infrared radiation, 259
Inhalation and exhalation (breathing), 90f
Inlet shoulder, 157
Input, 23
Input and output (I/O), 2
Input test standard, 66
Installation of control systems, 302–319
    blocking procedure, 310–311, 310f
    cable and conductor labeling, 318, 318f
    gas service, 307, 307f
    grounding, 316, 316f
    intrinsic barrier, 316–317, 317f
    liquid service, 307, 307f
    manifolds, 309–310, 309f, 310f
    mounting, 304–306
    piping accessories, 311, 311f
    pressure ratings, 309, 309f
    pressure taps, 306f, 307
    raceway or conduit installation, 311–313
    rotating the housing, 305–306
    shield isolation, 319f
    steam service, 308, 308f
    tubing, 306–308
    wiring terminations, 313–318
Instrument and control (I&C) projects, 371–376
Instrument-based controllers, 281, 281f
Instrument bubble, 44f
Instrument calibration. *See* Calibration procedure and documentation
Instrument calibration sheet, 70f
Instrument contractor, 375
Instrument elevation and suppression errors, 128–129
Instrument errors, 73. *See also* Calibration errors
Instrument identification, 48–50
Instrument identification bubble, 48f, 49f
Instrument index, 56
Instrument index sheet, 56, 58f
Instrument installation. *See* Installation of control systems
Instrument letter chart, 49f
Instrument line types, 43f
Instrument management databases, 78
Instrument manifolds, 309–310, 309f, 310f
Instrument manufacturer, 345
Instrument project manager, 368
Instrument spec sheets. *See* Specification sheet (spec sheet)
Instrument to process connection, 44f
Instrumentation
    defined, 2
    purpose, 2
    symbols. *See* Instrumentation symbols and diagrams
Instrumentation documents, 53–58
Instrumentation projects, 376. *See also* Project management

Instrumentation symbols and diagrams, 40–59
    behind-panel instrument, 45f
    blind/indicating transmitter, 50f
    common instrument line types, 43f
    common instruments and their symbols, 47f
    equipment symbols, 52, 53f
    failure mode indications, 53, 53f
    first-letter identifiers, 48, 48f, 49, 49f
    instrument identification, 48–50
    instrument index sheet, 56, 58f
    instrument letter chart, 49f
    instrument to process connection, 44f
    instrumentation documents, 53–58
    lines and instruments, 43–48
    location and function legend, 46f
    loop number, 50–51, 51f
    loop sheet, 54–55, 54f
    panel-mounted instrument, 45f
    process and instrumentation drawing (P&ID), 42f, 45, 53–54
    process lines, 44, 44f
    redundant device tags, 52, 52f
    redundant loop tags, 52, 52f
    remote-mount vs. shared device, 46f
    schematic diagram, 55–56, 56f
    spec sheet, 56, 57f
    tag number, 45, 48f, 55
    transmitter connections, 51f
    valve symbols and actions, 52–53, 53f
    wiring diagram, 55–56, 55f
Instrumentation technologies, 2
Instrumentation tubing, 306–308
Instrumentation vocabulary, 10–12
Instruments and their symbols, 47f
Integral action, 296, 296f, 298f
Integral control, 295–296
Integral windup, 296
Integration, 296f
International Organization for Standardization (ISO), 42, 68, 369, 370
International Society of Automation (ISA), 42, 369
Interposing relay, 270, 270f
Intrinsic barrier, 316–317, 317f
Intrinsically safe, 8
Intrinsically safe (IS) wiring, 317, 317f
Ion chamber, 134
Ionic concentration and pH, 244f
IS wiring. *See* Intrinsically safe (IS) wiring
ISA. *See* International Society of Automation (ISA)
ISA-RP60.8, 369
ISO. *See* International Organization for Standardization (ISO)
ISO 9001 certification, 370, 370f
ISO 9000 family of quality management standards, 369
ISO 9001 standard, 68, 370, 371
Isolating diaphragm, 311f
Isothermal block, 179f
Isotopes, 133
IT, 48
ITS-90 standard, 172

J
J, 48f, 49f
Japanese Industrial Standard (JIS), 369
Java, 282, 282f
JIS. *See* Japanese Industrial Standard (JIS)

K
K, 48f, 49f
K_{Pu}. *See* Ultimate gain (K_{Pu})
Kelvin, Lord, 172
Kelvin *(K)* temperature scale, 172, 172f
Kilo-Pascal (kPa), 106
Kinetic energy, 144, 170
kPa, 106

L
L, 48f, 49f
Ladder capacity ratings, 6f
Ladder logic, 30f
Ladder logic programming, 282f, 283
Ladder safety, 6, 7f
Laminar flow, 148, 148f
Law of conservation of energy, 145–148
Law of continuity, 145–146
Leak testing, 332
Legacy devices, 190
Legionella pneumophila, 246
LEL. *See* Lower explosion limit (LEL)
Letter codes, 48f, 49f
Level, 23. *See also* Principles of level
Level control loop, 12f
Level switch measurement, 112
Libraries, 368
Life safety oxygen measurement, 253
Limit switch, 231, 231f
Line breaking, 4
Linear cage, 216, 216f
Linear variable displacement transformer (LVDT), 103–104, 104f, 232–233, 233f
Linearity, 75
Lines and instruments, 43–48
Liquid filled impulse lines, 307f
Liquid head measurement, 116f
Liquid head pressure, 115f
Liquid level measuring devices, 112
Local, 45
Location and function legend, 46f
Lockout/tagout, 3–4
Logbooks, 267
Loop, 30
Loop calibrator, 326, 327f
Loop checking, 322–326. *See also* Process control loop checking
Loop-checking sheet, 323, 324f, 327
Loop number, 50–51, 51f
Loop sheet, 21, 21f, 54–55, 54f
Loop tuning, 38, 336–337
Low-damping values, 203
Low-gain controller settings, 295f
Low point drain ports, 313
Low pressure switch, 100
Low-viscosity fluid, 148
Lower explosion limit (LEL), 257, 257f
Lower-range value (LRV), 64
LRV. *See* Lower-range value (LRV)
LT, 48
LVDT. *See* Linear variable displacement transformer (LVDT)

M
M, 48f, 49f
Machine language, 282, 282f
Magnetic bond method, 122
Magnetic level indicator, 122f
Making, 27
Man-machine interface (MMI), 362
Management of change (MOC), 375
Managing projects, 368. *See also* Project management

Manifold blocking procedure, 310f
Manifolds, 309–310, 309f, 310f
Manometer, 96f, 97–98, 97f, 114–115
Manual and visual level instruments, 117–120
Manual control, 21–22
Manual loading stations, 334
Maple syrup manufacturing, 256
Marshalling cabinet, 27f
Mass, 88, 113
Mass flow meter, 161
Mathematical skills, 9–10
Matter, 88
Maximum working pressure (MWP), 309f
Measurement (input) standard, 66–67
Measuring junction, 178
Measuring section, 156
Mechanical level measurement instruments, 120–124
Mechanical signal, 28
Mechanical switching processes, 271
Mechanical switchover time, 271f
Mega-Pascal, 106
Membrane-covered amperometric sensor, 254
Meniscus, 97, 97f
Menu tree, 196, 196f
Mercury, 114, 115f
Mercury thermometer, 174, 174f
Metering runs, 156
Methane, 257f, 258
Methane and oxygen combustion, 249f
Metrology laboratory, 65
Microfarad, 131
Microprocessor controller, 264. *See also* Process controller
Microsiemens per centimeter, 243
Millisiemens per centimeter, 243
MMI. *See* Man-machine interface (MMI)
MOC. *See* Management of change (MOC)
Modbus, 105, 192
Modifiers (first-letter identifiers), 48, 48f, 49, 49f
Modulating control, 294
Motor-operated value (MOV) actuator, 223–224, 224f
Motor rotation meter, 333, 333f
Mounting, 304–306
MOV actuator. *See* Motor-operated value (MOV) actuator
Multi-loop controller, 277, 278f
Multiple-batch process tanks, 355, 355f
Multivariate relationships, 106–108
   density and temperature relationship, 135f
   ideal gas law, 107
   pressure and temperature relationship, 107, 107f
   volume and pressure relationship, 108, 108f
   volume and temperature relationship, 108, 108f
   water density versus temperature, 136f
MWP. *See* Maximum working pressure (MWP)

N

N, 48f, 49f
N/m^2, 88
Naming and identification scheme, 318
National Fire Protection Agency (NFPA), 368
National Institute of Standards and Technology (NIST), 68, 131, 253
NATO alphabet. *See* North Atlantic Treaty Organization (NATO) alphabet
Natural gas, 248
Natural gas boiler, 251
Negative temperature coefficient thermistor, 185
Network architecture, 355, 355f
Network diagram chart, 372, 373f
Network topologies, 358, 358f
Newton, Isaac, 91
Newtons per square meter (N/m^2), 88
NFPA. *See* National Fire Protection Agency (NFPA)
NFPA 70 Article 210, 362
NFPA 70 Article 250, 363
NFPA 86, 257
NIST. *See* National Institute of Standards and Technology (NIST)
Nitrogen-purged vessels and pipelines, 333
Non-indicating transmitter, 50f
Noncontact level instruments, 132–134
Nonlinearity, 74f, 75
Normal operating pressure, 309
North Atlantic Treaty Organization (NATO) alphabet, 322
Nozzle-flapper mechanism, 227–228, 228f
NPN-style devices, 273
NPN-type transistor, 272f
NRC. *See* Nuclear Regulatory Commission (NRC)
Nuclear measurement, 133–134
Nuclear Regulatory Commission (NRC), 62
Nutating disc meter, 152

O

O, 48f, 49f
Occupational Safety and Health Administration (OSHA), 3, 284
Offline configuration, 195
Offset, 295, 295f
Ohm, 13
Ohm, George, 13
Ohm's law, 13–14, 16, 275
On/off control, 293–294, 294f
Online configuration, 195
Opacity, 249–250
Open channel, 163
Open-channel meter, 163–165
Open-loop control, 30f, 291–292, 292f
Open-loop time clock control, 292f
Open-loop tuning technique, 336–337
Open-path infrared detector, 259, 259f
Open tank, 53f
Open-tank differential pressure measurement, 126, 126f
Operator control station, 267f
Operator interface
   control devices, 29
   graphical user interface (GUI), 28–29, 29f
   shared control, 30
Optical signal, 28
Opto-isolator, 362, 362f
Orifice plate, 153–155
ORP. *See* Oxidation reduction potential (ORP)
Oscillating tube Coriolis density, 255
OSHA. *See* Occupational Safety and Health Administration (OSHA)
OSHA pictograms, 9f
Output, 23
Output test standard, 66–67
Over pressurizing, 96
Owner's manual, 364
Oxidation reduction potential (ORP), 246–247
Oxygen, 253
Oxygen concentration, 253–254
Oxygen concentration analyzer, 333
Oxygen-deficient area, 253

P

P, 48f, 49f
P_u. *See* Ultimate period (P_u)
Pa, 88, 106
Packing, 212, 212f
Paddle flow switch, 152f
Palmer-Bowlus flume, 164, 164f
Panel mount, 304f
Panel mounted, 45
Panel-mounted current-to-pressure instrument, 45f
Panel-mounted instrument, 45f
Panel-mounted temperature indicators, 45f
Paperless chart recorder, 347f
Parallel PID control, 294
Parshall flume, 163, 163f
Partial immersion thermometer, 174
Pascal (Pa), 88, 106
Pascal, Blaise, 106
Pascal's law, 210
PD, 48f, 49f
PDCA. *See* Plan, do, check, act (PDCA)
PDT, 48
Permanently mounted oxygen analyzer, 253
Personal oxygen monitor, 253
Personal protective equipment, 262, 326, 333, 345
Petroleum and chemical industries, 369
PFD. *See* Process flow diagram (PFD)
pH analyzer, 245f
pH analyzer error, 245
pH measurement, 243–246
pH-sensitive paper, 244
PHP, 282f
PI, 99
Picofarad, 131
P&ID. *See* Process and instrumentation drawing (P&ID)
PID algorithm, 294, 298
PID block. *See* Proportional-integral-derivative (PID) block
PID control. *See* Proportional-integral-derivative (PID) control
PID control action, 297, 297f
PID control modes, 294–298
P&ID design standards. *See* Process and instrumentation drawing (P&ID) design standards
P&ID walkdown, 332
Piezoelectric-based transducer, 103
Piezoelectric effect, 103
Piezoresistive device, 102–103
Piezoresistive effect, 102
Pigtail, 311f
Pilot device, 270
PIP. *See* Process Industry Practices (PIP)
Pipe cleaning pig, 332f
Pipe elbow, 156–157, 157f
Pipe flanges, 80
Pipe mounts, 304, 305f
Pipe tap, 156, 156f
Piping accessories, 311, 311f
Piping diagram, 56
PISH, 99
PISL, 99
Piston-type actuator, 221–222
Pitot tube, 158, 158f
Plan, do, check, act (PDCA), 370, 370f
Plant start-up process, 332. *See also* Start-up and tuning of process control loops

Index

Plate and frame heat exchanger, 53f
Plate style mounts, 304
Plate-type heat exchanger, 171, 171f
PLC. *See* Programmable logic controller (PLC)
Pneumatic actuator, 219–223
Pneumatic and hydraulic signal, 28
Pneumatic boosters and amplifiers, 230–231, 230f
Pneumatic calibrator, 66f
Pneumatic controller, 276–277
Pneumatic controls, 266f
Pneumatic flow booster, 230f
Pneumatic PID controller, 277f
Pneumatic positioner, 235–236, 236f
Pneumatic regulator, 226–227, 227f
Pneumatic safety valve, 269f
Pneumatic schematics, 56
Pneumatic signaling system, 210
Pneumatic square-root extractor, 277
Pneumatic timer, 276
Pneumatic volume booster, 230
Pneumatics, 210–211
PNP-type transistor, 272f
Point level measurement, 112, 124
Point temperature control, 175
Poise, 148
"Popping" of the ears, 91
Portable combustion analyzer, 250f
Portable handheld labeling machine, 318
Portable oxygen analyzer, 253f
Position-indicating devices, 231–233
Positioner, 210, 210f. *See also* Valve positioner
Positive temperature coefficient thermistor, 185
Potential energy, 143–144
Potentiometer, 232
Potentiometer valve position indicator, 232, 233f
Pounds per square inch (psi), 88, 105
Pounds per square inch absolute (PSIA), 88
Power supply, 361
Pressure, 88. *See also* Principles of pressure
Pressure amplifier, 230–231
Pressure and temperature relationship, 107, 107f
Pressure arrows, 94f
Pressure conversion table, 106f
Pressure devices, 96
Pressure formula, 88
Pressure gauge, 98–99, 99f
Pressure range, 96
Pressure ratings, 309, 309f
Pressure regulator, 227, 227f
Pressure relief valve spring tension, 268
Pressure scales, 91–96
Pressure switch, 99–101
Pressure tap, 155–156, 306f, 307
Pressure transmitter, 82f, 101–105
Pressure transmitter adapter, 326f
Pressure units of measure, 88, 105–106
Pressurizer, 136
Pressurizer level measurement, 136–137, 137f
Primary element, 11, 32–33, 32f
Primary (input) test standard, 66
Primary measuring devices, 163
Principles of flow, 140–167
    ambient temperature variations, 165
    Bernoulli's principle, 146–147, 146f
    beta ratio, 150
    component variables, 142–143
    Coriolis meter, 161
    density, 148
    differential pressure, 143, 147f, 153–159
    differential pressure meter. *See* Differential pressure flow meter
    displacement meter, 152–153
    electromagnetic flow meter, 162
    environmental considerations, 165
    flow characteristics, 148–151
    flow switch, 152, 152f
    fluids, 142
    flume, 163–164
    gear meter, 152, 153f
    hot-wire anemometer, 161
    humidity, 165
    laminar flow, 148, 148f
    law of conservation of energy, 145–148
    law of continuity, 145–146
    mass flow meter, 161
    methods of flow measurement, 151–165
    open-channel meter, 163–165
    process energy, 143–144
    Reynolds number (RN), 149
    rotameter, 160–161, 160f
    square root signal extraction, 147–148
    target meter, 159–160, 160f
    turbulent flow, 149, 149f
    typical control loop, 142f
    ultrasonic flow equipment, 162–163
    velocity meter, 161–163
    vena contracta, 150, 151f
    viscosity, 148
    volumetric flow equation, 144–145
    weir, 164–165
Principles of level, 110–139
    ambient temperature variations, 138
    Archimedes's law, 122, 123f
    ball float, 120–121, 121f
    bubblers, 129–130, 129f
    capacitance level sensor measurement, 130–131
    chain float, 121, 121f
    closed-tank, dry reference leg measurement, 126–127, 126f
    closed-tank, filled (wet) reference leg measurement, 127–128, 127f, 128f
    conductivity probe method, 124, 124f
    contact level instruments, 130–132
    continuous level measurement, 112
    density, 113
    density compensation, 134–138
    direct/inferred device, 112
    discrete and level switches, 124–125
    displacer, 122–124
    environmental considerations, 138
    fillage/ullage, 112
    gauge glass method, 117–120
    guided wave radar, 131–132, 131f
    humidity, 138
    hydrostatic pressure, 115–117
    hydrostatic pressure instruments, 126–130
    instrument elevation and suppression errors, 128–129
    ion chamber, 134
    magnetic bond method, 122, 122f
    manual and visual level instruments, 117–120
    mass, 113
    mechanical instruments, 120–124
    methods of level measurement, 117–134
    noncontact level instruments, 132–134
    nuclear measurement, 133–134
    open-tank measurement, 126, 126f
    physics of level measurement, 112–117
    point level measurement, 112, 124
    pressurizer level instruments, 136–137
    radar level measurement, 133
    reference leg temperature considerations, 135–136
    rotating paddle level switch, 125, 125f
    scintillation tube, 134
    slurry, 138
    specific gravity, 114–115, 114f
    specific volume, 134–135
    steam generator level instruments, 137–138, 137f
    strapping tables, 112
    time-of-flight radar measurement, 132f
    ultrasonic level measurement, 132–133
    vibrating fork level switch, 124–125, 125f
Principles of pressure, 86–109
    absolute pressure, 94–95, 94f
    atmospheric pressure, 88, 89f, 90f, 91
    bellows-style pressure-sensing devices, 98–99
    Bourdon tube-type detectors, 98
    capacitance-based transducer, 102
    differential pressure, 94, 94f
    equivalent pressures (different units of measure), 105f, 106
    everyday situations requiring pressure measurements, 86
    gauge pressure, 95–96, 95f
    human breathing, 90, 90f
    inch-of-water measurement, 92–93, 93f
    linear variable displacement transformer (LVDT), 103–104
    manometer, 96f, 97–98, 97f
    multivariate relationships, 106–108
    over pressurizing, 96
    piezoelectric-based transducer, 103
    piezoresistive device, 102–103
    pressure, defined, 88
    pressure and temperature relationship, 107, 107f
    pressure arrows, 94f
    pressure conversion table, 106f
    pressure devices, 96
    pressure formula, 88
    pressure gauge, 98–99, 99f
    pressure range, 96
    pressure scales, 91–96
    pressure switch, 99–101
    pressure transmitter, 101–105
    static pressure, 91
    strain gauge, 102–103, 103f
    units of measure, 88, 105–106
    vacuum pressure, 96, 96f
    volume and pressure relationship, 108, 108f
    volume and temperature relationship, 108, 108f
Principles of temperature, 168–189
    continuous measurement, 176
    ITS-90 standard, 172
    methods of temperature measurement, 173–188
    molecular movement, 170f
    purpose of temperature measurement, 170–171
    resistance temperature detector (RTD), 180–184
    temperature scales, 171–172
    temperature sensors, 176–186
    temperature switch, 175, 175f
    thermistor, 184–186
    thermocouple, 176–180
    thermometer, 173–175
    thermowell, 188, 188f
    transmitter, 186–188

Problem statement, 342
Process, 23
Process action diagram, 334–335, 335f
Process and instrumentation drawing (P&ID), 34f, 35, 42f, 45, 53–54, 323
Process and instrumentation drawing (P&ID) design standards, 369
Process capacity, 290, 290f
Process control, 18, 23
Process control components, 24f
Process control loop checking, 320–329
    ANSI/ISA 62382 standard, 322
    documentation inspection, 323
    example (temperature transmitter loop checking), 326–327
    five-point test of analog systems, 326, 328
    forcing the process variable, 325–326
    functional inspection, 325–326
    handwriting, 322
    loop calibrator, 326, 327f
    loop-checking sheet, 323, 324f, 327
    NATO alphabet, 322
    stimulating the process variable, 326
    underlying principle, 322
    visual inspection, 325, 325f
Process control system start-up. *See* Start-up and tuning of process control loops
Process control valve, 36–37, 37f
Process controller, 264–287. *See also* Fundamentals of control
    analog I/O, 274–275
    auto-tune feature, 290
    binary input, 268–269
    binary output, 270–271
    control program, 282–284
    dedicated loop controller, 277–278
    digital communication mapping, 282
    distributed I/O, 279, 279f
    emergency shutdown (ESD) system, 284–285
    field I/O signals, 267–276
    gray-scale imaging, 285
    HART input cards, 275, 275f
    HMI graphics, 285f
    human-machine interface, 285
    instrument-based controllers, 281, 281f
    pneumatic controller, 276–277
    programmable logic controller (PLC), 278–281
    resistance temperature detector (RTD) input cards, 275–276
    sink and source devices, 271–274
    sinking inputs, 273, 273f
    sinking outputs, 273, 273f
    sourcing inputs, 273–274, 274f
    sourcing outputs, 274, 274f
    thermocouple input cards, 275–276
Process controller algorithm, 290
Process energy, 143–144
Process flow diagram (PFD), 42
Process hazard analysis, 8
Process hazards, 8–9
Process Industry Practices (PIP), 369
Process lag, 291, 291f
Process line break, 4–5
Process lines, 44, 44f
Process loop, 37–38
Process regulation, 36, 36f
Process tank. *See* Nuclear 63f
Process time constants, 291, 291f
Process variables, 3, 23
Profibus, 192, 236
Program, 282

Programmable logic controller (PLC), 26, 278–281
Programming language hierarchy, 282f
Project, 371
Project library, 368
Project life cycle, 371, 371f, 375f
Project management, 366–377
    changes in project design, 375, 375f
    closing the project, 375–376
    critical path items, 373
    defined, 371
    design standards, 368–369
    document standards, 369
    Gantt chart, 373, 373f
    instrument and control (I&C) projects, 371–376
    instrument contractor, 375
    ISO 9001 standard, 370, 371
    management of change (MOC), 375
    network diagram chart, 372, 373f
    project library, 368
    project life cycle, 371, 371f, 375f
    quality standards, 369–371
    recommended practices (RPs), 369
    request for information (RFI), 374f, 375
    request for proposal (RFP), 374
    request for quotation (RFQ), 375
    scope of work (SOW), 373
    user acceptance test (UAT), 376
    work breakdown structure (WBS), 372, 372f
Project milestones, 323f
Proof pressure, 309
Propane, 258
Proper zero procedure, 105f
Proportional action, 298f
Proportional control, 294–295
Proportional-integral-derivative (PID) block, 193, 193f
Proportional-integral-derivative (PID) control, 277
Proportional offset, 295, 295f
Proprietary protocol, 192
Protocol, 192–194, 357
Proving meter, 255
PSDH, 100
PSDL, 100
psi, 88, 105
PSIA, 88
PSL, 100
PT, 48
Pull-down resistor, 272
Pull-up resistor, 272
Python, 282, 282f

Q
Q, 48f, 49f
Q. *See* Volumetric flow rate *(Q)*
Quality standards, 369–371
Quarter-turn valve, 216, 217f
Quick-opening cage, 216, 216f
Quinhydrone, 246

R
R, 48f, 49f
R_T, 15, 16
Raceway or conduit installation, 311–313
Rack and pinion actuator, 222–223, 223f
Rack-mounted PLC, 279, 279f
Radar level measurement, 133
Radio or wireless signal, 28
Radius tap, 156
Range, 63
Rankine *(R)* temperature scale, 172, 172f

RATA. *See* Relative accuracy test audit (RATA)
Ratio control, 299, 299f
Real-time data highways, 358
Recommended practices (RPs), 369
Rectangular weir, 164
Redundancy, 358
Redundant bus data highway, 359f
Redundant communication lines, 25
Redundant controllers, 355
Redundant device tags, 52, 52f
Redundant loop tags, 52, 52f
Reed switch, 231–232, 232f
Reference junction, 178
Reference leg, 127, 127f
Reference leg temperature considerations, 135–136
Reflection gauge gas, 119–120, 119f
Refraction gauge glass, 120, 120f
Refractive index, 358
Regulator, 226–227, 227f
Relative accuracy test audit (RATA), 248
Relay output card, 271f
Relief valve, 269f
Remote control station, 267f
Remote diaphragm seal transmitter, 138f
Remote-mount vs. shared device, 46f
Remote seal, 129
Remote temperature well, 184f
Repeatability, 72
Repeatability vs. accuracy, 73f
Request for information (RFI), 374f, 375
Request for proposal (RFP), 374
Request for quotation (RFQ), 375
Reservoir manometer, 96f
Resistance, 12–13
Resistance temperature detector (RTD), 83f, 180–184
Resistance temperature detector (RTD) input cards, 275–276
Resolution, 67
Reverse-acting actuator, 221
Reverse-acting valve bodies, 221
Reverse-biased current, 272f
Reynolds, Osborne, 149
Reynolds number (RN), 149
RFI. *See* Request for information (RFI)
RFP. *See* Request for proposal (RFP)
RFQ. *See* Request for quotation (RFQ)
Ring network design, 358, 359f
Ring topology, 358f
Ringelmann, Maximilien, 249
Ringelmann chart opacity measurement, 249f
RJ45 connector, 357f
RJ45 Ethernet ports, 360f
RN. *See* Reynolds number (RN)
Rotameter, 160–161, 160f
Rotary valve, 216–218
Rotating paddle level switch, 125, 125f
Rotating the housing, 305–306
Rotational direction of motor, 333, 333f
RPs. *See* Recommended practices (RPs)
RS-232, 280
RS-485, 280
RTD. *See* Resistance temperature detector (RTD)
RTD input cards. *See* Resistance temperature detector (RTD) input cards

S
S, 48f, 49f
Safety data sheet (SDS), 5, 8, 305
Safety showers, 333
Salmonella, 246

Index

Sampling, 242
Scan, 30, 282
Scan time, 282
Schematic diagram, 55–56, 56f
Scintillation tube, 134
Scope of work (SOW), 373
Screw-in type thermowell, 188
SDS. *See* Safety data sheet (SDS)
Seal-off fittings, 312, 313f
Sealed capillary tubes, 138
Sealed tank, 53f
Seat ring, 212, 212f
Seebeck, Thomas, 176
Segmental orifice plate, 154, 154f
Self-regulating process, 2
Sensing element, 31, 32
Sensor trim, 77
Servo motor, 224
Set point, 11
Shared control, 30
Shared controller, 30
Shared device, 46f
Shield isolation, 319f
Shielded twisted pair (STP), 24–25
Shielded wiring, 268, 268f
Siemens per centimeter, 243
Signal, 11
Signal conditioner, 275
Signal conductors/power conductors, 363
Signal conversion, 36
Signal medium, 268f
Signal types
    analog signal, 27
    binary signal, 27–28
    digital signal, 28
    miscellaneous signals, 28
    wireless signal, 28
Signal wiring, 315, 316
Signaling devices, 38
Silica, 247
Silica analyzer, 247
Silica photometric measurement method, 247f
Simple process trainer, 44f
Single-acting hydraulic actuator, 224f, 225
Single-acting piston actuator, 222, 222f
Single-loop controller, 277, 277f
Single-point leak detection system, 259
Single-seated globe valve, 214, 214f
Sink and source devices, 271–274
Sinking inputs, 273, 273f
Sinking outputs, 273, 273f
Situational awareness graphics, 285
Skull and crossbones (HCS pictogram), 9f
Slant-body valve, 213f
Sliding-stem valve, 212f, 213–216
Slurry, 138
Small-batch process systems, 352
Smart communicator, 194–198
Smart device troubleshooting, 345–347
Smart device troubleshooting chart, 346f
Smart instrument, 76–77, 190–217
    advantages, 192
    calibration. *See* Smart instrument calibration
    device description (DD), 195
    fail high, 200, 200f, 201f
    fail low, 200, 200f, 201f
    fail-mode switch, 200, 200f
    fast key sequences, 197, 197f
    Foundation Fieldbus, 193–194, 194f
    functional operation of HART device, 199f
    HART, 192–193
    menu tree, 196, 196f
    microprocessor, 192
    online/offline configuration, 195
    operational range of typical HART device, 201f
    protocols, 192–194
    security switch, 200, 200f
    smart communicators, 194–198
    warning messages, 198f
    zero and span buttons, 199–200, 200f, 203
Smart instrument calibration, 76–77, 201–205
    AO trim, 204
    configure the conversion section, 203
    connect HART communicator, 201–202, 201f
    damping, 203
    full trim, 202
    minimum of 250 ohms of resistance, 202, 202f
    process flowchart, 205f
    step-by-step approach to bench calibration, 204
    verification of AO section, 203–204
    verification of sensor input section, 202
    zero-only trim, 202
Smart positioner, 236–237, 237f
Snubber, 272, 311f
Snubber diode, 273
Sodium hypochlorite, 246
Software, 22
Solderless wiring connectors, 317
Solenoid, 226, 226f
Solenoid air valve, 226
Solenoid spool valve, 226
Solid-state input devices, 273f
Solution conductivity, 242
Soot, 249
Sound waves, 132
"Sour" gas, 260
Sourcing inputs, 273–274, 274f
Sourcing outputs, 270, 274, 274f
SOW. *See* Scope of work (SOW)
Span, 63, 64–65
Span adjustment screws, 76, 76f
Span shift, 73, 73f, 76
Specific gravity, 114–115, 114f
Specific volume, 134–135
Specification sheet (spec sheet), 56, 57f, 80
    control valve, 81f
    pressure transmitter, 82f
    purpose, 80
    RTD sensor, 83f
    typical contents, 80
Spring force (F_S), 219
Square, 143f
Square root signal extraction, 147–148
Stack gas flowmeter, 248
Staffing needs, 375f
Standard developing bodies, 368
Standard for Electrical Safety in the Workplace (NFPA 70E), 3
Standard-setting organizations, 68
Standards, 368
    design, 368–369
    documentation, 369
    P&ID design, 369
    quality, 369–371
Star topology, 358f
Start-up and tuning of process control loops, 330–339
    cause-and-effect matrix, 334, 334f
    closed-loop technique, 337
    flushing the pipelines, 332–333
    leak testing, 332
    loop tuning, 336–337
    main control room/local control stations, 334
    open-loop technique, 336–337
    P&ID walkdown, 332
    pointers/tips, 335
    process action diagram, 334–335, 335f
    Ziegler-Nichols method, 336–337
Statement of work, 373
Static electricity, 278
Static pressure, 91
Steady state, 35
Steady-state error, 35
Steady state of process control, 35
Steam drum level measurement, 136, 136f
Steam flow nozzle, 159f
Steam generator, 136
Steam generator level system, 137–138, 137f
Steam measurements, 308f
Stem, 212, 212f
Stem sealing device, 212
Stepper motor, 224
Stimulating the process variable, 326
Stoichiometric condition, 249
STP. *See* Shielded twisted pair (STP)
Strain gauge, 102–103, 103f
Stranded conductor, 317f
Strapping tables, 112
Stripping tools and techniques, 318
Structured text, 284, 284f
Substitution method of troubleshooting, 344
Suppressed transmitter installation, 128f
Suppressed zero installation, 128
Suppressed zero shift, 73
Switched input, 269f
Symbols. *See* Instrumentation symbols and diagrams
Syntax, 284
Syphon, 311f
System, 21

T

T, 48f, 49f
Tag number, 45, 48f, 55
Target meter, 159–160, 160f
TD, 48f, 49f
TE. *See* Temperature element (TE)
TE 1004, 51f, 52
TE 1004a/TE 1004b, 52f
TE 1004b-1/TE 1004b-2, 52f
Temperature, 23. *See also* Principles of temperature
Temperature element (TE), 47f, 173
Temperature input controller card, 276f
Temperature measurement standards, 172
Temperature scales, 171–172
Temperature sensors, 176–186
Temperature switch, 175, 175f
Temperature transmitter (TT), 173, 184f, 186–188
Temperature transmitter loop checking, 326–327
Temperature transmitter styles, 186f
Temperature transmitter wiring, 185f
Temperature transmitter with RTD sensor, 187f
Temperature transmitter with thermocouple sensor, 187f
Temperature well, 184f
Terminology (instrumentation vocabulary), 10–12
Test equipment, 68–69, 69f

Test points, 69
Test terminals, 314–315, 315f, 347, 347f
Thermistor, 184–186
Thermocouple, 176–180
Thermocouple color coding and operating ranges, 177f
Thermocouple input cards, 275–276
Thermocouple reference chart, 177f
Thermocouple signals, 179f
Thermometer, 47f, 173–175
Thermostat-operated furnace, 267f
Thermowell, 188, 188f
Three-slot PLC controller, 26, 26f
Three-valve manifold, 309, 309f
TIA/EIA Recommended Standard (RS) 232, 280
TIA/EIA Recommended Standard (RS) 485, 280
Time constant, 291, 291f
Time-cycle control, 294
Time of flight, 133
Time-of-flight radar measurement, 32f, 132f
Tolerance, 67–68
Traceability, 68
Training manuals, 364
Transducer, 31, 35–36, 38
    capacitance-based, 102
    E/P, 228
    I/P, 228–230
    piezoelectric-based, 103
Transient regulation, 35
Transient response, 35
Transistor, 271–272, 272f
Transit-time meter, 163
Transmissometer, 250, 250f
Transmitter, 11, 33–34, 33f
Transmitter connections, 51f
Transmitter security switch, 200, 200f
Transmitter test terminals, 347f
Transparent tube level measurement, 118f
Trapezoidal weir, 164
Triangular weir, 164
Trim, 215
Troubleshooting algorithm, 342f
Troubleshooting equipment, 347–348
Troubleshooting heuristic, 343
Troubleshooting procedures, 340–351
    active process control systems, 344–345
    chart recorder, 347, 347f
    clamp-on milliammeter, 346f, 347
    define the problem, 342
    device history/device-specific data, 345
    divide-and-conquer method, 343, 343f
    new systems, 344
    role of technician, 349
    sample HART troubleshooting flowchart, 346f
    smart devices, 345–347
    substitution method, 344
    test terminals, 347, 347f
    troubleshooting algorithm, 342f
    troubleshooting equipment, 347–348
TT, 48. *See also* Temperature transmitter (TT)
TT 1004, 51, 51f
TT 1004 tag, 50f
TT 1004a/TT 1004b, 52f
TT 1004b-1/TT 1004b-2, 52f

Tubing, 306–308
Turbulence, 149
Turbulent flow, 149, 149f
Twisted-pair wiring, 24
Two-position control, 293–294
Type J thermocouple reference table excerpt, 178
Typical working environment, 2f, 3

U
U, 48f, 49f
U-tube manometer, 96f, 97, 97f
UAT. *See* User acceptance test (UAT)
UEL. *See* Upper explosion limit (UEL)
UL 508A, 271
Ullage, 112
Ultimate gain (K_{Pu}), 334
Ultimate period (P_u), 337, 337f
Ultra-low-sulfur diesel, 260
Ultrasonic flow equipment, 162–163
Ultrasonic flow meter, 162f
Ultrasonic level measurement, 132–133
Ultrasonic waves, 132
Ultraviolet signal, 28
Underwater scuba gear, 91
Uninterruptible power supply (UPS), 363
Unit conversions, 10, 106f, 107
United Nations Globally Harmonized System (GHS), 9
Units of volume, 112
Unshielded twisted pair (UTP), 24, 25
Upper explosion limit (UEL), 257, 257f
Upper-range value (URV), 64
UPS. *See* Uninterruptible power supply (UPS)
Upscale and downscale check, 70, 74
URV. *See* Upper-range value (URV)
U.S. Food and Drug Administration (FDA), 62
User acceptance test (UAT), 376
User interface, 362
UTP. *See* Unshielded twisted pair (UTP)

V
V, 48f, 49f
V-notch ball valve, 218, 218f
V-notch weir, 164
Vacuum, 96
Vacuum pressure, 96, 96f
Validation, 332
Valve actuator, 210, 210f, 218–225
    bench set, 220
    diaphragm, 210, 219–221
    direct-acting actuator, 221
    electric (motor-operated value) actuator, 223–224, 224f
    failure mode, 218–219
    hydraulic actuator, 224–225, 225f
    MOV actuator, 223–224, 224f
    piston-type actuator, 221–222
    pneumatic actuator, 219–223
    rack and pinion actuator, 222–223, 223f
    reverse-acting actuator, 221
Valve body, 212, 212f
Valve bonnet, 212, 212f
Valve bushing, 212
Valve diaphragm, 210, 219–221
Valve failure mode, 218–219

Valve positioner, 210, 210f, 233–238
Valve post, 212
Valve seat, 212, 212f
Valve stem limit switch, 231f
Valve symbols and actions, 52–53, 53f
Velocity, 142, 145
Velocity meter, 161–163
Vena contracta, 150, 151f
Venturi tube, 157, 157f
Vibrating fork level switch, 124–125, 125f
Vibronic density, 254, 255f
Viscosity, 148
Visual inspection, 325, 325f
Vocabulary/terminology, 10–12
Voltage, 12, 13
Voltage-based signals, 105
Voltage drop, 13, 13f, 14, 14f
Voltage-measurement process (battery), 245
Voltage output card, 270, 270f
Volume, 113
Volume and pressure relationship, 108, 108f
Volume and temperature relationship, 108, 108f
Volume booster, 230
Volume of cylindrical tank, 10, 10f, 113
Volumetric flow equation, 144–145
Volumetric flow rate *(Q)*, 144f
Vortex, 149
W, 48f, 49f
Water, 114, 115f, 148
Water density versus temperature, 136f
WBS. *See* Work breakdown structure (WBS)
WD, 48f, 49f
Wedge element flow meter, 158, 158f
Weir, 164–165
Well manometer, 96f
Wet reference leg, 127, 127f
"What gets measured, gets improved," 370
Windsock, 333, 333f
Wire cross-sectional area, 13f
Wireless signal, 28
Wiring diagram, 55–56, 55f
Wiring separation, 306f
Wiring terminations, 313–318
Word, 27
Work breakdown structure (WBS), 372, 372f
Working environment, 2f, 3
Workstation, 362

X
X, 48f, 49f

Y
Y, 48f, 49f

Z
Z, 48f, 49f
ZD, 48f, 49f
Zero and span buttons, 199–200, 200f, 203
Zero and span shift, 76
Zero-only trim, 202
Zero pressure reference, 94
Zero procedure, proper, 105f
Zero shift, 71f, 73, 75–76
Ziegler-Nichols method of loop tuning, 336–337